Agro-Waste Management and Valorization

Agro-Waste Management and Valorization

Pratibha S. Agrawal, Richa Tiwari, Pramod Belkhode, and Samuel Lalthazuala Rokhum

WILEY-VCH

Authors

Pratibha S. Agrawal
Laxminarayan Innovation Techn. Univ.
Nagpur
India

Richa Tiwari
Laxminarayan Innovation Techn. Univ.
Nagpur
India

Pramod Belkhode
Laxminarayan Innovation Techn. Univ.
Nagpur
India

Samuel Lalthazuala Rokhum
National Institute of Technology Silchar
Silchar
India

Cover Design: Wiley
Cover Image: © Kletr/Shutterstock

Typesetting: Lumina Datamatics, Madurai, India

Library of Congress Card No.: applied for

British Library Cataloguing-in-Publication Data
A catalogue record for this book is available from the British Library.

Bibliographic information published by the Deutsche Nationalbibliothek
The Deutsche Nationalbibliothek lists this publication in the Deutsche Nationalbibliografie; detailed bibliographic data are available on the Internet at <http://dnb.d-nb.de>.

Hardback ISBN: 9783527355754
ePDF ISBN: 9783527854219
ePub ISBN: 9783527854202
oBook ISBN: 9783527854226

Printed and bound by CPI Group (UK) Ltd, Croydon, CR0 4YY

C9783527355754_071225

Contents

Preface

The global imperative to transition toward sustainable development has intensified the focus on managing agricultural waste in ways that are both environmentally sound and economically viable. Agro-waste, once considered a liability, is now recognized as a valuable resource capable of supporting energy generation, soil health, and the circular economy. This book, *Agro-Waste Management and Valorization*, presents a comprehensive, multidisciplinary examination of the state-of-the-art technologies, scientific principles, and policy frameworks that underpin the valorization of agro-waste. By integrating fundamental concepts with applied research and case studies, the book aims to serve as an essential reference for researchers, practitioners, and policymakers committed to advancing sustainable agro-waste management.

Chapter 1: Waste-to-value – Opportunity and Challenges

This foundational chapter establishes the context for agro-waste valorization by exploring the spectrum of waste types, including hazardous and non-hazardous categories, and their respective management challenges. It systematically reviews current waste management practices – collection, consolidation, transportation, and disposal – highlighting operational inefficiencies and the persistent lack of awareness that impedes progress. The chapter delves into the economic competitiveness of waste-to-value initiatives, emphasizing the need for sustainable solutions that address feedstock availability, carbon sequestration, and integration with smart city infrastructure through IoT services. It provides a holistic overview of the environmental benefits and energy generation potential of agro-waste, setting the stage for subsequent technical discussions.

Chapter 2: A Perspective on the Emergence and Need for Alternate Fuels

Here, the limitations of conventional fossil fuels are scrutinized, particularly their environmental impact and finite nature. The chapter provides a technical analysis of alternative fuels derived from agro-waste, including ammonia, hydrogen, alcohol-based fuels, and biodiesel. It discusses the chemical and physical properties of these fuels, their application in dual-fuel modes and blends, and assesses their environmental and economic feasibility. Through a forward-looking lens, the chapter explores future prospects for alternative fuels, emphasizing the role of innovation in achieving energy security and sustainability.

Chapter 3: The Role of Waste in the Circular Economy, Policies, and Legislation

This chapter situates agro-waste management within the broader context of the circular economy. It articulates the principles of circularity – resource regeneration, systems thinking, and renewable energy adoption – and examines how agro-waste can be leveraged to close material loops. The chapter provides detailed case studies on biochar production, anaerobic digestion, and circular agriculture models. It also reviews national and international policy frameworks, including the EU Waste Framework Directive, US EPA regulations, and India's National Policy on Biofuels, highlighting both economic and technological challenges to implementation.

Chapter 4: Agro-waste Management

A technical overview of waste management systems is presented, encompassing the assessment of refuse-derived fuel (RDF) and solid recovered fuel (SRF), relevant legislation, and the classification of solid and liquid waste. The chapter details advanced processing methods – pelletization, incineration, and physico-chemical and biological treatments – alongside the management of e-waste. The environmental and health impacts of mismanaged waste are critically analyzed, and sustainable waste management strategies are proposed, supported by techno-economic analysis (TEA) and life cycle assessment (LCA).

Chapter 5: Waste Biorefinery

This chapter explores the science and engineering of biological conversion processes, focusing on composting and anaerobic digestion. It discusses the microbial ecology and biochemical pathways involved, as well as reactor design and process optimization. The chapter also surveys emerging biological conversion technologies, emphasizing their potential to transform agro-waste into high-value products while minimizing environmental impacts.

Chapter 6: Algal Biorefinery

Thermochemical processes such as pyrolysis, alkaline hydrolysis, and transesterification are dissected in this chapter. The mechanisms of bio-oil, biochar, and syngas production are elucidated, with attention to process parameters, feedstock characteristics, and reactor configurations. The chapter also incorporates techno-economic and LCA methodologies to evaluate the sustainability of these conversion routes, and discusses emerging trends that promise to further enhance process efficiency and product quality.

Chapter 7: Waste-to-bio-additive

This chapter introduces the concept of the waste biorefinery, where diverse feedstocks undergo integrated conversion processes to yield fuels, chemicals, and materials. It covers kinetic analysis of biomass, thermochemical and biochemical conversion pathways, and the unique challenges of food, municipal, and lignocellulosic waste biorefineries. The economic aspects of waste-to-energy biorefineries are critically assessed, providing a roadmap for scaling up and commercialization.

Chapter 8: Agro-waste to Compost

Focusing on algae as a versatile feedstock, this chapter examines cultivation systems, strain selection, and nutrient optimization – often leveraging agro-waste as a nutrient source. It details advanced processing techniques for extracting lipids, proteins, and carbohydrates, and discusses the production of biofuels and bioproducts through integrated biorefinery approaches. The chapter also addresses genetic engineering, environmental sustainability, and the regulatory landscape, positioning algal biorefining as a promising avenue for agro-waste valorization.

Chapter 9: Glycerol: From Abundant Byproduct to Valuable Bio-oil

This chapter reviews the production of bio-additives from various waste streams using biological, thermochemical, and physiochemical processes. It highlights the applications of bio-additives in enhancing fuel properties or as soil amendments, and provides a rigorous life cycle assessment of these technologies. Case studies illustrate successful projects and outline future directions for research and commercialization.

Chapter 10: Production of Biochar

The science of composting is explored in detail, from feedstock categorization and pretreatment to the biochemical and microbial processes that drive decomposition. The chapter compares traditional and advanced composting methodologies, such as vermicomposting and in-vessel composting, and discusses factors influencing compost quality and efficiency. The role of compost in sustainable agriculture and soil health is emphasized, supported by recent advances in process monitoring and control.

Chapter 11: Waste Valorization for Biogas Production: A Pathway to a Circular Economy

Addressing the challenge of managing glycerol, a byproduct of biodiesel production, this chapter investigates catalytic and non-catalytic conversion technologies for producing bio-oil. It explores the chemistry of glycerol transformation, reactor design, and process optimization, along with the characterization and application of the resulting bio-oil. The chapter also surveys commercial-scale plants and emerging trends in glycerol valorization.

Chapter 12: Digitalization for Agro-waste Management

This chapter provides a comprehensive analysis of biochar production from agro-waste through pyrolysis, gasification, and hydrothermal carbonization. It discusses feedstock selection, pretreatment, and the influence of process parameters on biochar properties. The environmental applications of biochar-including soil amendment, carbon sequestration, and pollution remediation-are examined, alongside comparative analyses and future research directions.

By systematically addressing the spectrum of agro-waste valorization pathways from biological and thermochemical conversions to integrated biorefinery concepts-this book equips researchers with a deep understanding of both the scientific fundamentals and practical applications. Each chapter integrates current research, techno-economic analysis, and policy considerations, offering a holistic perspective that is critical for driving innovation. Researchers will find in these pages not only a synthesis of the latest advances but also a roadmap for future inquiry, collaboration, and technology transfer. The book thus serves as a catalyst for interdisciplinary research, supporting the development of sustainable solutions that bridge the gap between waste management and resource recovery in the agricultural sector.

1

Waste-to-value Opportunity and Challenges

1.1 Introduction

Rapid economic growth and urbanization worldwide have resulted in a substantial increase in resource consumption and waste generation. Disposing of this growing volume of solid waste, particularly in landfills, is often neither economically feasible nor sustainable in urban environments. Energy recovery from waste, resource recovery, and waste volume reduction represent attractive and viable alternatives for waste management. Extracting energy from waste minimizes the need for energy produced from finite resources. Therefore, there is an urgent need to develop sustainable strategies for waste management. This study explores the latest advancements in sustainable waste-to-energy (WtE) technologies, resource recovery methods, and low-carbon biotechnologies and bioenergy options. Recent research highlights the growing importance of the circular economy as a sustainable waste management approach [1, 2]. WtE alternatives are crucial in the waste management sector. This chapter presents an innovative approach focused on addressing the challenges of managing organic waste by integrating WtE technologies. Globally, there is increasing emphasis on sustainable waste management practices, as utilizing organic waste for WtE purposes is a viable option. Identifying and prioritizing WtE alternatives is key, given the large quantities of waste in the management system. The goal is to reduce the environmental and logistical impacts of organic waste disposal, contributing to a more efficient and sustainable waste management system aimed at minimizing environmental effects and producing renewable energy. Ludlow et al. [3] emphasize that achieving net-zero greenhouse gas (GHG) emissions by 2050 fundamentally requires a shift in waste management practices. Organic waste disposal frequently presents significant challenges, including insufficient collection infrastructure, limited public awareness, and operating difficulties, particularly in urban areas. Daily per capita waste generation varies considerably globally, ranging from 0.11 to 4.54 kg [4].

Managing waste involves complexity and costs, which differ depending on local resource allocation and population density. However, as waste volume and disposal costs rise, developing countries face significant challenges in managing waste

Agro-Waste Management and Valorization, First Edition. Pratibha S. Agrawal, et al.

effectively [5]. Despite these challenges, alternative waste management options, such as anaerobic digestion (AD), pyrolysis, and gasification, are gaining traction. Implementing effective waste management practices and technologies is critical for mitigating negative environmental and health impacts [6]. Developing nations urgently need to adopt sustainable waste management practices.

1.1.1 Waste-to-energy

WtE technologies are commonly employed for sustainable waste management, facilitating simultaneous energy generation and effective disposal [7]. Among WtE methods, thermal conversion is the most widely adopted approach for large-scale waste-processing. Incineration, a thermal technology, is still under development; while pyrolysis remains experimental, it is not yet commercially viable for managing municipal solid waste (MSW) globally. Biological conversion methods like AD are also used, particularly for bio-methanization. AD is frequently utilized for processing organic-rich biodegradable waste like manure and sewage sludge [8]. Using landfills for waste disposal is less favored due to drawbacks such as unpleasant odors, GHG emissions from methane (CH_4), leachate production, and associated health risks. Evidence suggests that WtE technologies implemented in Latin American and Caribbean (LAC) countries offer long-term advantages, including economic development, environmental benefits, and improved human well-being [9]. Existing literature provides a detailed look at the scale of waste challenges in developing nations and explores potential methods to promote sustainable waste management practices in the future.

This chapter helps fill knowledge gaps by analyzing treatment methods used in developing countries and their potential to support continuous sustainability by aligning with renewable energy sources such as hydroelectricity, which are environmentally friendly and non-emission-generating. For example, South Africa's electricity mix in 2021 heavily relied on coal-fired plants (84%), resulting in substantial CO_2 emissions of 396 g CO_2/kWh. Energy consumption also surged sharply in New Zealand (66 kg CO_2/kWh, primarily powered by geothermics [57%], hydro [17%], and thermal sources ([9%]) [10, 11]. The rapid urbanization in many developing countries intensifies waste management challenges.

Nonetheless, these challenges also present significant opportunities. Treating waste as a valuable resource rather than a problem allows different nations to adopt sustainable waste management techniques and achieve future development goals. Studies conducted in Korea demonstrate how WtE operations reduced GHG emissions by 16,061 tCO_2eq annually; this reduction is projected to increase further to 26,477 tCO_2eq per year by 2021 [12]. The environmental advantages of WtE technologies (like incineration, pyrolysis, gasification, and incineration) vary depending on the specific method used [13]. The purpose of this chapter is to provide a systematic evaluation and highlight the benefits and challenges of WtE technology, especially in developing countries. This includes assessing the characteristics and volume of waste produced, current treatment methods, the opportunities and obstacles within the waste management sector, and the feasibility of implementing or expanding WtE solutions domestically.

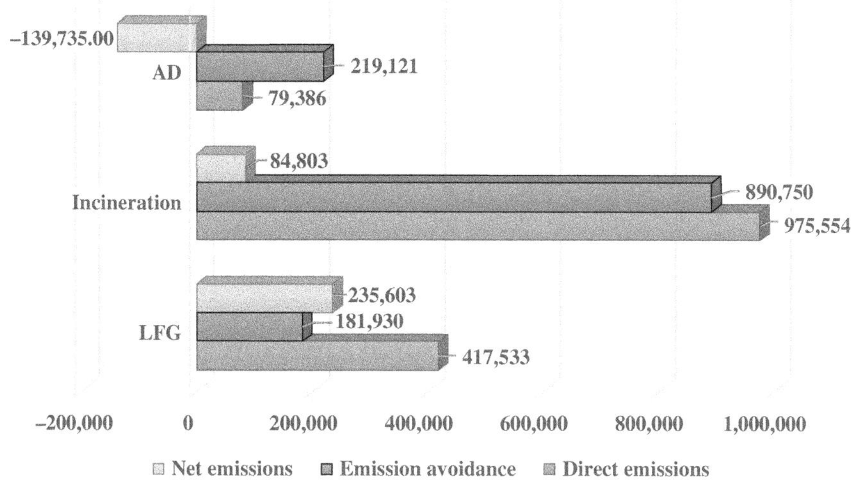

Figure 1.1 GHG emissions and emission saving from WtE technologies (tCO_2-eq/year). Adapted from [16].

1.1.2 Environmental Benefits

Figure 1.1 illustrates that incineration notably influences the global warming potential (GWP), contributing between 975,554 and 830,750 tCO_2-eq/year due to the large quantity of processed waste. In thermal technologies, emissions are a primary contributor to the GWP [14]. The environmental impact of incineration is largely dictated by its energy conversion efficiency; higher efficiency leads to lower emission rates. For instance, an incineration plant operating at 25% efficiency emits roughly 1,007 kg CO_2-eq per MWh of combustion, while increasing efficiency by just 5% (to 30%) reduces emissions by approximately 17% due to lower GHG emissions [15]. The significant emission reductions from incineration can be attributed to the characteristics of the burned materials. Plastic waste, comprising over 20% of facility input, has the most notable impact on GWP due to its chemical properties and low moisture content, resulting in greater emission contributions [16]. Moreover, the common practice of disposing of MSW without source separation leads to high moisture levels [17]. Achieving greater environmental and economic benefits from WtE technologies requires continuous improvements.

1.1.3 Energy Generation Potential

Uncollected CH_4 (25%) from landfill gas (LFG) is a major factor contributing to GWP, amounting to 417,533 tCO_2-eq/year of GHG emissions [18]. In contrast, AD methods produce the lowest emissions, achieving 99% efficiency in capturing CH_4, leaving only 5% to be released into the atmosphere. The environmental performance of WtE technologies compared to coal-based electricity generation was also examined [19]. Research indicates that LFG recovery results in the highest net GWP, estimated between 137,439 and 323,604 tCO_2-eq annually. Incineration shows a lower GWP, ranging from 219,121 to 84,803 tCO_2-eq/year. Incineration with energy recovery surpasses simple incineration in reducing emissions. Energy generation from WtE can offset some of the benefits gained from electricity production but significantly counteracts global warming effects. AD technology has substantially reduced impact on GWP, resulting in an avoidance of 95,321–195,395 tCO_2-eq/year of GHGs. From an environmental standpoint, AD technology is arguably the most suitable option due to its lower emissions.

1.1.4 Economic Potential

The global waste recycling services market is projected to reach US$99.2 billion by 2032, with an average annual growth rate of 5.1% from 2023 to 2032 [20]. In 2022, the Asia-Pacific region's waste recycling services market constituted over 80% of the total market share. The US Environmental Protection Agency (EPA) reports that the recycling industry employs 534,000 people and contributes US$13.2 billion in wages and benefits [15]. An analysis indicates that in 2020, only 9% of electronic waste (e-waste)was properly collected and recycled [21]. According to the EPA, the United States generated 292.4 million tons of MSW in 2018, of which 69 million tons were recycled [22]. The increasing adoption of advanced WtE technology offers substantial potential for adding value to waste and generating energy [23]. The precise potential varies based on factors such as the type and volume of available waste, technological readiness, environmental considerations, and government policies. Figure 1.2 illustrates potential applications of waste recycling and processing technologies that significantly influence the waste recycling services market value.

WtE presents considerable potential, but its viability is dependent on various factors including public acceptance, regulatory frameworks, technological maturity, and economic feasibility. WtE initiatives should be driven by efforts to generate waste, recover resources, and explore renewable energy sources to establish a truly sustainable waste management system.

Comparing WtE options involves evaluating different technologies and strategies for converting waste into energy. Figure 1.3 provides a comparative analysis of common WtE approaches.

The selection of a WtE option is influenced by factors such as waste composition, local regulations, infrastructure availability, energy demand, environmental impacts, and economic viability. Often, a combination of technologies is utilized to maximize resource recovery and energy generation while minimizing environmental effects. While WtE has shown notable advancements in waste management and energy production, recognizing that the most suitable approach is dependent on local circumstances, waste characteristics, technological readiness, and environmental factors is important. Therefore, the reasons for favoring WtE are often multifaceted:

1. **Waste Reduction:** WtE facilities substantially decrease the volume of waste sent to landfills by converting it into energy, thereby prolonging the lifespan of existing landfill sites and lessening the need for new ones [26].
2. **Energy Production:** WtE plants generate electricity or heat from burning MSW, biomass, or other waste sources, diversifying energy supply and meeting local energy demands.

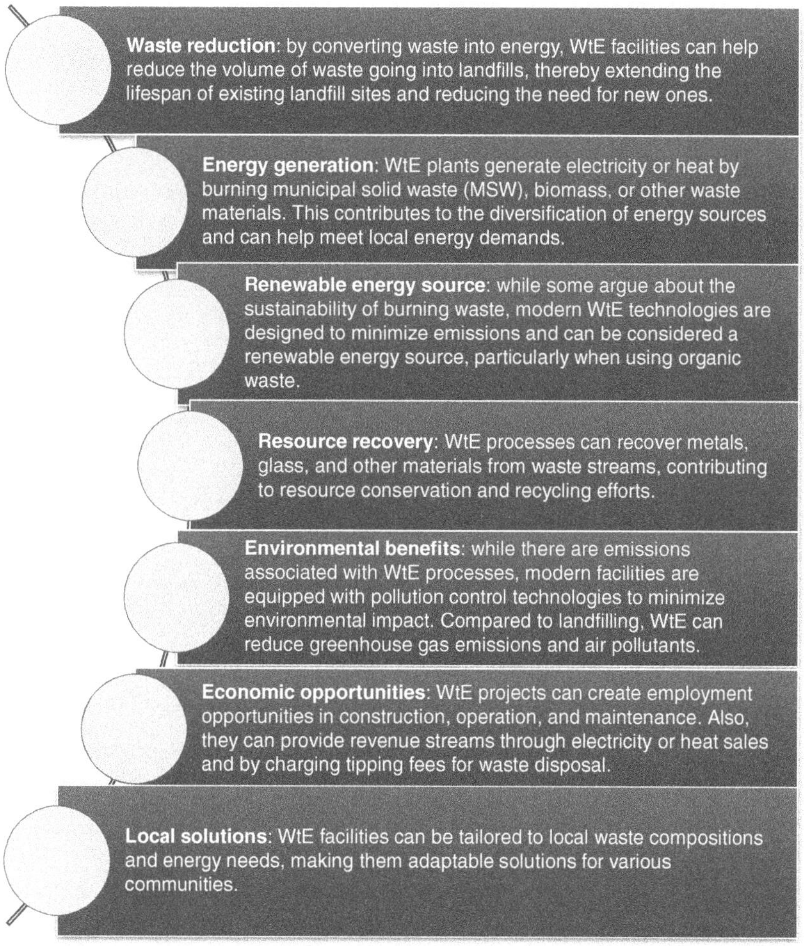

Figure 1.2 Potential of waste-to-energy (WtE) technology. Adapted from [24].

3. **Resource Recovery:** WtE processes can recover valuable materials like metals and glass from waste, supporting resource conservation and recycling efforts.
4. **Environmental Benefits:** WtE facilities are equipped with pollution control technologies to minimize harmful emissions. Compared to landfilling, WtE can reduce GHG emissions and air pollutants.
5. **Renewable Energy Source:** Depending on the type of waste processed, WtE can be considered a renewable energy source, especially when using organic waste.
6. **Job Creation:** WtE facilities create employment in construction, operation, and maintenance, benefiting local economies.
7. **Local Solutions:** WtE facilities can be customized to local waste streams and energy needs, providing flexible options for different communities.

However, acknowledging the complexity of waste management and the lack of a single solution for every situation is crucial. Effective implementation of sustainable waste management strategies, such as sorting, recycling, composting, and other

Incineration:
- **Process:** involves combusting waste at high temperatures in a controlled environment.
- **Energy output:** generates heat, which can be used for electricity generation or district heating.
- **Advantages:** mature technology, high energy recovery efficiency, reduces waste volume, can handle various types of waste.
- **Challenges:** concerns about air emissions (though modern facilities have advanced pollution control systems), high initial investment costs, public perception, and acceptance issues.

Gasification:
- **Process:** converts solid waste into synthetic gas (syngas) by reacting it with a controlled amount of oxygen and/or steam at high temperatures.
- **Energy output:** syngas can be used to produce electricity, heat, or biofuels.
- **Advantages:** can handle diverse feedstocks, potentially higher energy efficiency than incineration, lower emissions.
- **Challenges:** technically complex, requires preprocessing of waste, higher capital costs compared to incineration.

Pyrolysis:
- **Process:** involves heating waste in the absence of oxygen to produce bio-oil, gas, and char.
- **Energy output:** bio-oil can be used as a fuel, gas can be burned for energy, and char can be used for various applications.
- **Advantages:** can handle various types of waste, produces bio-oil suitable for transportation fuels or chemical feedstocks, potentially lower emissions.
- **Challenges:** technical challenges in scaling up, high capital costs, energy-intensive process.

Anaerobic digestion:
- **Process:** breaks down organic waste in the absence of oxygen by microorganisms to produce biogas (methane and carbon dioxide) and digestate (organic fertilizer).
- **Energy output:** biogas can be used for electricity generation, heat production, or vehicle fuel.
- **Advantages:** suitable for organic waste, produces renewable energy and organic fertilizer, reduces greenhouse gas emissions.
- **Challenges:** limited to organic waste streams, requires a steady supply of feedstock, biogas composition may vary.

Landfill gas recovery:
- **Process:** captures methane generated by the decomposition of organic waste in landfills and utilizes it for energy production.
- **Energy output:** methane can be used for electricity generation, heat production, or upgraded to natural gas quality.
- **Advantages:** utilizes existing landfill sites, reduces greenhouse gas emissions, relatively low capital costs.
- **Challenges:** dependent on landfill lifespan, requires significant infrastructure for gas capture and treatment.

Figure 1.3 Comparative analysis of some commonly used WtE techniques. Adapted from [25].

waste management methods, is essential. Furthermore, advancements in technology and continuous process optimization are necessary to determine the most appropriate waste management strategy for a given region or municipality.

1.2 Classification of Waste

Waste materials are classified based on various factors such as their composition, potential risks to human health and the environment, and relevant regulatory guidelines, as detailed below.

1.2.1 Hazardous Waste

Hazardous waste is defined as waste that poses significant risks to public health or the environment due to its inherent chemical, biological, or physical characteristics [6]. Waste materials are categorized as hazardous based on specific properties or origins, and regulations govern their handling. Key characteristics defined by regulatory bodies are as follows:

- **Ignitability:** Waste that can easily ignite, such as certain solvents, paints, and some types of oils.

- **Corrosivity:** Waste that is highly acidic or alkaline and can corrode containers, such as battery acid or specific cleaning agents.
- **Reactivity:** Waste that is unstable and can undergo violent reactions, release toxic gases, or explode under certain conditions, such as certain cyanide-containing wastes.
- **Toxicity:** Waste that contains substances harmful to human health or the environment, including heavy metals (like lead, mercury), pesticides, or certain industrial chemicals.

Specific hazardous wastes by Akpan et al. [27] are shown in Figure 1.4.

- **F-list:** Wastes from common manufacturing and industrial processes, such as specific solvents and wastewater.
- **K-list:** Wastes from specific industries (e.g., chemical manufacturing, pulp and paper industry), including sludges and residues from metal production.
- **P- and U-list:** Discarded commercial chemical products that are unused, off-specification, or spilled, including certain pesticides and laboratory chemicals.
- **Mixed Waste:** Waste containing both hazardous and radioactive components, often found in certain medical or research laboratories.
- **Medical Waste:** Waste generated by healthcare facilities, including contaminated sharps, materials soaked in blood, and expired medications.
- **Radioactive Waste:** Waste contaminated with radioactive materials, such as those from nuclear power generation, specific medical treatments, and research activities.
- **E-waste:** Discarded electronic devices containing hazardous substances like lead, mercury, and brominated flame retardants.

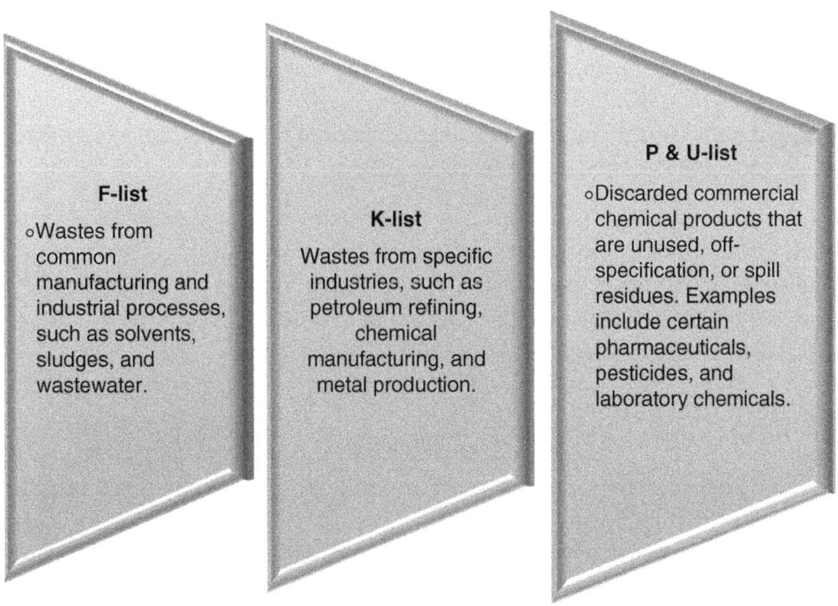

Figure 1.4 Category of hazardous wastes according to US EPA. Adapted from [27].

Table 1.1 Impact of hazardous waste on human health. Adapted from [28].

Toxic materials present in hazardous waste	Source of origin	Impact on human health
Cadmium	Battery waste/mining/fertilizer industries	Pulmonary diseases, liver and kidney damages, carcinogenic, skeletal/cardiovascular disorder
Manganese	Mining region	Neuropsychiatric disorder/respiratory diseases
Nickel	Mining/metal refining	Gastrointestinal system damage/lung or nasal cancer
Hydrocarbons	Industries of petrochemical	Nausea, bone marrow damage, headaches, nausea
PCBs	E-waste/fluorescent light/hydraulic fluid	Gastrointestinal/skin damage, carcinogenic
Pesticides	Insecticides	Stillbirths, carcinogenic, immune system disorder

Exposto et al. [28] documented adverse health effects in humans caused by exposure to toxic compounds in hazardous wastes, as presented in Table 1.1.

According to Saleh et al. [29], hazardous wastes are further categorized into special classes under specific regulations, even if they don't meet standard characteristics or appear on lists. Examples include medical waste, radioactive waste, and e-waste.

- **Medical Waste:** Waste generated by healthcare facilities, including contaminated sharps, materials soaked in blood, and expired medications.
- **Radioactive Waste:** Waste contaminated with radioactive materials, such as waste from nuclear power plants, specific medical treatments, and research materials.
- **E-waste:** Discarded electronic devices containing hazardous substances like lead, mercury, and brominated flame retardants.

Proper management of hazardous waste is crucial to prevent environmental contamination and protect human health. Regulations govern its generation, storage, transportation, treatment, and disposal to minimize risks. This often involves labeling, secure storage, specialized handling procedures, and disposal at authorized facilities equipped to manage hazardous materials safely.

1.2.2 Non-hazardous Waste

Non-hazardous waste is waste that does not pose significant risks to human health, or the environment based on its chemical, biological, or physical properties. This category encompasses various materials that can be safely disposed of in standard landfills or recycled without special handling procedures [30]. Some common examples and categories of non-hazardous waste are general non-toxic waste (paper, cardboard, plastic packaging, food scraps, wood, concrete, bricks, and textiles), recyclable waste such as paper, cardboard, glass, metals (aluminum, steel), e-waste and certain types of plastics, biodegradable organic waste like food scraps, yard waste (leaves, grass clippings), and other organic materials.

Regulatory bodies, such as the EPA in the United States or the European Environment Agency (EEA) in Europe, provide detailed criteria and guidelines for waste classification. These classifications are crucial for the correct handling, storage, transportation, and disposal of waste to minimize environmental and public health risks. Effective management of non-hazardous waste typically involves strategies like source reduction, recycling, composting, and landfill disposal in compliance with local regulations. Recycling and composting help reduce the volume of waste sent to landfills and conserve natural resources by recovering valuable materials for reuse. Additionally, initiatives promoting waste reduction and sustainable practices aim to minimize the overall environmental impact of waste generation.

1.3 Current Status in Waste Management

1.3.1 Waste Collection

As of early 2022, waste management methods and technologies continue to advance, influenced by increasing environmental awareness, regulatory demands, and progress in waste treatment and recycling. Waste collection methods vary considerably depending on location, infrastructure, regulations, and environmental factors. Usual waste collection practices [31] are as follows:

1. **Curbside Pickup:** A widely used method, especially in urban and suburban areas. Residents place separated waste (recyclables, organic waste, general trash) in designated containers or bags for collection on scheduled days. Waste management trucks collect and transport the waste to processing facilities.
2. **Drop-off Centers:** In some regions, residents transport their waste to central drop-off points. These centers often provide separate containers for different waste types, facilitating recycling and proper disposal.
3. **Composting Programs:** Many municipalities promote composting by offering residents compost bins or composting services. Organic waste (food scraps, yard trimmings) can be collected separately and processed into compost for gardening or landscaping use.
4. **Special Collection Services:** Some areas offer dedicated services for bulky items like furniture, appliances, or electronics that don't fit standard bins. Residents can arrange pickups or drop items off at specific locations.
5. **Hazardous Waste Collection:** Hazardous materials (batteries, paint, chemicals, e-waste) require special handling due to environmental and health risks. Many communities organize programs or events for the safe disposal of these items.
6. **Commercial Waste Collection:** Services for businesses, industrial facilities, and institutions differ from residential services. Commercial waste often requires specialized handling and uses larger bins or dumpsters.

7. **Recycling Programs:** Recycling collection is typically part of regular waste collection. Residents are encouraged to separate recyclables (paper, plastic, glass, metal) from general waste for collection and transport to recycling facilities.
8. **Public Awareness and Education:** There is growing emphasis on engaging the community and educating the public to promote waste reduction, recycling, and proper disposal. Effective collection practices include campaigns to inform residents about correct disposal methods, the importance of recycling, and strategies to reduce waste through composting and reuse initiatives. These efforts aim to raise awareness about waste generation's environmental impacts and encourage behavioral changes to minimize waste production and increase recycling rates.
9. **Technological Innovations:** Some waste management systems integrate technology, such as RFID tags on bins for tracking collection routes and optimizing efficiency, or smart sensors to monitor bin fill levels and schedule collections dynamically.
10. **Informal Waste Collection:** In some areas, informal waste collectors (scavengers or waste pickers) play a significant role by collecting recyclable materials from streets, dumps, or landfills for sale or reuse.

These practices aim to minimize environmental impact, promote sustainability, and efficiently manage waste to protect public health and the environment. The integration of technology with smart practices enhances efficiency and reduces collection burdens. Abubaker et al. [32] highlighted that many solid waste management practices involve combining household waste with commercial waste, which can contain hazardous substances during collection and handling, negatively impacting the environment. This process involves waste material collection, transportation, and disposal, which presents cost concerns globally. Literature from 1970 onwards shows that investing in waste collection equipment has resolved most technical problems related to SWM. Daily waste collection practices in high-, middle-, and low-income countries are compared in Table 1.2.

Furthermore, waste collection is a critical component of waste management worldwide. Various approaches and expenditures are adopted in developed and developing nations, as detailed in Table 1.2. Thus, the current state of waste collection involves:

1. **Efficiency and Optimization:** Waste collection services are increasingly adopting data-driven methods and technologies to optimize routes, schedules, and resource allocation. This includes using Geographic Information Systems (GISs), route optimization software, and real-time monitoring systems to improve efficiency and reduce costs.
2. **Smart Waste Management:** Smart waste management solutions, incorporating sensors, Internet of Things (IoT) devices, and data analytics are being deployed to monitor waste bin fill levels, optimize collection routes, and schedule pickups more effectively. These technologies help reduce unnecessary collections, minimize overflow, and enhance overall service quality.
3. **Separation at Source:** Many waste management programs emphasize source separation, where residents or businesses are encouraged to separate recyclable

Table 1.2 MSW collection practices commonly by country's economic development.

Activity	High-income countries	Low-income countries	Middle-income countries
Waste collection	Collection rate >90%. Transfer stations, compactors, and motorized trucks are commonly available. In design of system, volume of waste and aging collection workers is considered	Generally, it serves wealthy, business, and high-visibility areas. Overall collection <50%. Inefficient and infrequent	In residential areas, there is an improvement in collection and transportation. Mainly 50–80% collection, mechanization, large vehicle fleet, and transfer stations
Cost	Cost of collection <10% of budget, while treatment of intermediate waste is a large budget allocation. Community participation reduces costs and increases options for planners	Cost of waste collection is 80–90% of SWM budget. Government regulation of waste fees exists, but collection is insufficient, and a small budget portion is for disposal	Cost of collection is 50–80% of MSWM budget. Some national and local governments require waste fees and show innovation in collection methods. More collection fleets, mechanization, and disposal expenditure

materials, organic waste, and non-recyclable waste at the point of generation. Source separation facilitates more efficient processing and recycling downstream, reducing contamination and maximizing resource recovery.

4. **Specialized Collection Programs:** Municipalities and waste management companies are implementing specialized collection programs for hazardous waste, e-waste, bulky items, and other materials that require separate handling and disposal due to their potential environmental and health impacts.
5. **Integration of Sustainable Practices:** Eco-friendly practices such as composting, AD, and material recovery facilities (MRFs) are becoming more integrated into waste management systems to divert organic waste from landfills, recover valuable resources, and reduce GHG emissions.
6. **Innovations in Collection Vehicles:** Waste collection vehicles are evolving to incorporate cleaner and more efficient technologies, such as electric and hybrid propulsion systems, compressed natural gas (CNG), or biodiesel engines. These vehicles help reduce emissions, noise pollution, and dependency on fossil fuels [33].

According to Keerthana et al. [34], land pollution is primarily caused by garbage spillage. They proposed various existing methods for monitoring garbage. One such method is an advanced waste collection system that separates waste into wet, dry, and metallic categories, as shown in Figure 1.5. This system uses sub-bins for different waste types and a dual motor/tray mechanism. Waste is initially placed in a common tray, where an IR sensor detects its presence. A moisture sensor then activates, and if the moisture reading is high, the waste is classified as wet; otherwise, it's dry. The three bins are positioned below, and an ultrasonic sensor checks the

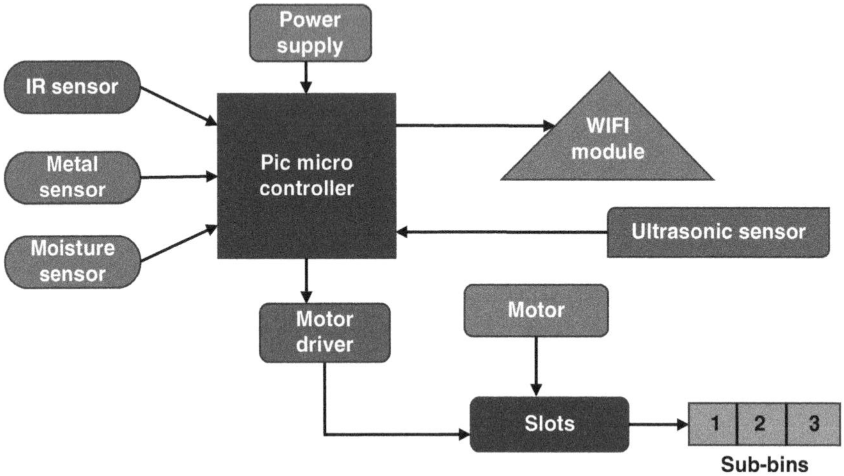

Figure 1.5 Block diagram of waste collection. Adapted from [34].

garbage level. The sensed data is transmitted via a PIC microcontroller. This entire process operates using an IoT application.

Overall, waste collection practices are increasingly focused on efficiency, sustainability, and innovation to address the challenges of waste management in an environmentally responsible manner while meeting the needs of growing urban populations.

1.3.2 Waste Consolidation and Transportation

Waste transportation involves moving collected waste from its origin (homes, businesses, or drop-off centers) to various destinations for processing, recycling, or disposal. Kumar et al. [35] state that India faces primary challenges in waste management related to waste generation, inefficient collection, transportation, treatment, and disposal, with waste volumes increasing daily. As illustrated in Figure 1.6, it is estimated that by 2041, total waste generation will reach 160.95×10^3 tons per year, with the population projected to be approximately 595.4×10^6. This suggests a per capita waste generation of around 0.741 kg/day. Since waste generation comes from various sources, including households, construction, organic matter, and agriculture, their collection, storage, and transportation are crucial aspects of waste management practices.

Here's a scenario illustrating waste transportation practices:

- **Consolidation:** After collection, waste is transported to a central transfer station or depot, where different types of waste are consolidated and loaded onto larger transport vehicles for efficient transport to processing facilities.
- **Recyclables:** Reusable materials (paper, plastic, glass, metal) are loaded onto separate trucks and transported to recycling facilities. These facilities may be located within or near the city, depending on contracts and logistics.

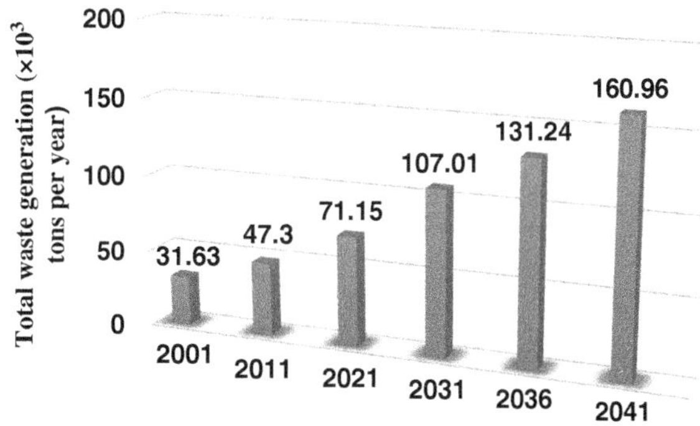

Figure 1.6 Total waste generation per year. Adapted from [35].

- **Organic Waste:** Organic waste from residential and commercial sources is transported to a composting facility. Specialized trucks are used, designed to handle organic material without odor or leakage during transit.
- **General Trash:** Non-recyclable and non-organic waste (residual household and commercial waste) is transported to a landfill or WtE plant. Trucks equipped with compaction mechanisms ensure efficient loading and transport to the designated facility.
- **Special Handling:** Hazardous materials from households and businesses (batteries, paints, chemicals) are transported separately to specialized treatment or disposal facilities equipped to safely handle hazardous waste.
- **Monitoring and Compliance:** Throughout transport, waste management authorities monitor vehicle routes, adherence to safety rules, and environmental standards. GPS tracking, electronic manifests, and regular inspections help ensure proper waste handling and transport.
- **Continuous Improvement:** Waste management authorities regularly evaluate and optimize transportation routes, vehicle fleet efficiency, and logistics to minimize costs, reduce environmental impact, and improve service delivery.

This scenario highlights the coordinated efforts involved in moving different types of waste from collection points to appropriate processing or disposal sites, aiming for effective waste management and environmental sustainability [36].

1.3.3 Waste Disposal Practices

A comprehensive approach is needed in order to manage large quantity of waste disposal in an energy efficient and environmentally friendly manner. Waste disposal practices involve the final stage of waste management, where waste is permanently removed from the environment through various methods [37]. Common waste disposal practices include:

1. **Landfilling:** Engineered landfills represent a primary method of waste disposal, designed with specific containment areas where waste is systematically deposited. Within these sites, the waste is typically compacted to reduce volume and subsequently covered with soil, a process often carried out daily or periodically. Contemporary landfill designs incorporate advanced environmental safeguards,

including impermeable liners and integrated systems for collecting leachate, the liquid that drains from waste, to prevent the contamination of surrounding groundwater and soil. Additionally, the anaerobic decomposition of organic matter within the contained waste mass produces LFG, which is primarily CH_4. Specialized collection systems are installed to capture this gas, which can then be either utilized for energy generation or safely flared, thus mitigating GHG emissions.

2. **Incineration:** Waste incineration serves as a thermal treatment process involving the combustion of waste materials at elevated temperatures within controlled facilities. The primary objectives of this method are to significantly reduce the volume and mass of the waste stream. Furthermore, modern energy recovery facilities are often integrated with incinerators to capture the heat generated during this combustion process. This heat is typically used to produce steam, which subsequently drives turbines for electricity generation. To mitigate the environmental impact, these facilities are equipped with sophisticated air pollution control systems, such as scrubbers and filters, which are designed to minimize the emission of harmful substances including particulate matter, heavy metals, and dioxins into the atmosphere.
3. **Composting:** Biological decomposition, specifically through composting, offers a viable treatment method for organic waste materials, including food scraps, yard trimmings, and paper products. This process is typically carried out in controlled environments where parameters such as moisture content, aeration, and temperature are carefully managed to optimize the microbial breakdown of organic matter. The outcome is the production of nutrient-rich compost. This valuable finished product can then be effectively applied as a soil amendment in various settings, including agriculture, landscaping, and erosion control initiatives, thereby enhancing soil health and promoting fertility.
4. **Recycling:** This process constitutes a crucial waste management strategy focused on recovering valuable materials from discarded products. This process entails collecting and processing waste materials such as paper, plastics, glass, and metals to convert them into new products, thereby facilitating resource conservation and reducing the demand for virgin raw materials. Dedicated recycling facilities are equipped to sort, clean, and process these recyclable streams, transforming them into secondary raw materials or feedstocks suitable for various manufacturing industries. Furthermore, policy instruments like Extended Producer Responsibility (EPR) programs play a vital role by establishing manufacturer accountability for the collection and recycling of their products at the conclusion of their intended use phase.
5. **Biological Treatment:** Bio-treatment methods leverage the metabolic activity of microorganisms to decompose organic waste, yielding valuable byproducts. Key techniques within this category include AD and aerobic digestion, which are employed to break down biodegradable materials. AD, conducted in the absence of oxygen, results in the production of biogas, primarily composed of CH_4, which can be harnessed for energy generation. In contrast, aerobic digestion utilizes oxygen to facilitate the biological decomposition of organic matter, typically yielding compost or other stable organic residues.
6. **Hazardous Waste Treatment:** Hazardous waste, a category encompassing materials such as chemicals, solvents, heavy metals, and e-waste, necessitates

specialized treatment protocols due to the inherent and significant environmental and health risks associated with improper handling. The primary objective of hazardous waste treatment is to minimize these risks. Employed methods are diverse and may include chemical neutralization to alter hazardous properties, thermal processes like high-temperature incineration, or stabilization and immobilization techniques designed to render the waste less harmful or inert before final disposition.

7. **Waste-to-energy:** WtE facilities represent systems that convert waste materials into usable energy forms, including heat, electricity, or fuel. This conversion is achieved through a variety of thermal and biochemical processes. Advanced WtE technologies, such as gasification, pyrolysis, and plasma arc gasification, are capable of converting solid waste into syngas, a versatile intermediate product. This syngas can then be utilized directly for energy production or as a valuable chemical feedstock for further industrial processing.
8. **Secure Disposal:** Secure waste handling is a critical practice designated for specific waste streams, including medical waste, pharmaceuticals, and confidential documents, to ensure the prevention of unauthorized access and minimize the risk of environmental contamination. The methods employed for secure disposal are designed to guarantee safe handling and containment. These techniques may incorporate physical destruction processes such as shredding, biological neutralization through sterilization, or physical containment methods like encapsulation within specialized containers specifically designed for secure handling and ultimate disposition.

Agamuthu and Babel et al. [38] studied about SWM practices in Asia and pointed out that from last five decades, it was ranged over in Asia and identification of different challenges/issues were carried out. Since there is an increment in waste generation worldwide and concerning their disposal sites, most of the adopted disposal techniques are open dumps/landfills, which have become more widespread in low- or middle-income countries (79% south, 64% southeast, 51.1% south and central Asia) as summarized in Table 1.3.

Table 1.3 Distribution (%) of waste disposal techniques by different income level. Adapted from [38].

Waste management type	High-income counties (%)	Upper-middle-income countries (%)	Lower-middle-income countries (%)	Low-income countries (%)
Open dump	2	30	66	93
Landfill	39	54	18	3
Incineration	22	10	—	—
Composting	6	2	10	—
Other advanced methods	2	—	—	—
Recycling	29	4	6	3.7

Effective waste disposal practices aim to minimize environmental impact, conserve resources, and protect public health while meeting regulatory requirements and societal expectations for sustainable waste management.

1.3.4 Lack of Awareness

Limited awareness about waste treatment is a significant barrier to implementing sustainable waste management practices. In accordance with Hasan et al. [39], current human health problems and environmental degradation are due to lack of awareness and improper handling of waste. Various reasons for this lack of awareness and its potential consequences can be limited education, as communities may lack awareness of available waste treatment options and their importance. In some cultures or communities, there may be a lack of emphasis on proper waste management, leading to a mindset of disposability and indifference toward waste treatment practices. People may be unaware of the types of waste that can be treated and recycled, as well as the facilities and services available for waste treatment in their area. Without access to clear and accurate information, individuals may resort to improper disposal methods out of convenience or ignorance. Some individuals may perceive waste treatment options as too costly or inconvenient, particularly if they are accustomed to traditional disposal methods like landfilling or open burning. This perception can discourage people from seeking out and utilizing more sustainable waste treatment practices. Weak enforcement of waste management regulations can contribute to a culture of non-compliance and disregard for proper waste treatment practices. When there are no consequences for improper disposal, individuals may be less motivated to adopt responsible behaviors. Misinformation or misconceptions about waste treatment methods, such as the belief that certain materials cannot be recycled or composted, can lead to confusion and apathy among the public [40].

- **Limited Access to Infrastructure**: May be another reason, as in some regions, especially in rural or underserved areas, there may be a lack of infrastructure for waste treatment, such as recycling facilities or composting centers. Without access to these facilities, communities may resort to dumping or burning waste, exacerbating environmental pollution.

Addressing the lack of awareness for waste treatment requires a multi-faceted approach, including education, outreach, infrastructure development, regulatory enforcement, and community engagement. By raising awareness, providing accessible information, and promoting the benefits of sustainable waste management practices, individuals and communities can make informed choices and contribute to a cleaner and healthier environment.

1.4 Problems Encountered in Waste Handling

Technological bottlenecks in waste handling refer to challenges or limitations in the application of technology to effectively manage and process waste. These bottlenecks can arise at various stages of waste handling, from collection and sorting to treatment and disposal, which are shown in Figure 1.7.

Figure 1.7 Common technological challenges associated with waste handling. Adapted from [41].

Kumar and Agrawal et al. [41] reported that municipalities face problems in MSWM related to collection efficiency, treatment, and other factors. Inefficient collection systems, stemming from outdated routes or inadequate scheduling, can lead to missed pickups, uneven service distribution, and higher costs. Furthermore, limited integration of technology, such as GPS tracking or route optimization software, can hinder the optimization of collection routes and real-time monitoring.

Manual sorting at recycling facilities is labor-intensive and time-consuming, resulting in inefficiencies and inconsistencies in material recovery. Automated sorting technologies (optical sorters, magnetic separators) struggle to accurately identify and separate mixed or contaminated waste streams. Regarding quality control and contamination, it is estimated that contamination of recyclables with non-recyclables or hazardous substances lowers the quality and value of recovered materials [42, 43]. Existing technologies for contamination detection and quality control may have limited ability to effectively identify and remove contaminants. Processing capacity and efficiency also impact MSWM; limited processing capacity and outdated equipment in waste treatment facilities can create bottlenecks in waste-processing and disposal. Similarly, technological constraints, such as inefficient energy recovery systems or inadequate throughput, can impede the efficient conversion of waste into energy, compost, or recyclables.

Moreover, resource recovery and recycling are the crucial parts of MSWM, yet these are also responsible for waste disposal management such as technological limitations in recycling processes, such as the inability to effectively recycle certain types of plastics or composite materials, can restrict the recovery of valuable resources from waste streams, and lack of investment in research and development of innovative recycling technologies may impede progress toward achieving higher recycling rates and closing material loops [44]. In terms of data management and analytics, the limited use of data-driven technologies in waste management can hinder decision-making, resource allocation, and performance monitoring. Challenges in collecting, integrating, and analyzing data across different waste management systems and stakeholders can impede efforts to optimize waste handling. Additionally, infrastructure development plays a key role; insufficient investment in facilities for waste handling (recycling centers, composting plants, WtE facilities) can limit the adoption and scalability of advanced technologies. Technological bottlenecks may arise from delays or setbacks in planning, permitting, and constructing new infrastructure projects. Thus, addressing technological bottlenecks in waste handling requires investment in research, development, deployment of advanced technologies, capacity building, and collaboration among stakeholders. Overcoming these challenges can improve the efficiency, sustainability, and resilience of waste management systems.

1.5 Economic Competitiveness

Waste is a byproduct of economic actions and related to it by government, household as well as suggested by Balasubramanian et al. [45] and further pointed out that in terms of energy recovery/material, waste is also an input to economic activity. Economic competitiveness in waste management refers to the ability of waste management systems, technologies, and practices to deliver cost-effective solutions while meeting environmental and regulatory requirements. Several factors contribute to the economic competitiveness of waste management, which are given in Table 1.4:

Table 1.4 Factors contributing to economic competitiveness of waste management. Adapted from [45].

Cost efficiency	Resource recovery	Energy generation	Market demand for recyclables
• Waste management operations must be cost-efficient to remain competitive • This includes factors such as collection, transportation, processing, and disposal costs • Investments in technologies and infrastructure that improve operational efficiency can help reduce costs over the long term	• Waste management systems that prioritize resource recovery and recycling can enhance economic competitiveness by capturing the value of materials that can be reused or repurposed • Recycling and recovery facilities can generate revenue from the sale of recovered materials, offsetting some of the costs associated with waste management	• Waste-to-energy technologies, such as incineration or anaerobic digestion, can contribute to economic competitiveness by generating electricity, heat, or biogas from waste materials • Revenue generated from the sale of energy can offset operational costs and provide additional income streams	• Economic competitiveness in waste management is influenced by market demand for recyclable materials • Fluctuations in commodity prices can impact the profitability of recycling operations • Diversification of markets and development of domestic recycling infrastructure can help mitigate market volatility

Blagoeva et al. [46] suggested that waste management contributes to economic growth and development through innovation and technology. Investments in innovation and technology can drive economic competitiveness by improving sorting and processing technologies, automation, and data analytics, thereby enhancing efficiency, reducing operational costs, and improving the quality of recovered materials. Compliance with environmental regulations and waste management standards is essential for economic competitiveness. Non-compliance penalties can lead to additional costs and damage reputation. Proactive measures to meet or exceed regulations can help minimize risks and ensure long-term viability.

Moreover, transitioning to a circular economy, where materials are reused, recycled, and regenerated, can boost economic competitiveness in waste management. Circular economy principles promote resource efficiency, reduce waste generation, and create opportunities for innovation and value creation throughout the supply chain. Collaboration between public and private sectors can enhance economic competitiveness by combining resources, expertise, and innovation. Public–private partnerships can facilitate infrastructure investment, technology deployment, and service delivery, leading to cost savings and better outcomes. Thus, by prioritizing cost-effectiveness, resource recovery, innovation, and collaboration, waste management systems can improve economic competitiveness while advancing environmental sustainability and meeting societal needs.

1.6 Sustainable Challenges

Sustainable challenges in waste management systems encompass a range of issues related to environmental, social, and economic sustainability. These challenges vary depending on factors such as geography, population density, waste composition, and available resources. Common sustainable challenges as adumbrated by Srivastava et al. [47] in waste management are:

1. **Waste Generation and Composition:** Rapid population growth, urbanization, and industrialization contribute to increased waste generation and changing waste composition, posing challenges for waste management systems. The proliferation of single-use plastics and packaging materials exacerbates waste management challenges due to their persistence in the environment and difficulties in recycling.
2. **Waste Minimization and Prevention:** Promoting waste minimization and prevention strategies, such as source reduction, reuse, and repair, is essential for sustainable waste management. Encouraging behavioral changes and consumer awareness to reduce consumption, adopt eco-friendly alternatives, and embrace a circular economy mindset can help mitigate waste generation.
3. **Resource Recovery and Recycling:** Maximizing resource recovery and recycling rates is critical for sustainable waste management. However, challenges such as contamination of recyclable materials, limited markets for recycled products, and inefficient recycling infrastructure hinder progress. Investment in recycling technology, market development, and public education is needed

to improve recycling rates and ensure the economic viability of recycling programs.

4. **Waste Collection and Transportation:** Inefficient waste collection and transportation systems lead to increased energy consumption, GHG emissions, and operational costs. Optimizing collection routes, investing in alternative fuel vehicles, and implementing smart technologies for real-time monitoring and route optimization can improve the efficiency and sustainability of waste collection and transportation.
5. **Waste Treatment and Disposal:** Sustainable waste treatment and disposal methods are essential to minimize environmental pollution, conserve natural resources, and protect public health. Challenges include limited capacity and outdated infrastructure for waste treatment, inadequate regulatory enforcement, and concerns about the environmental and social impacts of incineration and landfilling. Investing in advanced WtE technologies, composting facilities, and hazardous waste treatment plants can help address these challenges and promote sustainable waste management practices.
6. **Community Engagement and Participation:** Engaging communities in waste management initiatives and fostering stakeholder participation is crucial for sustainable waste management. Lack of awareness, cultural attitudes, and socio-economic disparities can hinder community participation and support for waste reduction and recycling programs. Implementing education campaigns, outreach activities, and incentive programs to promote behavior change and community involvement can enhance the sustainability of waste management efforts.
7. **Policy and Governance:** Strong policy frameworks and effective governance structures are essential for promoting sustainable waste management practices. Challenges include inadequate regulation and enforcement, lack of coordination among government agencies and stakeholders, and insufficient funding for waste management initiatives. Strengthening waste management policies, enhancing regulatory compliance, and fostering multi-stakeholder partnerships are key strategies for addressing these challenges and advancing sustainable waste management [48].

Addressing sustainable challenges in waste management requires a holistic approach that integrates environmental, social, and economic considerations. By adopting innovative technologies, engaging stakeholders, promoting waste reduction and recycling, and strengthening policy and governance frameworks, communities can build more resilient and sustainable waste management systems.

1.6.1 Feedstock Availability

Feedstock availability is a crucial consideration in waste management techniques, particularly in processes that involve the conversion of waste materials into valuable products or energy. The availability and composition of feedstock can vary depending on factors such as waste generation rates, waste streams, and regional characteristics [49]. Here's how feedstock availability influences various waste management techniques:

1. **Waste-to-energy:** WtE technologies, such as incineration and gasification, rely on the combustion or thermal conversion of waste materials to generate heat or electricity. Feedstock availability is critical for maintaining consistent energy output and operational efficiency in WtE facilities. The right investment in institutional and technological changes contributing in conversion of waste into sources that take part in socio-economic development [50]. MSW, biomass, and certain industrial wastes can serve as feedstock for WtE processes, but the composition and calorific value of the feedstock may vary, affecting process performance. Jamilatun et al. [51] proposed that composition, characteristics, and amount of waste differ in every country based on growth rate of population, income, and urbanization, and variety of alternative treatments can be carried out depending upon the waste composition such as paper, wood, plastics, cloth, etc., and transformed into energy via WtE technology as shown in Figure 1.8.

   Thus, feedstock availability in the form of waste is an important part of WtE technologies, and in the same manner, these technologies even require consideration of financial, environmental aspects and energy efficiency.
2. **Anaerobic Digestion:** This is a biological process that converts organic waste into biogas (primarily CH_4) and digestate through microbial activity in the absence of oxygen. Feedstock for AD includes organic waste streams such as food waste, agricultural residues, sewage sludge, and animal manure. The availability and quality of organic feedstock influence biogas production rates, CH_4 content, and nutrient content in the digestate. Haque et al. [52] indicated that AD is a widely used technology for converting organic waste into valuable products due to its cost-effectiveness and applicability. However, its success depends on operating conditions and feedstock properties, which affect the quantity and quality of biogas and digestate produced. Supplementary processes like pyrolysis, hydrothermal liquefaction, and pretreatment methods can enhance biogas

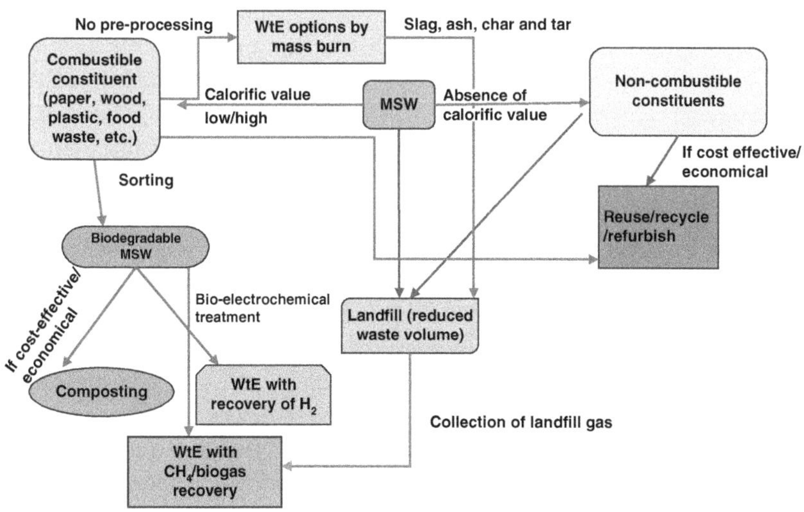

Figure 1.8 Technologies of waste treatment. Adapted from [52].

production and yield valuable byproducts, although challenges related to high operating temperatures and feedstock composition remain. Furthermore, understanding the composition and generation rate of MSW feedstock characteristics is necessary for designing full-scale biodigesters for municipal use and assessing environmental impact [53].

3. **Composting:** Composting is a biological process where organic waste decomposes into nutrient-rich compost through microbial activity. Feedstock for composting includes organic materials like food scraps, yard waste, paper products, and agricultural residues. A proper balance of carbon (C) and nitrogen (N) ratios, moisture content, and particle size in the feedstock is essential for efficient composting and high-quality compost production. Cao et al. [54] proposed that feedstock characteristics (seasonality, maturation regimes) and processing technologies are key factors that improve the composition of MSW compost and the production of good-quality compost by influencing feedstock selection/screening. Additionally, industrialization and urbanization lead to complex MSW composition with more diverse feedstock and larger logistical networks, further contributing to the variety of compost properties.
4. **Recycling:** Recycling processes involve collecting, sorting, and processing recyclable materials into raw materials or products for reuse. Feedstock for recycling includes materials like paper, plastics, glass, metals, and e-waste [42–55]. The availability of high-quality recyclable feedstock relies on effective waste collection and sorting systems, as well as consumer participation and material recovery rates [56].
5. **Biofuels Production:** Biofuels, such as biodiesel and ethanol, can be produced from organic feedstock like vegetable oils, animal fats, crop residues, and dedicated energy crops. Feedstock availability, composition, and logistics are crucial factors in biofuel production, influencing the process's economics and sustainability. Skaggs et al. [57] emphasized the importance of feedstock in biofuel production, as biofuels are recognized as sustainable renewable energy that can replace fossil fuels and offer environmental benefits. Biofuels are classified into different generations based on feedstock characteristics, and these feedstocks are considered readily available and accessible, contributing to socio-economic growth. Utilizing waste feedstocks helps mitigate controversies related to using food materials for biofuels. Waste materials (including animal manures) contain various types of feedstocks that can be converted into diverse biofuels. Sharma et al. [58] suggested that organic waste (waste feedstock) is an efficient feedstock for biofuel production, offering opportunities across various manufacturing processes and potential for biofuels. Organic waste (food waste, farming waste, paper, corporate garbage, forestry, etc.) is inexpensive and readily available, thereby boosting the economic potential of biofuels and reducing pollution [23, 59].
6. **Advanced Recycling Technologies:** Advanced recycling technologies, such as pyrolysis and chemical recycling, convert plastic waste into valuable products or feedstock for producing new plastics, chemicals, or fuels. Feedstock availability, feedstock quality, and feedstock consistency are key considerations in developing and scaling up advanced recycling processes [60].

In summary, feedstock availability plays a critical role in determining the feasibility, efficiency, and sustainability of various waste management techniques. Ensuring a reliable and consistent supply of suitable feedstock is essential for optimizing process performance and maximizing resource recovery from waste streams. Effective waste management strategies should consider feedstock availability as a fundamental aspect of planning, design, and implementation.

1.6.2 Operational Challenges

Operational challenges in waste management can arise at various stages of the waste management process, from collection and transportation to treatment and disposal. These challenges can impact efficiency, cost-effectiveness, and environmental performance. Some common operational challenges in waste management as suggested by Kwakye et al. [61] are given in Table 1.5:

Table 1.5 Challenges associated with waste management. Adapted from [61].

Common operational challenges	
Inefficient waste collection	• Poor route planning and scheduling can lead to inefficient collection, resulting in missed pickups, uneven service coverage, and higher operational costs • Inconsistent collection frequencies and service levels may cause overflowing bins, illegal dumping, and community dissatisfaction
Contamination of waste streams	• Contamination of waste streams with non-recyclable materials, hazardous substances, or inappropriate waste items can compromise the quality of recyclable materials and increase processing costs • Lack of public awareness, improper sorting practices, and inadequate enforcement of recycling guidelines contribute to contamination challenges
Limited access to collection services	• Rural and remote areas may face challenges in accessing waste collection services due to logistical constraints, long distances, and dispersed populations • Providing equitable access to collection services for underserved communities while addressing cost and operational barriers presents a challenge for waste management providers
Technological constraints	• Outdated or inadequate waste management infrastructure and equipment can hinder operational efficiency and performance • Limited use of technology, such as GPS tracking, route optimization software, and smart sensors, may impede real-time monitoring, data analysis, and decision-making in waste management operations
Waste transportation challenges	• Inefficient waste transportation systems, including outdated fleets, congested roadways, and long travel distances, can increase fuel consumption, emissions, and operational costs • Addressing logistical challenges, optimizing transportation routes, and reducing vehicle idle times are key strategies to improve the efficiency and sustainability of waste transportation

(Continued)

Table 1.5 (Continued)

Common operational challenges	
Capacity constraints in waste treatment facilities	• Limited capacity in waste treatment facilities, such as landfills, incinerators, and recycling plants, can lead to bottlenecks in waste-processing and disposal • Balancing waste generation rates with treatment capacity, implementing diversion strategies, and investing in infrastructure upgrades are necessary to address capacity constraints
Regulatory compliance and permitting	• Compliance with environmental regulations, permitting requirements, and operational standards presents challenges for waste management operators • Navigating complex regulatory frameworks, obtaining permits, and maintaining compliance with evolving regulations require dedicated resources and expertise
Financial sustainability	• Funding constraints, revenue fluctuations, and budgetary pressures pose challenges for achieving financial sustainability in waste management • Balancing service affordability with revenue generation, optimizing cost structures, and exploring alternative funding mechanisms are critical for ensuring long-term financial viability

Addressing operational challenges in waste management requires a comprehensive approach that integrates technology, infrastructure development, regulatory compliance, community engagement, and financial planning. Collaboration among stakeholders, innovation, and continuous improvement are essential for overcoming these challenges and advancing sustainable waste management practices.

1.7 Carbon Sequestration for Green and Sustainable Environment

Carbon sequestration in WtE processes involves capturing and storing carbon dioxide (CO_2) emissions generated during the combustion or thermal conversion of waste. While WtE technologies generate energy, they also release CO_2 and other GHGs, contributing to climate change [62]. Carbon sequestration aims to mitigate these emissions by capturing CO_2 and preventing its release into the atmosphere [63]. Carbon sequestration can be incorporated into WtE processes as follows:

- **Carbon Capture and Storage (CCS):** CCS technologies are specifically designed to intercept CO_2 emissions generated at WtE facilities before they are vented into the atmosphere. Post-combustion capture represents one primary approach, focusing on separating CO_2 from the flue gases downstream of the combustion process, often employing techniques such as absorption or adsorption. Alternatively, pre-combustion capture can be integrated with processes like waste gasification or pyrolysis, where waste is converted into a synthesis gas (syngas), and the CO_2 is removed from this syngas prior to its combustion.

Table 1.6 Processing conditions of CCS Technology. Adapted from [64].

Technology of CCS	Pressure (bar)	Processing temperature (K)	Concentration of CO_2 (%)
Direct air capture	1.0	298	~0.05
Pre-combustion	14.0–70.0	473–723	16–61
Oxy-combustion	1.0	218	17–71
Post-combustion	1.0	313–333	4–21

The captured CO_2 is then typically transported and permanently stored in secure geological formations, such as depleted oil and gas reservoirs or deep saline aquifers, effectively preventing its re-entry into the carbon cycle. Table 1.6 provides an overview of typical operating conditions associated with various CCS technologies [64]. Table 1.6 summarizes the operating conditions of CCS technology.

- **Carbon Capture and Utilization:** In parallel with storage, Carbon Capture and Utilization (CCU) technologies offer an avenue to convert captured CO_2 into commercially valuable products or industrial feedstocks. This approach provides an economic incentive that can support the implementation of carbon capture systems. One significant application of captured CO_2 is in mineralization processes, where it is reacted with minerals to form stable, long-term carbonate compounds. Beyond mineralization, captured CO_2 can also be leveraged in various other applications, including the synthesis of fuels, production of chemicals, polymerization processes, and manufacturing of building materials, effectively integrating waste-derived carbon into new product value chains.
- **Bioenergy with Carbon Capture and Storage (BECCS):** BECCS represents a synergistic approach that couples the generation of energy from biomass with carbon capture and geological storage. This method utilizes biomass feedstocks, which can include organic waste, agricultural residues, or dedicated energy crops. A key advantage of using biomass is that it absorbs CO_2 from the atmosphere during its growth phase. When this biomass is subsequently converted into energy and the resulting CO_2 emissions are captured and stored, the overall process can potentially achieve net negative CO_2 emissions, meaning more CO_2 is removed from the atmosphere than is released. Consequently, BECCS holds significant potential to offsct CO_2 emissions generated not only by WtE processes but also from other sources, thereby contributing substantially to climate change mitigation efforts.
- **Enhanced Landfill CH_4 Capture:** Landfills constitute a notable source of CH_4 emissions, which is a potent GHG significantly contributing to climate change. This CH_4 is primarily produced through the anaerobic decomposition of organic waste materials buried within landfill sites. Enhanced LFG capture systems are specifically designed to collect these CH_4 emissions effectively. These systems can be readily coupled with WtE facilities, enabling the captured CH_4 to be utilized for energy production. By capturing and converting this potent GHG into usable energy, enhanced landfill CH_4 capture plays a

vital role in reducing overall GHG emissions and advancing climate change mitigation.

- **Carbon Offsetting and Carbon Credits:** Market-based mechanisms, such as carbon offsetting and carbon credits, provide financial frameworks to support emission reduction initiatives. WtE facilities that successfully implement carbon sequestration technologies may be eligible to generate carbon offset credits or participate in carbon trading schemes. These credits offer a financial incentive that can enhance the economic feasibility of emission reduction projects. Carbon offsetting, facilitated by these credits, allows organizations to compensate for their unavoidable emissions by investing in projects elsewhere that reduce or sequester an equivalent amount of CO_2, such as those implemented at WtE facilities equipped with CCS capabilities.

Integrating carbon sequestration technologies into WtE processes can help mitigate GHG emissions, reduce the carbon footprint of waste management systems, and contribute to climate change mitigation efforts. However, challenges related to technological feasibility, cost-effectiveness, and regulatory requirements must be addressed to fully realize the potential of carbon sequestration in WtE operations.

1.8 IoT Services

According to Tamakloe et al. [65], the integration of IoT technologies in WtE techniques offers several benefits, including improved efficiency, real-time monitoring and data-driven decision-making, and proposed smart waste management design as shown in Figure 1.9.

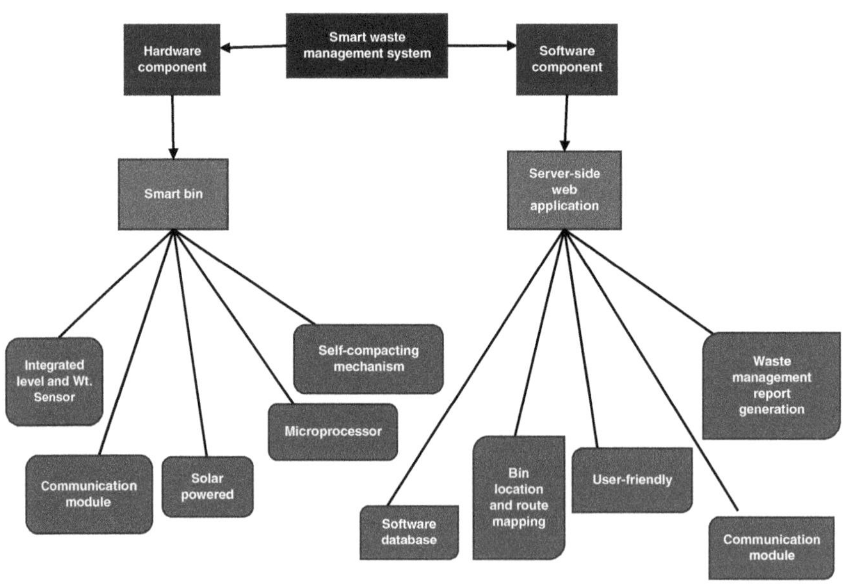

Figure 1.9 Design component structure of waste management. Adapted from [65].

As shown in Figure 1.9, the hardware design implementation involves efficient coordination of sensors and microprocessors for effective system function. The software design utilizes a developed web application based on the Adonis JS web framework, representing a modern waste management system. As recommended by Khan et al. [66], the integration of IoT technologies within waste management significantly enhances operational efficiency across collection, routing, and equipment monitoring. At the level of individual bins, IoT sensors provide real-time data on fill levels, temperature, and other parameters. This sensor data is leveraged to dynamically optimize waste collection routes and schedule pickups precisely according to necessity, thereby preventing both overflowing bins in dense areas and inefficient trips to underutilized containers. Extending this capability to transportation, IoT-enabled fleet management systems offer real-time tracking of waste collection vehicles, monitoring their location, operational status, and performance. By synthesizing data from bins, vehicles, and traffic conditions, sophisticated route optimization algorithms can be applied to enhance logistical efficiency, resulting in reduced fuel consumption and minimized associated emissions. Furthermore, IoT facilitates the remote monitoring of waste-processing equipment and facilities, including incinerators, gasifiers, and digesters. Sensors placed on these assets continuously monitor operational parameters and performance metrics, providing essential data for performance assessment and proactive maintenance planning. This remote oversight enables prompt detection of equipment anomalies, identification of inefficiencies, and scheduling of preventive repairs, thereby minimizing unplanned downtime and optimizing facility throughput.

Beyond direct operational improvements, IoT applications contribute significantly to energy management, environmental oversight, and predictive capabilities within waste management systems. IoT-based energy management systems offer comprehensive monitoring of energy consumption, generation, and distribution at WtE facilities. Utilizing smart meters and sensors, these systems enable the optimization of energy usage, allow for adaptive production adjustments based on demand, and pinpoint opportunities for efficiency enhancements. Environmental monitoring and regulatory compliance are also bolstered by IoT integration. Sensors measure key parameters such as air quality, GHG emissions, and other environmental indicators. Continuous Emissions Monitoring Systems (CEMSs), often IoT-enabled, track pollutant levels, detect non-compliance anomalies, and trigger alerts for necessary corrective actions. Operational data, including historical trends, is analyzed using IoT data analytics platforms to identify performance patterns and predict potential equipment failures, enabling a shift toward predictive maintenance strategies. These algorithms prioritize maintenance tasks and schedule interventions proactively, further reducing unscheduled downtime. Finally, IoT-enabled WtE systems can integrate with smart grids and energy markets. This connectivity allows for the optimization of energy production, storage, and distribution in response to real-time grid conditions, market prices, and renewable energy availability, contributing to grid stability and potentially increasing revenue streams.

By leveraging IoT services, WtE techniques can achieve greater efficiency, reliability, and sustainability while reducing operational costs and environmental impact. However, challenges such as data security, interoperability, and scalability need to be addressed to realize the full potential of IoT in WtE applications.

1.9 Smart City Infrastructure

WtE plays a significant role in smart city infrastructure by contributing to sustainability, resilience, and resource efficiency [67]. WtE technologies occupy a significant and multifaceted role within the development of smart urban infrastructure, contributing fundamentally to sustainability, resilience, and overall resource efficiency [60]. A primary contribution lies in energy production and security; WtE facilities convert organic waste into various energy carriers – electricity, heat, or biofuels – decreasing reliance on fossil fuels and enhancing energy independence. Furthermore, their capacity for distributed energy generation and integration with smart grids allows for optimized production, distribution, and consumption, bolstering the reliability and efficiency of the urban energy network and contributing to grid stability by balancing supply and demand. Beyond energy, WtE drives significant advancements in waste management optimization. Providing a vital alternative to traditional landfilling and open burning, these technologies minimize environmental impacts and substantially reduce landfill CH_4 emissions. This aligns with circular economy principles, as WtE facilities facilitate resource recovery by extracting energy and valuable materials such as metals, organic matter, and plastics from waste streams. When integrated with smart waste management systems, WtE enables real-time monitoring, optimized collection routes, and more efficient utilization of waste resources, supporting conservation efforts within the smart city framework. The environmental and public health benefits are also profound; by treating waste that would otherwise decompose and release CH_4 or be combusted inefficiently, WtE processes contribute directly to climate change mitigation by reducing GHG emissions, with advanced techniques like carbon capture and utilization/storage (CCUS) potentially amplifying these benefits. Moreover, modern WtE facilities employ sophisticated pollution control technologies that significantly minimize emissions of air pollutants, particulate matter, and hazardous substances, leading to improved air quality and contributing positively to public health and overall environmental sustainability within urban areas. The implementation of smart WtE systems fosters data-driven decision-making by leveraging sensors, advanced data analytics, and predictive algorithms to optimize operational performance, maintenance scheduling, and energy efficiency, providing necessary insights for proactive management. Finally, successful WtE initiatives actively promote community engagement and public education, raising awareness about waste management, energy recovery, and sustainability, which cultivates a sense of shared ownership and responsibility essential for achieving sustainable development goals in future smart cities.

By leveraging WtE technologies and integrating them with smart city infrastructure, urban areas can achieve greater sustainability, resilience, and livability while addressing challenges related to waste management, energy security, and climate change mitigation.

1.10 Conclusion and Discussion

In conclusion, integrating WtE technologies into smart city infrastructure offers significant benefits across multiple dimensions of sustainability, resilience, and resource efficiency. By converting organic waste into electricity, heat, or biofuels, WtE facilities contribute to energy security, reduce GHG emissions, and promote the transition to a circular economy. Furthermore, WtE plays a key role in optimizing waste management processes, enhancing resource recovery, and mitigating environmental pollution in urban areas. The synergy between WtE and smart technology initiatives facilitates data-driven decision-making, real-time monitoring, and optimization of energy production, waste collection, and grid integration. By harnessing IoT technologies, advanced analytics, and smart grid capabilities, WtE systems improve operational efficiency, reduce environmental impacts, and encourage community engagement in sustainable waste management practices. Overall, the integration of WtE represents a holistic approach to urban sustainability, tackling interconnected challenges in waste management, energy production, climate change mitigation, and public health. Through cooperation among government bodies, private sector stakeholders, and local communities, cities can utilize WtE technologies to build more resilient, livable, and sustainable urban environments for present and future generations.

References

1 Das, A., Li, H., Kataki, R., Agrawal, P.S., Moyon, N.S., Gurunathan, B., et al. Terminalia arjuna bark – A highly efficient renewable heterogeneous base catalyst for biodiesel production. *Renew. Energy*. 2023. https://doi.org/10.1016/j.renene.2023.05.066.

2 Saikia, K., Rajkumari, K., Moyon, N.S., Basumatary, S., Halder, G., Rashid, U., et al. Sulphonated biomass-based catalyst for solketal synthesis by acetalization of glycerol – A byproduct of biodiesel production. *Fuel Process. Technol.* vol. 238, no.15, pp. 107482, Dec. 2022. https://doi.org/10.1016/J.FUPROC.2022.107482.

3 Ludlow, J., Jalil-Vega, F., Rivera, X.S., Garrido, R.A., Hawkes, A., Staffell, I., et al. Organic waste to energy: resource potential and barriers to uptake in Chile. *Sustain. Prod. Consum.* 2021. https://doi.org/10.1016/j.spc.2021.08.017.

4 Maalouf, A. and Mavropoulos, A. Re-assessing global municipal solid waste generation. *Waste Manag. Res.* 2023. https://doi.org/10.1177/0734242X221074116.

5 Chadar, S.N. Solid waste pollution: a hazard to environment. *Recent. Adv. Petrochemical. Sci.* 2017. https://doi.org/10.19080/rapsci.2017.02.555586.

6 Zolnikov, T.R., Furio, F., Cruvinel, V. and Richards, J. A systematic review on informal waste picking: occupational hazards and health outcomes. *Waste. Manag.* 2021. https://doi.org/10.1016/j.wasman.2021.03.006.

7 Nanda, M.A., Sugandi, W., Wijayanto, A.K., Imantho, H., Sutawijaya, A., Nelwan, L.O., et al. The Waste-to-Energy (WtE) technology to support alternative fuels for agriculture in the context of effective solid waste management in the Jabodetabek Area, Indonesia. *Energies*. 2023. https://doi.org/10.3390/en16247980.

8 Bywater, A., Heaven, S., Zhang, Y. and Banks C. J. Potential for biomethanisation of CO2 from anaerobic digestion of organic wastes in the United Kingdom. *Processes*. 2022. https://doi.org/10.3390/pr10061202.

9 Zhang, Y. Heaven, S. and Banks, C. J. Validation of two theoretically derived equations for predicting pH in CO2 biomethanisation. *Processes*. 2023. https://doi.org/10.3390/pr11010113.

10 Makgato, S.S. Roadmap towards clean coal technology for sustainable power sector in South Africa. *Chem. Eng. Trans*. 2023. https://doi.org/10.3303/CET2399038.

11 Skubko, Y. ESKOM and the problems of South African energy. *Uchenie Zap. Instituta Afriki RAN*. 2023. https://doi.org/10.31132/2412-5717-2023-65-4-31-43.

12 Lim, H., Kwon, I.S., Lee, H. and Park, J.W. Environmental impact assessment using a GSR tool for a landfarming case in South Korea. *Environ. Monit. Assess*. 2016. https://doi.org/10.1007/s10661-016-5243-1.

13 Adar, E., Karatop, B., Ince, M. and Bilgili, M.S. Comparison of methods for sustainable energy management with sewage sludge in Turkey based on SWOT-FAHP analysis. *Renew. Sustain. Energy Rev*. 2016. https://doi.org/10.1016/j.rser.2016.05.007.

14 Tian, H., Wang, X., Lim, E.Y., Lee, J.T., Ee, A.W., Zhang, J., et al. Life cycle assessment of food waste to energy and resources: centralized and decentralized anaerobic digestion with different downstream biogas utilization. *Renew. Sustain. Energy Rev*. 2021. https://doi.org/10.1016/j.rser.2021.111489.

15 Wang, Y., Pan, S., Yin, J., Feng, H., Wang, M. and T. Chen. Resource potential and global warming potential of fruit and vegetable waste in China based on different treatment strategies. *Waste Manag*. 2022. https://doi.org/10.1016/j.wasman.2021.11.016.

16 Pheakdey, D.V., Van Quan, N. and Xuan. D. Economic and environmental benefits of energy recovery from municipal solid saste in Phnom Penh Municipality, Cambodia. *Energies*. 2023. https://doi.org/10.3390/en16073234.

17 Kashid, M.N., Kiwi-Minsker, L., Gehr, R., Chen, D., Moreau, M., Kikutani, Y., et al. Peracetic acid (CAS No. 79-21-0) and its equilibrium solutions. *Chem. Eng. Sci*. vol. 2, no. 1, pp. 1–6, 2011.

18 Sulaiman, A., Othman, N., Baharuddin, A.S., Mokhtar, M.N. and Tabatabaei, M. Enhancing the halal food industry by utilizing food wastes to produce value-added bioproducts. *Procedia - Soc. Behav. Sci*. 2014. https://doi.org/10.1016/j.sbspro.2014.01.1106.

19 Slorach, P.C., Jeswani, H.K., Cuéllar-Franca, R. and Azapagic, A. Assessing the economic and environmental sustainability of household food waste management in the UK: current situation and future scenarios. *Sci. Total Environ*. 2020. https://doi.org/10.1016/j.scitotenv.2019.135580.

20 Senanayake, D., Reitemeier, M., Thiel, F. and Drechsel. P. Business models for urban food waste prevention, redistribution, recovery and recycling. *Resour. Recover. Reuse*. 2021. https://doi.org/10.5337/2021.208.

21 Jungmeier, G. The biorefinery fact sheet. *Int. J. Life Cycle Assess*. 2017. vol. 23, pp. 1788–805.

22 Wang, C., Liu, D., Li, Y., Wang, L. and Gu, W. A multidisciplinary perspective on the evolution of municipal waste management through text-mining: a mini-review. *Waste Manag. Res.* 2021. https://doi.org/10.1177/0734242X20962841.

23 Laskar, I.B., Deshmukhya, T., Biswas, A., Paul, B., Changmai, B., Gupta, R., et al. Utilization of biowaste-derived catalysts for biodiesel production: process optimization using response surface methodology and particle swarm optimization method. *Energy Adv.* 2022. https://doi.org/10.1039/d2ya00011c.

24 AlQattan, N., Acheampong, M., Jaward, F.M., Ertem, F.C., Vijayakumar, N. and Bello, T. Reviewing the potential of Waste-to-Energy (WTE) technologies for Sustainable Development Goal (SDG) numbers seven and eleven. *Renew. Energy Focus.* 2018. https://doi.org/10.1016/j.ref.2018.09.005.

25 Alao, M.A., Popoola, O.M. and Ayodele, T.R. Waste-to-energy nexus: An overview of technologies and implementation for sustainable development. *Cleaner Energy Systems.* 2022. https://doi.org/10.1016/j.cles.2022.100034.

26 Tiwari, R., Agrawal, P.S., Belkhode, P.N., Brijpuriya, D., Devase, A. and Samarth, P. Utilization of waste transformer oil for the synthesis of biodiesel and glyptal. In: Yadav, K.D., Jariwala, N.D., Kumar, A., Sinha, A. (eds) *Recent Advances in Sustainable Waste Management Practices. SWMP 2023. Lecture Notes in Civil Engineering*, vol 430. Springer, Singapore. 2024. https://doi.org/10.1007/978-981-99-4186-5_7.

27 Akpan, V.E. and Olukanni, D.O. Hazardous waste management: an African overview. *Recycling.* 2020. https://doi.org/10.3390/recycling5030015.

28 Exposto L.A.S.M. and Sujaya, I.N. The impacts of hazardous and toxic waste management: a systematic review. *Interdiscip. Soc. Stud.* 2021. https://doi.org/10.55324/iss.v1i2.20.

29 Saleh, H.E.-D.M. Introductory Chapter: introduction to hazardous waste management. In: *Management of Hazardous Wastes.* IntechOpen, 2016. https://doi.org/10.5772/64245.

30 Kumar, A., Thakur, A.K., Gaurav, G.K., Klemeš, J.J., Sandhwar, V.K., Pant, K.K., et al. A critical review on sustainable hazardous waste management strategies: a step towards a circular economy. *Environ. Sci. Pollut. Res.* 2023. https://doi.org/10.1007/s11356-023-29511-8.

31 Mwanza, B.G., Mbohwa, C. and Telukdarie, A. The influence of waste collection systems on resource recovery: a review. *Procedia Manuf.* 2018. https://doi.org/10.1016/j.promfg.2018.02.192.

32 Abubakar, I.R., Maniruzzaman, K.M., Dano, U.L., AlShihri, F.S., AlShammari, M.S., Ahmed, S.M.S., et al. Environmental sustainability impacts of solid waste management practices in the global south. *Int J Environ Res Public Health.* 2022. https://doi.org/10.3390/ijerph191912717.

33 Mousavi, S., Hosseinzadeh, A. and Golzary, A. Challenges, recent development, and opportunities of smart waste collection: a review. *Sci. Total. Environ.* 2023. https://doi.org/10.1016/j.scitotenv.2023.163925.

34 Keerthana, S., Kiruthika, B., Lokeshvaran, R., Midhunchakkaravarthi, B. and Dhivyasri, G.A. review on smart waste collection and disposal system. *J. Phys. Conf. Ser.* 2021. https://doi.org/10.1088/1742-6596/1969/1/012029.

35 Kumar, S., Smith, S.R., Fowler, G., Velis, C., Kumar, S.J., Arya, S., et al. Challenges and opportunities associated with waste management in India. *R. Soc. Open Sci.* 2017. https://doi.org/10.1098/rsos.160764.

36 Hemidat, S., Achouri, O., El Fels, L., Elagroudy, S., Hafidi, M., Chaouki, B., et al. Solid waste management in the context of a circular economy in the MENA region. *Sustain.* 2022. https://doi.org/10.3390/su14010480.

37 Fakorede, C.O., Anguruwa, G.T., Ajayi, O.B. and Odega, C.A. Waste disposal practices and its environmental implications. *Niger. J. Environ. Sci. Technol.* 2021. https://doi.org/10.36263/nijest.2021.02.0284.

38 Agamuthu, P. and Babel, S. Waste management developments in the last five decades: Asian perspective. *Waste Manag. Res.* 2023. https://doi.org/10.1177/0734242X231199938.

39 Hasan, M.J.A., Hanafiah, M.M. and Satchet, M.S. Public awareness on solid waste management: a case study in Al-Nassyriah City, Iraq. *AIP Conf. Proc.* 2019. https://doi.org/10.1063/1.5111275.

40 Debrah, J.K., Vidal, D.G. and Dinis, M.A.P. Raising awareness on solid waste management through formal education for sustainability: a developing countries evidence review. *Recycling*. 2021. https://doi.org/10.3390/recycling6010006.

41 Kumar, A. and Agrawal, A. Recent trends in solid waste management status, challenges, and potential for the future Indian cities – A review. *Curr Res Environ Sust.* 2020. https://doi.org/10.1016/j.crsust.2020.100011.

42 Agrawal, P.S., Belkhode, P.N. and Rokhum, S.L. Agro-waste-produced biofuels. In: *Sustainability in Biofuel Production Technology*. 2022. https://doi.org/10.1002/9781119888864.ch7.

43 Laskar, I.B., Gupta, R., Chatterjee, S., Vanlalveni, C. and Rokhum, S.L. Taming waste: waste Mangifera indica peel as a sustainable catalyst for biodiesel production at room temperature. *Renew. Energy*. 2020. https://doi.org/10.1016/j.renene.2020.07.061.

44 Abdel-Shafy, H.I. and Mansour, M.S.M. Solid waste issue: sources, composition, disposal, recycling, and valorization. *Egypt. J. Pet.* 2018. https://doi.org/10.1016/j.ejpe.2018.07.003.

45 Balasubramanian, M. Economics of solid waste management: a review. In: *Strategies of Sustainable Solid Waste Management*. 2021. https://doi.org/10.5772/intechopen.95343.

46 Blagoeva, N., Georgieva, V. and Dimova, D. Relationship between GDP and municipal waste: regional disparities and implication for waste management policies. *Sustain.* 2023. https://doi.org/10.3390/su152115193.

47 Srivastava, R.R., Rajak, D.K., Ilyas, S., Kim, H. and Pathak, P. Challenges, regulations, and case studies on sustainable management of industrial waste. *Minerals*. 2023. https://doi.org/10.3390/min13010051.

48 Fidje, E., Haddara, M. and Langseth, M. Sustainable smart waste management adoption challenges in developing countries. *Procedia Comput. Sci.* 2023. https://doi.org/10.1016/j.procs.2023.01.366.

49 Shah, K.J., Tripathi, S., Hussain, T. and You, Z. Feedstock: a solution to energy and environment sustainability. *Curr. Chinese Sci.* 2021. https://doi.org/10.2174/2210298102666211230103252.

50 Alam, S., Rahman, K.S., Rokonuzzaman, M., Salam, P.A., Miah, M.S., Das, N., et al. Selection of waste to energy technologies for municipal solid waste management—towards achieving sustainable development goals. *Sustain.* 2022. https://doi.org/10.3390/su141911913.

51 Jamilatun, S. Pitoyo, J. and Setyawan. M. Technical, economic, and environmental review of waste to energy technologies from municipal solid waste. *J. Ilmu Lingkung.* 2023. https://doi.org/10.14710/jil.21.3.581-593.

52 Haque, F. Fan, C. and Lee, Y.Y. From waste to value: addressing the relevance of waste recovery to agricultural sector in line with circular economy. *J. Clean. Prod.* 2023. https://doi.org/10.1016/j.jclepro.2023.137873.

53 Agbede, O.O., Aworanti, O.A., Osuolale, F.N., Adebayo, A.O., Ogunleye, O.O., Agarry, S.E., et al. Anaerobic conversion of biodegradable municipal solid waste to biogas: a review. *LAUTECH J. Civ. Environ. Stud.* 2019. https://doi.org/10.36108/laujoces/9102/20(0230).

54 Cao, X., Williams, P.N., Zhan, Y., Coughlin, S.A., McGrath, J.W., Chin, J.P., et al. Municipal solid waste compost: global trends and biogeochemical cycling. *Soil Environ Health.* 2023. https://doi.org/10.1016/j.seh.2023.100038.

55 Anli Dino, A., Kishore, G., Rokhum, S.L. and Baskar, G. Bioplastic production using waste macroalgal biomass: a holistic review on challenges, prospects, economic viability and sustainability analysis. *J. Environ. Chem. Eng.* vol. 13, no. 2, pp. 116108, Apr. 2025. https://doi.org/10.1016/j.jece.2025.116108.

56 Peng, X., Jiang, Y., Chen, Z., Osman, A.I., Farghali, M., Rooney, D.W., et al. Recycling municipal, agricultural and industrial waste into energy, fertilizers, food and construction materials, and economic feasibility: a review. *Environ. Chem. Lett.* 2023. https://doi.org/10.1007/s10311-022-01551-5.

57 Skaggs, R.L., Coleman, A.M., Seiple, T.E. and Milbrandt, A.R. Waste-to-Energy biofuel production potential for selected feedstocks in the conterminous United States. *Renew. Sustain. Energy Rev.* 2018. https://doi.org/10.1016/j.rser.2017.09.107.

58 Sharma, A., Srikavya, B.P., Urade, A.D., Joshi, A., Narain, R.S., Dwarakanath, V., et al. Economic and environmental impacts of biofuels in Indian context. *Mater. Today Proc.* 2023. https://doi.org/10.1016/j.matpr.2023.05.118.

59 Kishor, N. and Agrawal, P.S. Nitration of glycerol-A byproduct of biodiesel production. *Mater. Today Proc.* 2020. https://doi.org/10.1016/j.matpr.2020.03.065.

60 Kunlere, I.O. and Shah, K.U. A recycling technology selection framework for evaluating the effectiveness of plastic recycling technologies for circular economy advancement. *Circ. Econ.*, 2023. https://doi.org/10.1016/j.cec.2023.100066.

61 Kwakye, S.O., Amuah, E.E.Y., Ankoma, K.A., Agyemang, E.B. and Owusu, B.G. Understanding the performance and challenges of solid waste management in an emerging megacity: insights from the developing world. *Environ. Challenges.* 2024. https://doi.org/10.1016/j.envc.2023.100805.

62 Tiefenbacher, A., Sandén, T., Haslmayr, H.P., Miloczki, J., Wenzel, W. and Spiegel, H. Optimizing carbon sequestration in croplands: a synthesis. *Agronomy.* 2021. https://doi.org/10.3390/agronomy11050882.

63 Zhang, Y., Zhao, S., Li, L., Feng, J., Li, K., Huang, Z., et al. Integrated CO2 capture and utilization: a review of the synergistic effects of dual function materials. *Catal. Sci. Technol.* 2024. https://doi.org/10.1039/d3cy01289a.

64 Lai, J.Y., Ngu, L.H. and Hashim, S.S. A review of CO2 adsorbents performance for different carbon capture technology processes conditions. *Greenhouse Gases: Sci. Technol.* 2021. https://doi.org/10.1002/ghg.2112.

65 Tamakloe, C.N. and Rosca, E.V. Smart systems and the Internet of Things (IOT) for waste management. In: *2020 IEEE International Conference on Computational Intelligence and Virtual Environments for Measurement Systems and Applications (CIVEMSA)*. 2020. https://doi.org/10.1109/CIVEMSA48639.2020.9132968.

66 Khan, R., Kumar, S., Srivastava, A.K., Dhingra, N., Gupta, M., Bhati, N., et al. Machine learning and IoT-Based waste management model. *Comput. Intell. Neurosci.* 2021. https://doi.org/10.1155/2021/5942574.

67 Octavianthy, D. and Purwanto, W.W. Designing smart energy system for smart city through municipal solid waste to wlectricity: techno-economic analysis. In: *E3S web of conferences*. 2018. https://doi.org/10.1051/e3sconf/20186702044.

2

A Perspective on the Emergence and Need for Alternate Fuels

2.1 Introduction

Recent years have seen a significant increase in the demand for both vehicles and related services, which has consequently driven up the demand for fuel to power these vehicles and meet energy requirements using conventional fuels. The transportation sector is a major contributor to global greenhouse gas (GHG) emissions, accounting for approximately 30–35% of the total, primarily due to the combustion of fossil fuels releasing carbon dioxide (CO_2). CO_2 exacerbates the greenhouse effect (solar heat trapped in the atmosphere). Other harmful gases emitted, such as methane (CH_4) and NOx, also contribute to global warming and climate severity.

The economic development of countries is generally influenced by energy-related expenses. Various approaches have been explored to generate energy, with synthetic fuels being evaluated for their utility [1, 2]. Relying solely on 100% renewable energy presents complex economic and technical challenges. However, achieving certain pre-established targets is possible through energy consumption savings, improvements in energy efficiency, and the development of diverse renewable energy sources [3]. Global oil reserves are depleting at an estimated rate of 4 billion tons annually, suggesting exhaustion by 2060 if this rate continues. Therefore, the current global objective is to reduce the carbon footprint and emissions from fossil fuels, with countries committed to transitioning to sustainable alternative fuels (AFs) by 2030–2050. The transportation sector, being a major source of emissions and pollution, is specifically targeted, with the European Union having already proposed an action plan to cut GHG emissions from this sector.

The move toward AFs is driven by the combined force of environmental, economic, and social imperatives. By transitioning from traditional fossil fuels to cleaner, more sustainable options, we can help mitigate climate change, enhance energy security, foster social equity, and build more prosperous and resilient societies. However, realizing this vision requires coordinated efforts and investments from governments, businesses, and individuals to accelerate the shift toward a low-carbon future. The urgency for this transition away from conventional fossil fuels is underscored by several interconnected factors [4].

Agro-Waste Management and Valorization, First Edition. Pratibha S. Agrawal, et al.

Firstly, environmental concerns are paramount. The combustion of fossil fuels – such as coal, oil, and natural gas – releases substantial amounts of GHGs into the atmosphere, particularly CO_2, a primary driver of climate change. Unchecked climate change leads to consequences like rising global temperatures, melting polar ice caps, more frequent and intense weather events, and disruptions to ecosystems and biodiversity. AFs provide a path to reduce these emissions, thereby lessening the severe impacts of climate change and preserving the planet's habitability for future generations.

Secondly, the economic arguments for AFs are compelling. Traditional fossil fuels are finite resources prone to price volatility and geopolitical tensions. As global energy demand rises, especially in rapidly industrializing nations, the pressure on fossil fuel reserves increases, leading to price fluctuations and supply interruptions. AFs, including solar, wind, hydroelectric, and biofuels, offer a more sustainable and diverse energy mix, decreasing reliance on imported oil and improving energy security. Furthermore, investing in renewable energy technologies stimulates innovation and job creation, boosting economic growth and global competitiveness [5].

Thirdly, the social drivers for AFs arise from concerns about energy access, fairness, and public health. In many areas globally, especially in rural and underserved communities, access to reliable and affordable energy remains limited. By decentralizing energy production and harnessing local renewable resources, AFs can extend energy access to those in need, empowering communities and enhancing quality of life. Additionally, burning fossil fuels contributes to air and water pollution, posing significant risks to public health, particularly in urban environments. Adopting cleaner AFs can alleviate these health burdens, reducing respiratory illnesses, cardiovascular diseases, and premature deaths linked to pollution exposure.

Furthermore, the increasing adoption of AFs aligns with broader societal trends toward sustainability and environmental stewardship. Consumers are increasingly seeking eco-friendly products and services, prompting businesses and governments to adopt greener practices and policies. AFs offer a tangible solution to meet these demands, allowing individuals and organizations to lower their carbon footprint and contribute to a more sustainable future [6]. Whether through electric vehicles, renewable energy grids, or biodegradable biofuels, AFs present versatile and scalable solutions for the complex challenges of the twenty-first century. Consequently, this chapter examines the difficulties associated with fossil fuels, their environmental impact, the emergence and various types of AFs, their application in engines, and their environmental and economic considerations.

2.2 Challenges Associated with Conventional Fuels

Fossil fuels are currently crucial for global energy needs, fulfilling approximately 75–81% of worldwide energy consumption. The development of many countries also relies heavily on their energy systems. Fossil fuels are hydrocarbons formed from the ancient remains of plants and animals, found within the Earth's crust as non-renewable resources. Oil, coal, and natural gas are types of fossil fuels originating from organic matter over millions of years, extracted through drilling or mining. However, the use of fossil fuels leads to significant environmental issues such as air pollution and global warming, negatively impacting living organisms and presenting considerable environmental and economic challenges.

These challenges include the urgent need to reduce GHG emissions to mitigate climate change. Burning fossil fuels pollutes the air by releasing particulate matter, polycyclic aromatic hydrocarbons (PAHs), sulfur oxides, and nitrogen oxides. Nitrogen oxides in the air contribute to acid rain and smog formation, and their excessive accumulation on land and entry into water bodies leads to water pollution, promoting the growth of toxic algal blooms and oxygen-depleted aquatic zones, which harm aquatic life. Moreover, human activities, primarily through energy use, release around 30–36 billion metric tons of carbon annually, making carbon emission reduction a significant challenge. Beyond health and environmental issues, the uneven distribution of fossil fuels raises concerns about energy security due to their non-renewable nature and limited future availability. Market price unpredictability is another major disadvantage, negatively affecting economies [7].

Awareness regarding the need to reduce dependence on fossil fuels has been increasing, putting pressure on civil society organizations, governments, and businesses to work toward a transition to renewable energy sources and lower GHG emissions. Conventional fuels, predominantly coal, oil, and natural gas, face numerous challenges in the modern era, arising from environmental, economic, and geopolitical factors. Environmental concerns are prominent; burning fossil fuels releases GHGs like CO_2, CH_4, and nitrous oxide (N_2O) into the atmosphere, intensifying climate change and its effects. Rising global temperatures, melting ice caps, more frequent severe weather events, and ecosystem disruptions highlight the urgent need to reduce reliance on conventional fuels. Furthermore, the extraction and transport of conventional fuels can cause environmental degradation, habitat destruction, and air, water, and soil pollution, compounding ecological harm [8].

Secondly, conventional fuels face economic difficulties, including price volatility, supply constraints, and geopolitical tensions. As finite resources, fossil fuels are susceptible to fluctuations in global demand and supply, causing price spikes and market instability. Dependence on imported oil also exposes countries to geopolitical risks, as conflicts or disruptions can arise over resources, transit routes, and market access. Additionally, the long-term costs associated with conventional fuels, including subsidies, externalities, and environmental cleanup, can impose heavy economic burdens on societies, hindering competitiveness and sustainability.

Thirdly, conventional fuels present social challenges related to energy access, equity, and public health. In many regions, particularly developing countries, limited access to reliable and affordable energy persists, perpetuating poverty and inequality. Moreover, the detrimental health effects of conventional fuel use, such as air pollution, respiratory illnesses, and premature deaths, disproportionately affect vulnerable populations, including children, the elderly, and low-income communities. Reliance on conventional fuels also reinforces dependence on centralized energy systems, limiting opportunities for decentralized energy production, community empowerment, and energy justice.

Furthermore, conventional fuels encounter technological challenges, particularly as advancements in renewable energy, energy storage, and efficiency technologies offer increasingly viable alternatives. Renewable energy sources like solar, wind, hydroelectric, and geothermal power provide cleaner and more sustainable options compared to conventional fuels, with rapidly decreasing costs and expanding deployment globally. Innovations in energy storage systems, smart grids, and electric vehicle technology are transforming energy systems and transportation networks, lessening dependence on fossil fuels and facilitating greater integration of renewables. In summary, conventional fuels face a range of challenges, including environmental degradation, economic instability, geopolitical tensions, and social inequities [9]. As societies address the urgent need to mitigate climate change, enhance energy security, and promote social justice, the transition from conventional fuels to cleaner, more sustainable alternatives becomes imperative. However, overcoming these challenges necessitates coordinated efforts and investments from governments, businesses, and individuals to accelerate the shift toward a low-carbon, resilient, and equitable energy future.

2.3 Alternative Fuels

It is widely recognized that national progress is linked to energy expenditure and the availability of resources. The finite nature and unpredictable costs of conventional fuels threaten the development of all sectors, leading developing countries to seek alternative energy sources for a long-term solution to fossil fuel depletion. AFs are generally derived from renewable sources, and their properties vary depending on the production process and origin. However, they are typically produced through clean and sustainable methods with significantly lower CO_2 emissions compared to gasoline. AFs can be obtained in solid, liquid, or gaseous forms, depending on their production method and intended application. Solid fuels are commonly used in power plants for stationary applications, while gaseous and liquid fuels are utilized primarily for transportation. Over recent years, alternative energy and fuel sources like wind, solar, hydroelectric, and geothermal power have emerged as potential replacements for conventional, polluting energy sources. Research into AFs involves assessing their economic, technical, and social impacts, feedstock availability, consumption patterns, and environmental effects.

The primary focus regarding environmental impact is that AFs contribute to emission reduction, and the main challenge is replacing non-renewable fuels with renewable ones. Furthermore, due to limitations in consumption or production processes/technologies, most AFs are not yet widely commercialized, which affects their economic viability [10]. Significant research has been conducted on AFs for transportation, leading to the emergence of fuels like ethanol, biodiesel, dimethyl ether (DME), methanol (CH_3OH), bioethanol, synthetic natural gas (SNG), diethyl ether, hydrotreated vegetable oils (HVO), hydrogen (H_2), and Fischer–Tropsch diesel as potential substitutes for conventional fuels. AFs can be synthesized via two primary routes: thermochemical processes and direct electricity utilization. Variable renewable energy sources generating electricity superfluity are used to synthesize carbon-neutral fuels, such as electro-fuels in liquid or gas form. Carbon neutrality is achieved by capturing CO_2 directly from the air or from exhaust gases. CO_2 and H_2 are fundamental components for synthesizing electro-fuels, yielding products like

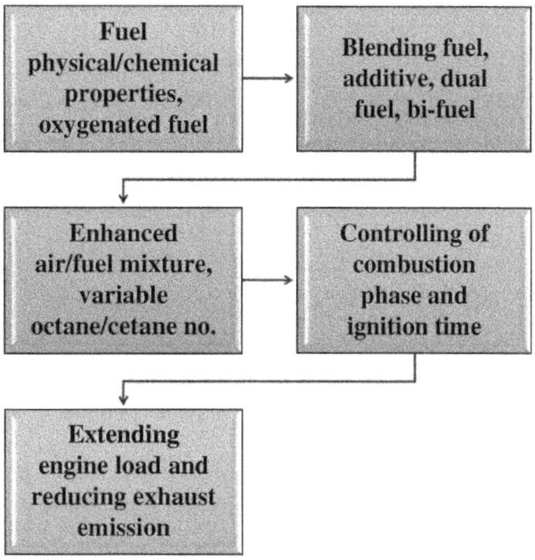

Figure 2.1 Fuel design principles for IC engine. Adapted from [12].

CH_4 or C_4H_{10} (butane), synthetic hydrocarbon gases/liquid alcohol fuels (CH_3OH). Another pathway for synthesizing AFs is the thermochemical method, where raw materials are thermally converted into AFs. Nowadays, the thermochemical process is of significant interest for converting waste materials into liquid or gaseous biofuels, which aids in waste management while addressing issues related to conventional fuels [11]. However, a key challenge in using AFs is reducing production costs. Once produced, AFs are generally easy to distribute and store. Production technologies and sources for AFs have been studied, and the principal fuel design considerations for internal combustion (IC) engines are illustrated in Figure 2.1.

Furthermore, the use of AFs/biofuels can enhance engine performance and reduce emissions, as the interaction between fuel and air is critical for combustion. This beneficial aspect of AFs has spurred extensive global research into their potential application in IC engines.

2.4 Types of AFs

Given the environmental concerns and the depletion of fossil fuel reserves, research into AFs for the transportation sector has become increasingly important. The global economy is encouraging the use of AFs to reduce emissions and mitigate climate change problems. Estimates suggest that oil reserves have only about 50–55 years remaining before depletion. Therefore, utilizing sustainable AFs is considered a better strategy to address these issues. Some of the AFs are discussed in the following sections.

2.4.1 Ammonia

The negative impacts of climate change have driven interest in transitioning to renewable energy, and in this context, ammonia (NH_3) is seen as a significant renewable clean transportation fuel with zero carbon emissions. It also serves as a medium for energy storage and is suitable for future use despite certain limitations. Furthermore, identifying chemicals capable of acting as H_2 carriers is essential because H_2 storage is expensive and complex. NH_3 demonstrates the capacity to store H_2, thereby lowering storage costs and offering an efficient and safe method. Various routes exist for storing H_2; one method involves converting it into nitrogen-based fuels, the simplest being NH_3. NH_3 is produced on a large scale, with 80–82% of production used in fertilizer manufacturing. Globally, steam reforming of natural gas yields the highest amount of NH_3 compared to other sources, as depicted in Figure 2.2 [13].

Scamardella et al. [14] suggested that NH_3 could be used as an energy carrier in specific fuel cells and investigated its potential as an AF in IC engines for cruise ships. They highlighted that NH_3, as a chemical compound, possesses properties such as density $(0.73\ kg/m^3)$, kinematic viscosity at 40 °C (0.08–0.11 mm^2/s), and has a CO_2-specific emission of 0 g/MJ. In contrast, CH_4/LNG, with a CO_2-specific emission of 54.88 g/MJ, is less suitable as an Internal Combustion Engine (ICE) fuel as it reduces engine efficiency, although it lowers CO_2 emissions. However, as a gas carrier, NH_3 is suitable for H_2 storage. Tornatore et al. [15] proposed that NH_3 holds potential as a sustainable fuel for both retrofitted and new engines and can be applied in stationary or marine engines. NH_3 is viewed as an excellent energy carrier and a carbon-free fuel for ICEs. Nevertheless, applying pure NH_3 requires increasing the pressure/compression ratio to compensate for its slow flame speed. In spark ignition (SI) engines, adding H_2 to NH_3 improves flame propagation and combustion, although exhaust modifications are needed to prevent elevated NOx and unburnt NH_3. It was also noted that NH_3 is produced industrially via the Haber–Bosch (H-B) process, which is not considered a green method. Consequently, a green route is

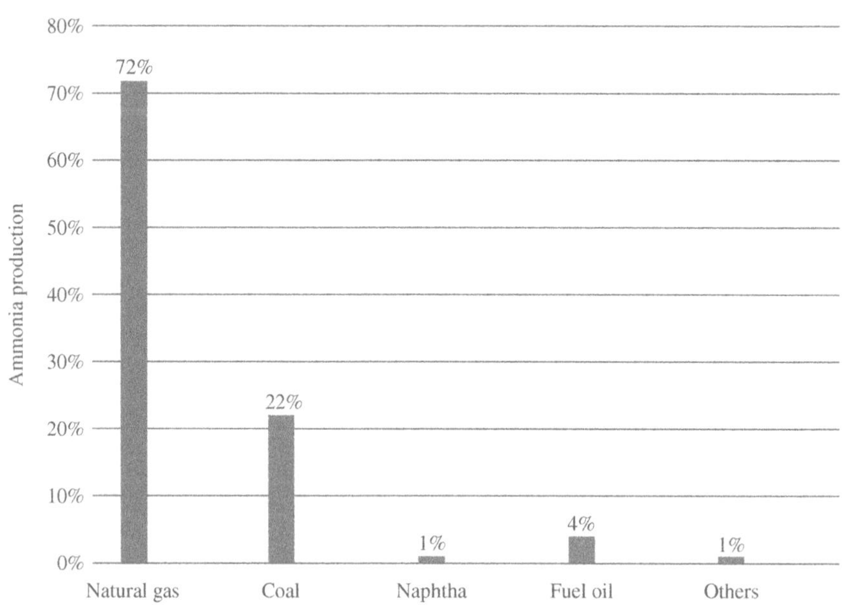

Figure 2.2 Ammonia content from different sources. Adapted from [13].

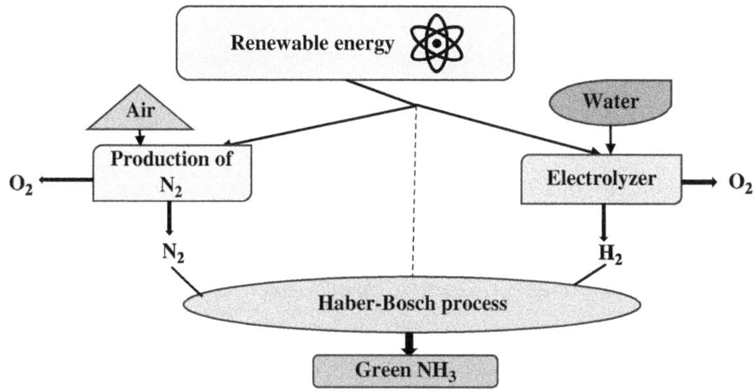

Figure 2.3 Production process of green ammonia. Adapted from [15].

suggested as shown in Figure 2.3. This route involves either using green H_2 produced by electrolysis with electricity in the H-B process or electrochemical synthesis of NH_3.

For IC engines, the application of NH_3 is gaining interest among stakeholders in the shipping industry. However, further analysis and development are necessary for large-scale production. Langella et al. [16] identified the main challenges for NH_3 application in IC engines as its low flame propagation speed and high ignition temperature (IT). This suggests it is best used in blended form with fuels that burn more readily, which also improves engine efficiency. Since NH_3 has lower production and storage costs compared to H_2, it is recommended as a means for H_2 transportation, from which H_2 can be extracted via reforming.

2.4.2 Hydrogen

In the effort to replace fossil fuels with renewable alternatives, H_2 emerges as a promising and environmentally sound option. It is abundant in plant substances and is derived from readily available water, which covers 68–70% of the Earth's surface. H_2 boasts a high calorific value (121–143 MJ/kg) compared to other conventional fuels. Chemical compounds containing H_2, such as H_2O and biomass (organic compounds), produce H_2O and heat upon combustion without toxic pollutants. H_2 is considered a clean alternative to fossil fuels because its combination with oxygen can be used in fuel cells (electrochemical reactions) to generate electricity for various uses, including refrigeration, information processing, transportation, and communication. H_2-based fuel cell vehicles and low-carbon fuels are expected to play a role in decarbonizing shipping, aviation, and mobile sectors [17]. H_2 is also being considered as a replacement for conventional fuels in transportation since its use in IC engines does not produce harmful gases. Wang et al. [18] investigated the performance of a diesel engine operating in dual-fuel mode with H_2. They found that hydrogen's fuel properties – density at 25 °C (0.083 kg/m^3), viscosity at 25 °C (8.916 × 10^{-6} Pa.s), octane number (>131), and low calorific value (120.2 MJ/kg) – differ significantly from diesel fuel, which has higher density, viscosity, and lower octane number. When the engine was tested with diesel-hydrogen fuel blends (0, 5, 10, and 15% H_2), in-cylinder pressure and temperature improved by 10.55 and 15.13%, respectively, with reductions in HC, CO, soot, and CO_2 exhaust emissions. However, an increase in NOx emissions was noted, which could potentially be reduced by optimizing in-cylinder temperature and pressure.

Kanna et al. [19] pointed out that fuel cells powered by H_2 can be used in a wide range of applications, including vehicles, household electrical appliances, and as battery replacements in portable devices. In the petroleum and chemical industries, the main sources for H_2 production are natural gas, electrolysis, oil, and coal, as shown in Figure 2.4.

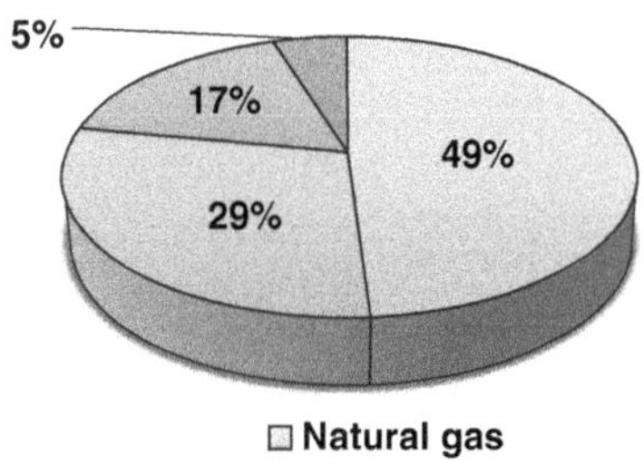

Figure 2.4 Worldwide hydrogen production sources. Adapted from [19].

IC engines running on H_2 require specific modifications to existing fuel cell car infrastructure. The high production cost of H_2 currently makes H_2 vehicles more expensive than gasoline vehicles. However, H_2 engines are 23–25% more efficient, making H_2 a viable AF. Various methods exist for producing H_2, as illustrated in Figure 2.5 [20]. H_2 can be obtained from renewable sources like lignocellulosic biomass or through water splitting. Water splitting can be achieved via water

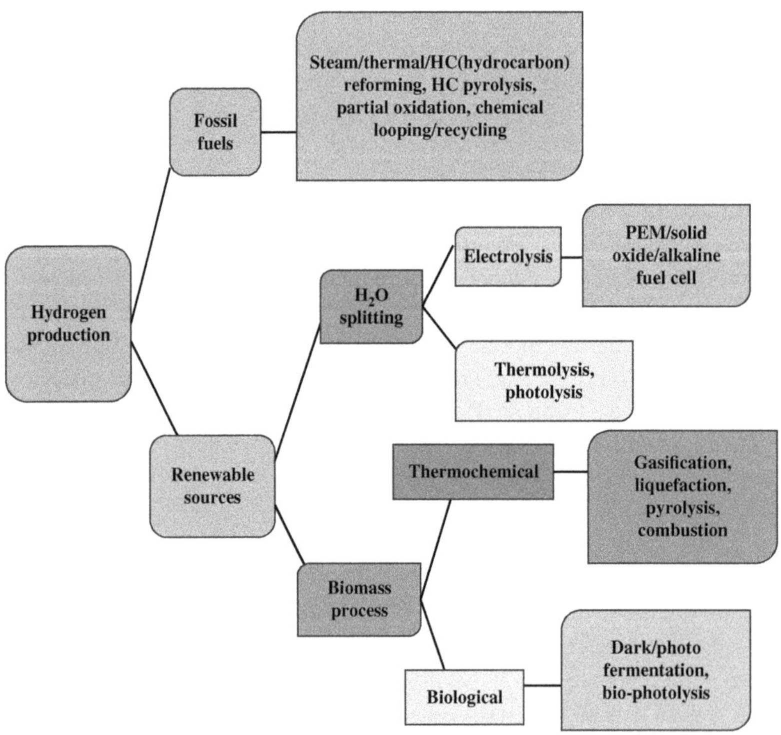

Figure 2.5 Production methods of hydrogen. Adapted from [20].

electrolysis, using fossil fuels, microorganisms, or chemical reforming. Chemical reforming, such as CH_4 catalysis at temperatures above 1073 K, produces H_2 and CO_2. Other methods for generating H_2 include thermal cracking, steam reforming, and partial oxidation.

2.4.3 Alcohol-derived Fuels

Alcohols have a long history of use as fuels in the automotive sector, burning cleanly with fewer emissions. Compared to fossil fuels, alcohols offer advantages: they can be produced from readily available, low-cost raw materials such as natural gas, coal, and biomass; they reduce GHG emissions and harmful air pollutants; they have a higher octane rating than gasoline, enhancing fuel and power efficiency in IC engines; alcohol combustion promotes higher pressure, improving thermal efficiency and power output compared to conventional fuels; they are miscible with water and easily removed if spilled; and their lower carbon content results in less ash upon combustion [21, 22]. Aliphatic alcohols like CH_3OH, C_2H_5OH, C_3H_7OH, C_4H_9OH, and DME are gaining interest as they can be synthesized chemically or biologically from renewable raw materials (biomass) and possess properties suitable for modern engines [23]. However, their use has some limitations, leading to their application primarily in blended forms (typically not exceeding 20% in diesel fuel). Some alcohol-based fuels are discussed in the following sections.

2.4.3.1 Methanol

CH_3OH has been recommended as an alternative transportation fuel for the past 40 years. It has a high-octane number and favorable combustion properties with low emissions. Its ability to be blended with conventional fuels has attracted engine researchers globally as a potential substitute for conventional fuels, especially with rising crude oil prices and limited reserves. Research on CH_3OH use was initiated in India in the 1980s. Different countries and markets have developed and implemented optimal strategies and infrastructure for flexible fuel vehicles and filling stations [24].

Valera et al. [25] suggested that CH_3OH could be directly used as fuel in SI engines instead of gasoline due to its high octane rating, though its use in compression ignition (CI) engines requires substantial effort. Table 2.1 summarizes the fuel properties of CH_3OH, highlighting its similarities to conventional fuels.

CH_3OH has a higher auto- IT than diesel fuel, making it suitable for transportation. However, its low cetane number (<6) can lead to challenges for smooth combustion. CH_3OH can be sourced from various origins, as discussed below:

- **From Coal:** This process involves four main steps: syngas production, purification, CH_3OH synthesis, and rectification. This is illustrated in Figure 2.6.

  Syngas (H_2, CO, CO_2, H_2S) is produced in a gasifier by mixing coal slurry with oxygen at high pressure, followed by purification to remove CO and H_2S. Hydrogen sulfide is recovered using amine and converted to sulfur (S). A shift reaction is then performed to achieve the required stoichiometric

Table 2.1 Comparative analysis of diesel, methanol, and gasoline fuel properties. Adapted from [25].

Fuel characteristics	Diesel	Methanol	Gasoline
Chemical formulation	C_8–C_{25}	CH_3OH	C_4–C_{12}
O-Content (wt.%)	—	50	—
C-Content (wt.%)	85.1	38.2	86.0
H-Content (wt.%)	15.0	13.0	14.0
Density (kg/m^3)	841	797.9	741
Kinematic viscosity (cSt at 20 °C)	2.5–3.1	0.75	0.60
Octane no.	15–26	108	90–101
Cetane no.	39–54	<6	—
Flash point (°C)	54	14	−46

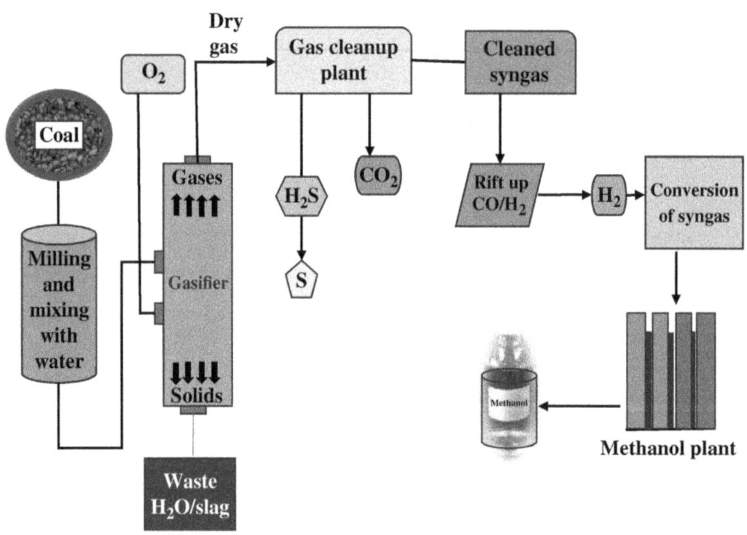

Figure 2.6 Flowchart of methanol production from coal. [25]/Springer Nature.

ratio of CO/H_2 in the syngas for CH_3OH synthesis. CH_3OH is synthesized using a Zn-Cr (zinc-chromium) catalyst at 573–673 K. The relevant reactions are as follows:

$$CO + 2H_2 \rightarrow CH_3OH \text{ (}CH_3OH\text{ synthesis reaction)}$$
$$H_2O + CO \rightarrow H_2 + CO_2 \text{ (water–gas shift (WGS) reaction)}$$
$$CO_2 + 3H_2 \rightarrow CH_3OH + H_2O \text{ (}CH_3OH\text{ synthesis reaction)}$$

- **From Natural Gas:** This process occurs in three stages: syngas production, conversion of the obtained syngas into crude CH_3OH, and subsequent purification via distillation. The reactions involved in syngas production are as follows:

$CH_4 + H_2O \rightarrow 3H_2 + CO$ (steam reforming reaction)
$2CO_2 + CH_4 \rightarrow CO_2 + 2H_2O$ (auto thermal reforming [ATR])
$H_2 + CO_2 \rightarrow H_2O + CO$ (WGS reaction)
$1/2\ O_2 + CH_4 \rightarrow 2H_2 + CO$ (partial oxidation reaction)

The syngas composition is represented by the stoichiometric number S_n, which depends on the raw material used and is defined as:

$$S_n = \frac{\textit{no. of moles of hydrogen} - \textit{no. of moles of carbon dioxide}}{\textit{no. of moles of carbon dioxide} + \textit{no. of moles of carbon monoxide}}$$

For CH_3OH production, the S_n value should be 2. The conversion of syngas to crude CH_3OH occurs at high temperatures (473–573 K) and pressures (49.3–98.7 atm). The overall reaction is:

$3H_2 + CO_2 \rightleftharpoons H_2O + CH_3OH$ (hydrogenation of CO_2)

This exothermic reaction requires substantial cooling.

- **From Biomass/Municipal Solid Waste (MSW):** CH_3OH production from biomass/MSW generally yields lower amounts and follows a similar process to natural gas conversion: syngas generation, conversion to crude CH_3OH, and purification. The process involves shredding the raw material, removing ferrous substances with a magnet, gasifying the biomass/MSW using steam/oxygen/flue gases, and then gas scrubbing to remove water. Subsequent processes mirror those used for CH_3OH production from coal/natural gas to yield the final product.

2.4.3.2 Ethanol

Ethanol is a widely adopted alcohol-based AF in the automotive sector globally, primarily used as an additive to conventional fuels in concentrations like E10 (10%), E20 (20%), E85 (85%), and E100 (100%). Low concentrations of ethanol do not significantly alter engine characteristics, but higher concentrations necessitate rigorous testing. Ethanol improves fuel thermal efficiency and anti-detonation properties. It is reported to reduce negative impacts on human health by approximately 1.4% and on ecosystems and natural resources by 1.45%, potentially reaching 13% [26].

According to Singh Bisht et al. [27], India's total ethanol production increased significantly (from 700 Cr L to 1500 Cr L) with the introduction of a 10% ethanol blend (E10) in 2022. The rollout of E20 began in 2023, with a target of blending 80% gasoline with 20% ethanol by 2030. Ethanol blends enhance engine performance and contribute to reducing pollution from conventional fuels, while also supporting agriculturalists by maintaining currency exchange rates and increasing the cost of sugarcane products. Testing ethanol blends at various loads and constant steady speeds showed that vehicles running on ethanol will be able to achieve higher speeds than conventional fuel vehicles by 2030. Table 2.2 summarizes

Table 2.2 The demand for ethanol alternative fuel by 2025. Adapted from [27].

	Ethanol blend	
Type of vehicles	**E10 (Cr L)**	**E20 (Cr L)**
Conservative (low EVs)	445.32	921.45
Business-as-usual (med EVs)	374.56	778.39
Low carbon (good EV uptake)	348.75	721.63

the projected demand for ethanol AF in India by 2025, influenced by the rise of electric vehicles.

It is suggested that ethanol blending in India supports sustainable development goals (SDGs) by improving economic, environmental, and social conditions. Jegan et al. [28] investigated the performance of SI engines using ethanol-gasoline blends of varying percentages. Ethanol's fuel properties – density (796 kg/m^3), viscosity (1.53 mm^2/s), and calorific value (26.42 MJ/kg) – differ from gasoline's (750.2 kg/m^3, 0.4–0.81 mm^2/s, 45.31). Test fuels prepared with different proportions (E10, E20, E25) were tested on a standard bike engine (TVS Apache 150, 4-stroke) with its highest torque (12.5 Nm @ 6000 rpm) and power (13.8 bhp @ 8500 rpm). The experiment concluded that E-blends resulted in lower total fuel consumption (TFC) compared to gasoline. While Specific Fuel Consumption (SFC) increased with higher E-blend percentages, the E-blends ultimately improved fuel economy. Given ethanol's richness in oxygen and higher octane number than gasoline, it is believed that with engine modifications, ethanol blends can help address the current fuel crisis.

2.4.3.3 Dimethyl Ether

Among various AFs, DME (CH_3OCH_3) has emerged as a promising option with potential positive societal impacts by reducing reliance on petroleum. DME can serve as an effective CI fuel, leading to reduced emissions of NOx, SOx, and particulate matter. Like other fuels, it doesn't pose issues related to production, transportation, infrastructure, or toxicity. DME is considered environmentally benign; its volatile organic compounds (VOCs) are non-poisonous, non-cancerous, and non-genotoxic [29].

Park et al. [30] reported on the application of DME as an AF in CI engines, examining its atomization/spray characteristics, fuel properties, exhaust emissions, and combustion characteristics. At mild conditions, DME is gaseous with a low C : H ratio and properties similar to liquid petroleum gas (LPG), which consists of C_4H_{10} and C_3H_8 (propane), making its handling similar to LPG. DME has a lower heating value (LHV 25.4 MJ/kg) compared to diesel fuel (LHV 42.6 MJ/kg) but higher than ethanol/CH_3OH. Although DME has a lower IT than diesel fuel, its IT is higher (623 K) than that of diesel fuel (523 K) at atmospheric temperature and pressure. DME's high cetane number facilitates its combustion in CI engines due to the presence of oxygen (approximately 38.9 wt.%), which lowers the auto- IT, resulting in nearly negligible smoke emissions. DME's kinematic viscosity (0.185–0.23 cSt) is lower than the recommended ASTM range, making it less lubricating and more viscous than diesel fuel, necessitating additives to improve its properties. Ardy et al. [31] demonstrated in their experiment that DME can be synthesized from CH_3OH via dehydration. Their experimental setup included a tubular reactor (20 mm diameter) electrically heated, 2.0 g of catalyst, a gas sampling point, a nitrogen bottle, a bubbling bottle, and a bubble soap meter. Purging with nitrogen flow (100 ml/min) for an hour at the required temperature removed gases from the reactor. Catalytic CH_3OH dehydration was performed between 513 and 563 K, with the reactor feed containing CH_3OH vapor in N_2. CH_3OH conversion was calculated using the following formula:

$$Y = \frac{\left(\frac{A}{An}\right)in - \left(\frac{A}{An}\right)out}{\left(\frac{A}{An}\right)in}$$

where Y is the MeOH conversion, *A* is the chromatogram area of MeOH, *An* is the chromatogram area of N_2, and subscripts *in*/*out* refer to inlet/outlet of the reactor sampling points.

The experiment concluded that CH_3OH conversion using γ-Al_2O_3 followed a first-order reaction, achieving a maximum CH_3OH conversion of approximately 72% with the γ-Al_2O_3 catalyst.

Thus, DME is environmentally favorable and shows potential as an AF for diesel engines. It has a good C/H ratio, producing water vapor upon combustion. The absence of a direct carbon–carbon bond in DME results in less particulate matter and allows for higher exhaust gas recirculation rates, leading to lower NOx emissions. Using DME as an AF also reduces unburned HC and CO emissions. DME can be produced from biomass sources such as wood chips/waste [32].

2.4.4 Biodiesel

Biodiesel, known as fatty acid alkyl esters, is produced by the transesterification of animal fats or vegetable oils with alcohol and has been used as a substitute for fossil fuel derivatives for decades. Biodiesel production varies regionally based on feedstock availability. In some areas, edible oils like soybean oil serve as raw materials, but their higher cost compared to diesel fuel makes waste vegetable oils, non-edible vegetable oils, or waste oils/fats more attractive and cost-effective feedstocks for biodiesel production. The composition of fatty acids in oils differs depending on their source, varying in chain length, chemical functional groups, and degree of unsaturation. The properties of biodiesel are heavily influenced by its fatty acid ester composition, so enriching biodiesel with specific types of fatty acid esters improves its fuel characteristics.

Knothe et al. [33] proposed that biodiesel, being composed of fatty acid esters, possesses significant fuel properties, which are enhanced by the individual fatty acid ester constituents. Transesterification is the process of converting one ester to another, used here to convert fatty acids in fats or oils into esters for biodiesel production. The standard conventional method for biodiesel production is shown in Figure 2.7.

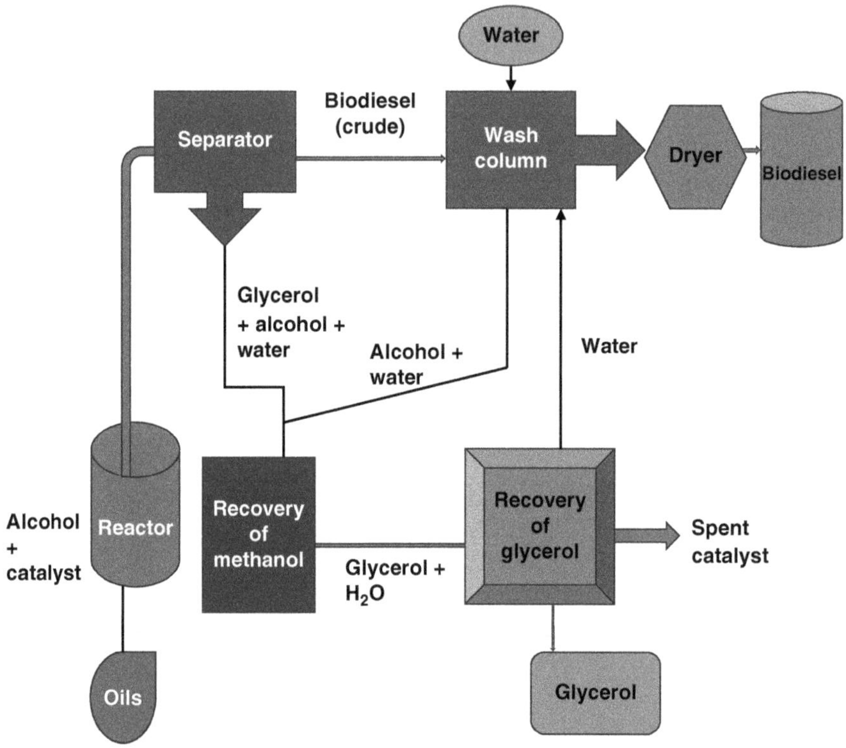

Figure 2.7 Common production manner of biodiesel from oils/fats. Adapted from [33].

The typical conventional procedure for industrial biodiesel production is shown in Figure 2.7. Standard transesterification conditions face limitations related to mass transfer, free fatty acid content, other components, and moisture. Analysis of biodiesel indicates that most types meet kinematic viscosity standards and have a higher bulk modulus than petro-diesel, enabling quicker pressure buildup and earlier injection, resulting in lower exhaust emissions. Abed et al. [34] converted palm oil, waste cooking oil (WCO), jatropha oil, and algae into biodiesel via transesterification and evaluated engine exhaust emissions. The process for algae involved grinding and drying (353 K, 20 minutes), oil extraction with 21 mL hexane, settling, filtering, and then transesterification with CH_3OH (24 mL) catalyzed by sulfuric acid. After the reaction, the product was filtered, biodiesel was separated from glycerol (byproduct), washed with 5% acid water, and dried. Similar transesterification was performed for palm oil, WCO, and jatropha oil. Biodiesel samples from these oils were blended with diesel fuel at 10–20% (B10 and B20 blends), and their fuel properties and engine performance were assessed. The density of B10 and B20 blends was slightly higher than diesel fuel. Kinematic viscosity of palm oil and WCO biodiesel blends (2.4–2.9 cSt) was close to that of diesel fuel, while jatropha and algae blends (3.3–5.3 cSt) were higher. Similar trends were observed for flash point. Exhaust emissions (CO_2, HC, smoke, and CO) were reduced for B10 and B20 blends, but NOx emissions increased. Adding specific additives enhanced the properties of the blends. Pasha et al. [35] noted various approaches used for biodiesel production

Table 2.3 Production sources of biodiesel and their conversion technologies. Adapted from [35].

Diversity of feedstock			Conversion process	Products
First generation (edible oils)	**Second-generation (non-edible oils and animal fats)**	**Third-generation microalgal/ waste oils)**	**Transesterification and pyrolysis**	
Soybean	Mahua	Bacteria	Alkali-catalyzed	Biodiesel and glycerol
Rice bran oil	Karanja	Various microalgae	Acid-catalyzed	
Coconut	Jatropha	Waste cooking oil	Heterogeneous acid-catalyzed	
Palm	Fish oil	Waste tire oil	Supercritical	
Sunflower	Pork lard	Waste plastic oil	Enzyme-catalyzed	

from different sources and environmental impact analyses conducted. It was estimated that producing in-biodiesel using second- and third-generation low-cost feedstocks offers a potential path to reduce biodiesel prices compared to those from first-generation feedstocks, as summarized in Table 2.3.

It was further estimated that the cost of biodiesel from edible oils (rapeseed oil US$1.16/L, palm oil US$2.3/L, and canola oil 1.0/L) is higher than that from WCO (0.703/L) and waste canola oil (US$0.632/L). However, biodiesel costs vary depending on the production method and catalyst used.

Transesterification is generally the preferred method for biodiesel production due to its advantages, lower cost, and simplicity compared to other methods. The resulting biodiesel typically exhibits reduced emissions, improved combustion efficiency, and a higher cetane number. Conventional methods can be time-consuming and energy-intensive, leading to the development of methods like ultrasound irradiation and microwave-assisted transesterification. A disadvantage of using animal fats for biodiesel is that the fuel can have higher cold flow/pour point properties and low oxidation stability. Blending the oils (raw materials) and converting the mixture into biodiesel is creating increased interest in the field and promoting the use of lower-grade oils.

2.5 Applications of AFs

2.5.1 Dual-fuel Mode

Dual-fuel combustion mode offers various advantages, and numerous researchers have investigated using AFs in dual-fuel diesel engines to enhance energy efficiency and reduce emissions in transportation and distribution. In this mode, AFs are introduced into the manifold, forming a homogeneous mixture with air. This mixture is ignited by a spark plug when the piston reaches top dead center. The quantity of AF injected is varied based on engine power output and operational conditions, while the amount of diesel fuel remains constant. Various AFs, including alcohols, DME, biodiesel, and natural gas, have been used in dual-fuel combustion mode. Studies in literature have shown that dual-fuel mode with AFs like alcohol and natural gas, using diesel as the pilot fuel, performs quite well. Compared to conventional fuels, dual-fuel mode is estimated to significantly reduce particulate matter and NOx emissions. However, dual-fuel combustion faces operational challenges under varying loads, particularly at medium/low load conditions where combustion efficiency can be low, leading to increased HC and CO emissions due to incomplete burning. More research is needed to thoroughly assess the impact of using dual-fuel combustion mode on engine performance [36].

2.5.2 Blend Form

Applying mixed or blended fuel modes in engines involves using a combination of different fuels to power an engine. This approach can provide several benefits, such as improved performance, enhanced fuel efficiency, reduced emissions, and greater flexibility in fuel sourcing.

A common example of mixed or blended fuel mode is seen in the automotive industry, particularly with hybrid vehicles [37]. These vehicles often combine gasoline and electricity to power the engine, offering better fuel efficiency and lower emissions compared to purely gasoline-powered vehicles. In hybrid systems, the engine can switch between using gasoline and electric power, or even utilize both simultaneously, depending on driving conditions and energy demands. Blended fuel modes are also used in conventional IC engines, where AFs are mixed with gasoline or diesel to create blends that meet specific performance and emission standards. For instance, ethanol, CH_3OH, biodiesel, or compressed natural gas (CNG) can be blended with gasoline or diesel to produce fuels with lower emissions and decreased reliance on fossil fuels. These blends can be used in existing engine designs with minimal modifications, providing a relatively straightforward path to reduce environmental impact and improve energy security [38].

In aviation, mixed fuel modes are being explored to reduce emissions and dependence on traditional jet fuels. Sustainable Aviation Fuels (SAFs), derived from sources like algae, waste oils, or agricultural residues, can be blended with conventional jet fuel [39]. These SAF blends can be used in existing aircraft engines without major modifications, offering a feasible way to reduce the carbon footprint of air travel.

Furthermore, mixed fuel modes can be applied in stationary power generation, such as electricity generation in power plants or heating in industrial facilities. By blending renewable fuels like biomass, biogas, or H_2 with conventional fuels such as coal or natural gas, these systems can lower emissions and increase overall efficiency while maintaining reliable and flexible energy production. Overall, the application of mixed or blended fuel modes in engines is a versatile and adaptable strategy to address challenges in energy security, environmental sustainability, and economic viability [40]. By combining different fuels in transportation, power generation, or other uses, significant reductions in emissions and dependence on fossil fuels can be achieved, while also improving overall performance and energy system resilience.

2.6 Environmental Impact and Economic Feasibility

The adoption of AFs profoundly affects the environment, primarily by reducing GHG emissions and other pollutants. Unlike conventional fossil fuels, many AFs, including biodiesel, ethanol, H_2, and electricity, produce lower or even zero emissions during combustion. Replacing fossil fuels with cleaner alternatives helps mitigate climate change, reduce air pollution, and protect ecosystems and biodiversity [41]. Additionally, AFs often have smaller lifecycle carbon footprints, considering their production, transport, and use, which further enhances their environmental benefits. However, it is essential to evaluate the environmental impacts of AF production processes themselves, such as land use changes, water consumption, and chemical inputs, to ensure overall sustainability and avoid unintended negative consequences.

The adoption of AFs also carries significant economic implications, influencing energy markets, investment decisions, and overall economic development trajectories. Shifting to AFs can decrease dependence on imported oil, bolster energy security, and mitigate economic risks associated with volatile fossil fuel prices. Moreover, AF industries, such as those producing renewable energy and biofuels, can stimulate economic growth, generate employment, and foster innovation and competitiveness in the global market [42]. However, the economic viability of AFs depends on various factors, including technological progress, government policies, market demand, and resource availability. Ensuring cost-effectiveness and scalability is crucial for the widespread adoption and sustainable growth of AF sectors.

Promoting the adoption of AFs necessitates developing and implementing environmental and economic policies to support their deployment and address potential trade-offs. Environmental policies, such as carbon pricing mechanisms, emission standards, renewable energy mandates, and sustainability criteria, can encourage the adoption of AFs and drive investment in cleaner energy technologies. Economic policies, such as tax incentives, subsidies, grants, and research funding, can further support the development and commercialization of AF technologies, making them more competitive with conventional fuels. However, policymakers must balance promoting AFs with addressing potential conflicts with other policy objectives, such as food security, land use, and social equity [43].

The transition to AFs presents both challenges and opportunities for stakeholders across different sectors. Technical challenges, such as developing energy storage solutions, building necessary infrastructure, and ensuring vehicle compatibility, must be overcome for widespread adoption. Market barriers, regulatory uncertainties, and socio-political factors can also impede the transition process and hinder progress toward sustainability goals. Nevertheless, overcoming these challenges can unlock numerous opportunities, including creating jobs, fostering technological innovation, achieving energy independence, and promoting environmental stewardship. Collaboration among governments, industries, academia, and civil society is vital for navigating the complexities of transitioning to AFs and realizing their full potential in addressing environmental and economic challenges [44].

2.7 Future Aspects of AFs

The outlook for AFs is highly promising as societies worldwide increasingly seek cleaner, more sustainable energy sources to tackle pressing environmental, economic, and social issues. Several ongoing trends and developments point toward continued growth and adoption of AFs in the coming years. A primary trend shaping the future of AFs is the escalating focus on decarbonization and climate change mitigation. Governments, businesses, and individuals are increasingly recognizing the urgent need to reduce GHG emissions and move away from fossil fuels. Alternative energy sources like solar, wind, hydroelectric power, and biofuels offer a viable route to achieving these goals by providing low-carbon or carbon-neutral options [45, 46] compared to conventional fossil fuels.

Technological advancements are also propelling the future of AFs, making them more efficient, cost-effective, and readily scalable. Innovations in areas such as energy storage solutions, biofuel production methods, H_2 fuel cells, and electric vehicle technologies are expanding the range of choices and lowering barriers to adoption. For instance, battery technology improvements are making electric vehicles increasingly competitive with IC engine vehicles, while progress in biofuel production processes is boosting yields and lowering costs. Furthermore, there is growing recognition of the importance of a diversified energy portfolio for enhancing energy security and resilience. AFs offer opportunities to reduce dependence on imported oil, mitigate geopolitical risks, and strengthen domestic energy production [47, 48]. By diversifying energy sources and decreasing reliance on finite fossil fuels, countries can improve their energy independence and reduce vulnerability to supply disruptions and price fluctuations.

Another significant factor driving the future of AFs is the increasing alignment of economic incentives with sustainability objectives. As the costs of renewable energy technologies continue to fall and the impacts of climate change become more evident, the economic benefits of transitioning to AFs are gaining wider recognition. Renewable energy industries are creating jobs, driving economic growth, and attracting investment, making them increasingly appealing from an economic perspective. In conclusion, the future of AFs appears bright, motivated by the necessity to address climate change, technological advancements, energy security concerns, and economic opportunities. Continued investment in research, innovation, and policy support will enable societies to accelerate the transition to cleaner, more sustainable energy sources and fully realize the potential of AFs in creating a more prosperous, resilient, and equitable future.

2.8 Conclusion

There is increasing acknowledgment of the significance of diversifying energy sources to enhance energy security and resilience. AFs present opportunities to lessen dependence on imported oil, reduce geopolitical risks, and boost domestic energy production. By diversifying energy portfolios and decreasing reliance on finite fossil fuels, countries can improve their energy independence and reduce susceptibility to supply disruptions and price volatility, thereby enhancing national security and economic stability. Economic incentives are also becoming more aligned with sustainability objectives, fostering investment and growth in the AF sectors. As the costs of renewable energy technologies continue to decline and the effects of climate change become more apparent, the economic benefits of transitioning to AFs are increasingly recognized. Renewable energy industries are generating employment, stimulating economic growth, and attracting investment, making them more attractive economically. Furthermore, the AF sector offers considerable economic opportunities for innovation and entrepreneurship, driving further investment and growth.

However, despite the promising outlook for AFs, challenges remain that must be addressed to fully realize their potential. These challenges include technological hurdles, policy uncertainties, market barriers, and issues related to social acceptance. Overcoming these challenges will require coordinated efforts from governments, industries, academia, and civil society to invest in research, innovation, and policy support, as well as to raise public awareness and build support for AFs. In conclusion, the future of AFs holds great promise as societies aim for cleaner, more sustainable energy sources to address climate change, improve energy security, and drive economic growth. By leveraging technological progress, aligning economic incentives, and tackling remaining challenges, we can accelerate the transition to a more sustainable energy future powered by AFs.

References

1 Chatterjee, S., Dhanurdhar and Rokhum, S.L. Extraction of a cardanol based liquid bio-fuel from waste natural resource and decarboxylation using a silver-based catalyst. *Renew. Sustain. Energy Rev.* 2017. https://doi.org/10.1016/j.rser.2017.01.035.

2 Agrawal, P.S., Belkhode, P.N. and Rokhum, S.L. Advanced biofuels – alternatives to biofuels. In: *Sustainability in Biofuel Production Technology*. Wiley-Blackwell, pp. 53–73. 2022. https://doi.org/10.1002/9781119888864.ch3.

3 Maradin, D. Advantages and disadvantages of renewable energy sources utilization. *Int. J. Energy Econ. Policy*. 2021. https://doi.org/10.32479/ijeep.11027.

4 Chiong, M.C., Kang, H.S., Shaharuddin, N.M.R., Mat, S., Quen, L.K., Ten, K.H., et al. Challenges and opportunities of marine propulsion with alternative fuels. *Renew. Sustain. Energy Rev.* 2021. https://doi.org/10.1016/j.rser.2021.111397.

5 Burchart-Korol, D., Gazda-Grzywacz, M. and Zarębska, K. Research and prospects for the development of alternative fuels in the transport sector in poland: a review. *Energies*. 2020. https://doi.org/10.3390/en13112988.

6 Law, L.C., Foscoli, B., Mastorakos, E. and Evans, S. A comparison of alternative fuels for shipping in terms of lifecycle energy and cost. *Energies*. 2021. https://doi.org/10.3390/en14248502.

7 Eweade, B.S., Karlilar, S., Pata, U.K., Adeshola, I. and Olaifa, J.O. Examining the asymmetric effects of fossil fuel consumption, foreign direct investment, and globalization on ecological footprint in Mexico. *Sustain. Dev.* 2023. https://doi.org/10.1002/sd.2825.

8 Armaroli, N. and Balzani, V. The legacy of fossil fuels. *Chem Asian J.* 2011. https://doi.org/10.1002/asia.201000797.

9 Gani, A. Fossil fuel energy and environmental performance in an extended STIRPAT model. *J. Clean. Prod.* 2021. https://doi.org/10.1016/j.jclepro.2021.126526.

10 Ashrafi, M., Lister, J. and Gillen, D. Toward a harmonization of sustainability criteria for alternative marine fuels. *Marit. Transp. Res.* 2022. https://doi.org/10.1016/j.martra.2022.100052.

11 Müller-Casseres, E., Carvalho, F., Nogueira, T., Fonte, C., Império, M., Poggio, M., et al. Production of alternative marine fuels in Brazil: an integrated assessment perspective. *Energy*. 2021. https://doi.org/10.1016/j.energy.2020.119444.

12 Lu, X., Han, D. and Huang, Z. Fuel design and management for the control of advanced compression-ignition combustion modes. *Prog. Energy. Combust. Sci.* 2011. https://doi.org/10.1016/j.pecs.2011.03.003.

13 Yapicioglu, A. and Dincer, I. A review on clean ammonia as a potential fuel for power generators. *Renew. Sustain. Energy Rev.* 2019. https://doi.org/10.1016/j.rser.2018.12.023.

14 Scamardella, F., Zamboni, G., Canepa, E., Gualeni, P. and MacOcco, A. Ammonia as an alternative fuel for large passenger ships: benefits and challenges. *Prog. Mar. Sci. Technol.* 2022. https://doi.org/10.3233/PMST220018.

15 Tornatore, C., Marchitto, L., Sabia, P. and De Joannon, M. Ammonia as green fuel in internal combustion engines: state-of-the-art and future perspectives. *Front. Mech. Eng.* 2022. https://doi.org/10.3389/fmech.2022.944201.

16 Langella, G., De Joannon, M., Sabia, P., Iodice, P. and Amoresano, A. Ammonia as a fuel for internal combustion engines: latest advances and future challenges. *J. Phys. Conf. Ser.* 2022. https://doi.org/10.1088/1742-6596/2385/1/012036.

17 Balasubramanian, R., Abishek, A., Gobinath, S. and Jaivignesh, K. Alternative fuel: hydrogen and its thermodynamic behaviour. *J. Human. Earth. Futur.* 2022. https://doi.org/10.28991/HEF-2022-03-02-05.

18 Wang, S., Li, Y., Lv, J., Liu, Z., Gao, S., Hu, J., et al. Evaluation of hydrogen addition on combustion and emission characteristics of dual-fuel diesel engines with different compression ratios. *Processes*. 2023. https://doi.org/10.3390/pr11092675.

19 Vinoth Kanna, I. and Paturu, P. A study of hydrogen as an alternative fuel. *Int. J. Ambient. Energy*. 2020. https://doi.org/10.1080/01430750.2018.1484803.

20 Qazi, U.Y. Future of hydrogen as an alternative fuel for next-generation industrial applications; challenges and expected opportunities. *Energies*. 2022. https://doi.org/10.3390/en15134741.

21 Ghosh, N., Rhithuparna, D., Khatoon, R., Rokhum, S.L. and Halder, G. Sulphonic-acid functionalized novel limonia acidissima carbonaceous catalyst for biodiesel synthesis from millettia pinnata oil: optimization, kinetics, thermodynamics and cost analysis. *J. Clean. Prod.* 2023. https://doi.org/10.1016/j.jclepro.2023.136362.

22 Karmakar, B., Lalthazuala Rokhum, S. and Halder, G. Injecting superheated C1 and C3 alcohol supports towards non-catalytic semi-continuous conversion of *Hevea brasiliensis* oil into biodiesel. *Fuel*. vol. 314, no. 15, pp. 122777, Apr. 2022. https://doi.org/10.1016/j.fuel.2021.122777.

23 Gawale, G.R., Srinivasulu, G.N. and Lingayat, A. Experimental investigation on diesel ignited dual fuel homogeneous charge compression ignition engine for the selection of suitable alcohol based fuel. *J. Clean. Prod.* 2022. https://doi.org/10.1016/j.jclepro.2022.135048.

24 Wouters, C., Burkardt, P., Steeger, F., Fleischmann, M. and Pischinger, S. Comprehensive assessment of methanol as an alternative fuel for spark-ignition engines. *Fuel.* 2023. https://doi.org/10.1016/j.fuel.2023.127627.

25 Valera, H. and Agarwal, A.K. Methanol as an alternative fuel for diesel engines. In: *Energy, Environment, and Sustainability*. Springer, 2019, pp. 9–33. https://doi.org/10.1007/978-981-13-3287-6_2.

26 Rao, R.N., Silitonga, A.S., Shamsuddin, A.H., Milano, J., Riayatsyah, T.M.I., Sebayang, A.H., et al. Effect of ethanol and gasoline blending on the performance of a stationary small single cylinder engine. *Arab J. Sci. Eng.* 2020. https://doi.org/10.1007/s13369-020-04567-7.

27 Singh Bisht, Y., Kumar Shah, S. and Singh Rana, V. Road map for ethanol blending & prospective scope towards sustainable development in Indian scenario. *Mater. Today. Proc.* 2023. https://doi.org/10.1016/j.matpr.2023.03.375.

28 Jegan, T.M.C., Chitra, R. and Glivin, G. Testing of ethanol as an alternative fuel for IC engine. *Energy. Res. J.* 2021. https://doi.org/10.3844/erjsp.2021.13.21.

29 Putrasari, Y. and Lim, O. Dimethyl ether as the next generation fuel to control nitrogen oxides and particulate matter emissions from internal combustion engines: a review. *ACS Omega*. 2022. https://doi.org/10.1021/acsomega.1c03885.

30 Park, S.H. and Lee, C.S. Applicability of dimethyl ether (DME) in a compression ignition engine as an alternative fuel. *Energy. Convers. Manag.* 2014. https://doi.org/10.1016/j.enconman.2014.06.051.

31 Ardy, A., Pohan, R.D.A., Rizkiana, J., Laniwati, M. and Susanto, H. Dehydration of methanol to dimethyl ether (DME): performance of three types of catalyst at atmospheric pressure. *AIP Conf. Proc.* 2019. https://doi.org/10.1063/1.5095042.

32 Palomo, J., Rodríguez-Cano, M.Á., Rodríguez-Mirasol, J. and Cordero, T. Biomass-derived activated carbon catalysts for the direct dimethyl ether synthesis from syngas. *Fuel.* 2024. https://doi.org/10.1016/j.fuel.2024.131264.

33 Knothe, G. and Razon, L.F. Biodiesel fuels. *Prog. Energy. Combust. Sci.* 2017. https://doi.org/10.1016/j.pecs.2016.08.001.

34 Abed, K.A., Gad, M.S., El Morsi, A.K., Sayed, M.M. and Elyazeed, S.A. Effect of biodiesel fuels on diesel engine emissions. *Egypt. J. Pet.* 2019. https://doi.org/10.1016/j.ejpe.2019.03.001.

35 Pasha, M.K., Dai, L., Liu, D., Guo, M. and Du, W. An overview to process design, simulation and sustainability evaluation of biodiesel production. *Biotechnol. Biofuels.* 2021. https://doi.org/10.1186/s13068-021-01977-z.

36 Ramesha, D.K., Bangari, A.S., Rathod, C.P. and Samartha, C.R. Experimental investigation of biogas-biodiesel dual fuel combustion in a diesel engine. *J. Middle. Eur. Constr. Des. Cars.* 2015. https://doi.org/10.1515/mecdc-2015-0003.

37 Tiwari, R., Agrawal, P.S., Belkhode, P.N., Ruatpuia, J.V.L. and Rokhum, S.L. Hazardous effects of waste transformer oil and its prevention: a review. *Next. Sustain.* 2024. https://doi.org/10.1016/j.nxsust.2024.100026.

38 Ghadikolaei, M.A., Cheung. C.S. and Yung, K.F. Study of combustion, performance and emissions of diesel engine fueled with diesel/biodiesel/alcohol blends having the same oxygen concentration. *Energy*. 2018. https://doi.org/10.1016/j.energy.2018.05.164.

39 Bhujade, R.S. and Agrawal, P. Algal biocrude as drop-in feedstock for green fuel and green chemicals. In: *The Production of Biodiesel and Related Fuel Additives*. Bentham Science Publishers, 2024, pp. 178–198.

40 Schifter, I., González, U., Díaz, L., Sánchez-Reyna, G., Mejía-Centeno, I. and González-Macías, C. Comparison of performance and emissions for gasoline-oxygenated blends up to 20 percent oxygen and implications for combustion on a spark-ignited engine. *Fuel*. 2017. https://doi.org/10.1016/j.fuel.2017.07.065.

41 Ao, S., Gouda, S.P., Selvaraj, M., Boddula, R., Al-Qahtani, N., Mohan, S., et al. Active sites engineered biomass-carbon as a catalyst for biodiesel production: process optimization using RSM and life cycle assessment. *Energy. Convers. Manag*. 2024. https://doi.org/10.1016/j.enconman.2023.117956.

42 Rather, R.A., Wani. A.W., Mumtaz. S., Padder, S.A., Khan, A.H., Almohana, A.I., et al. Bioenergy: a foundation to environmental sustainability in a changing global climate scenario. *J. King. Saud. Univ - Sci*. 2022. https://doi.org/10.1016/j.jksus.2021.101734.

43 Bongers, A. Energy mix, technological change, and the environment. *Environ. Econ. Policy. Stud*. 2022. https://doi.org/10.1007/s10018-021-00324-8.

44 Deniz, C. and Zincir, B. Environmental and economical assessment of alternative marine fuels. *J. Clean. Prod*. 2016. https://doi.org/10.1016/j.jclepro.2015.11.089.

45 Ghosh, N., Rhithuparna, D., Rokhum, S.L. and Halder, G. Ethical issues pertaining to sustainable biodiesel synthesis over trans/esterification process. *Sustain. Chem. Pharm*. 2023. https://doi.org/10.1016/j.scp.2023.101123.

46 Kumawat, M.K. and Rokhum, S.L. Biodiesel production from oleic acid using biomass-derived sulfonated orange peel catalyst. *Front. Catal*. vol. 2, pp. 1–12. Jun. 2022, https://doi.org/10.3389/fctls.2022.914670.

47 Agrawal, P.S., Belkhode, P.N. and Rokhum, S.L. Biofuel production technologies – an overview. Sustain. *Biofuel. Prod. Technol*. 2022. https://doi.org/10.1002/9781119888864.ch4.

48 Pathak, G., Das, D., Rajkumari, K. and Rokhum, L. Exploiting waste: towards a sustainable production of biodiesel using: musa acuminata peel ash as a heterogeneous catalyst. *Green. Chem*. vol. 20, no.10, pp. 2365–2373, Apr. 2018. https://doi.org/10.1039/c8gc00071a.

3

The Role of Waste in the Circular Economy, Policies, and Legislation

3.1 Introduction

Agricultural waste, a byproduct of farming activities, presents substantial environmental challenges if not managed effectively. However, embracing the principles of a circular economic system offers a sustainable pathway for effectively utilizing agro-waste. This chapter explores how the definition of an economy can be adapted to maximize the value derived from such waste. Agro-waste refers to residual organic material generated during agricultural operations, including crop residues, animal manure, and processing byproducts. Globally, agriculture produces large volumes of waste, which, if improperly handled, can lead to environmental degradation, contribute to greenhouse gas (GHG) emissions, and result in resource inefficiency. This situation underscores the need to investigate innovative methods for converting agro-waste into valuable resources. Agro-waste presents both obstacles and opportunities within the context of sustainability [1]. On one hand, inappropriate disposal can cause water and soil pollution, generate methane emissions, and lead to the loss of valuable nutrients. Conversely, these waste materials contain organic matter and nutrients that can be reintegrated into agricultural systems, improving soil health and reducing the need for synthetic fertilizers. Furthermore, agro-waste can be converted into energy using techniques like anaerobic digestion and biomass conversion [2, 3].

The circular economy is an economic model focused on eliminating waste and optimizing resource utilization. Its aim is to regenerate products and materials at the end of their service life to create a closed-loop system that minimizes waste and conserves resources. Key tenets include designing for longevity and recyclability, promoting the use of renewable energy, and encouraging a systemic approach to resource management [4].

The shift from an agriculture-based economy during the Industrial Revolution led to economies centered on large-scale industry, mechanized production, and the factory system. The third industrial revolution incorporated elements like resource scarcity and the need for new reserves, IoT and system integration, renewable energy alongside oil scarcity, robotics, and the circular economy. The circular economy poses

Agro-Waste Management and Valorization, First Edition. Pratibha S. Agrawal, et al.

a significant challenge for all industrial sectors. It is considered a primary strategy for managing resource scarcity and ensuring the efficient use of critical raw materials. The circular economy necessitates advanced control and the elimination of waste generated by industrial activities [5]. This concept is rooted in understanding the existing and potential circularity of materials before they become waste.

The circular economy focuses on the production and use of products, aiming to reuse, repair, and recycle materials and products as much as possible. This involves extending product lifespan and minimizing waste. Protecting the environment is a core goal. Reusing and recycling products can slow the depletion of natural resources, reduce habitat disruption, and help limit biodiversity loss. It also aims to decrease dependence on raw materials. The circular flow model illustrates how economic activity transforms production factors into goods and services, which are then exchanged with consumers for value, allowing businesses to continue producing and potentially increasing their output and profitability (Figure 3.1). Interest in the circular economy is growing among economic stakeholders, making it very important. Adopting innovative and more sustainable methods for producing and consuming goods is crucial. This approach helps to utilize resources efficiently, which enhances the competitiveness and sustainability of businesses. Managing resources effectively is the only way to address the needs of a growing population while resources are limited and controlling biocapacity, which is unevenly distributed globally [6].

A major advantage of the circular economy is its capacity to safeguard both people and the planet [3]. This economic model seeks to lessen negative impacts on natural systems and biodiversity while fostering sustainable development and creating wealth and jobs. The Ellen MacArthur Foundation states that the circular economy, particularly through recycling and reusing end-of-life products, could cut GHG emissions by roughly 39%. This could prevent billions of tons of CO_2 emissions and help limit global warming to 2 °C, aligning with the Paris Agreement's goals. In this regard, the circular economy is vital for the energy and ecological transition by

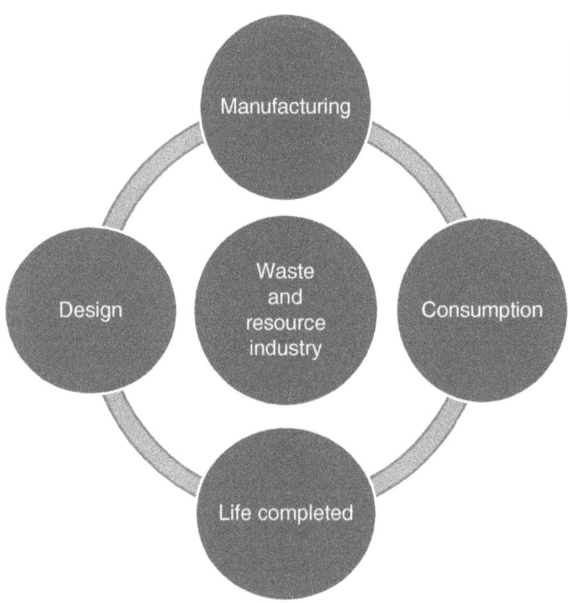

Figure 3.1 Third Industrial revolution, SWM industry, and circular economy.

reducing the environmental footprint of business operations [7, 8]. Ultimately, all the advantages of the circular economy are interconnected, opening up new avenues for a future regenerative economy. The key now is for companies to embrace these opportunities.

3.2 Crucial Reasons for Implementing a Circular Economy

The linear economy faces inherent limitations due to resource consumption, following a "take-make-dispose" model that generates enormous amounts of waste annually as products and materials are discarded before their maximum value is extracted. This practice puts excessive strain on the environment and risks exhausting finite natural resources (Figure 3.2). At both micro- and macro-economic levels, the circular economy offers significant operational and strategic development benefits, including considerable potential for new products, innovation, employment, and overall growth by maximizing value from goods [9].

The circular economy aims to restore or conserve the environment by limiting resource consumption, which leads to reduced waste [10, 11]. This highlights the importance of successfully implementing a circular economy by minimizing waste, reusing goods, and recycling materials. These are critical reasons why citizens, governments, businesses, and society are adopting the circular economy – to reduce waste and mitigate the environmental consequences associated with the linear economy [12].

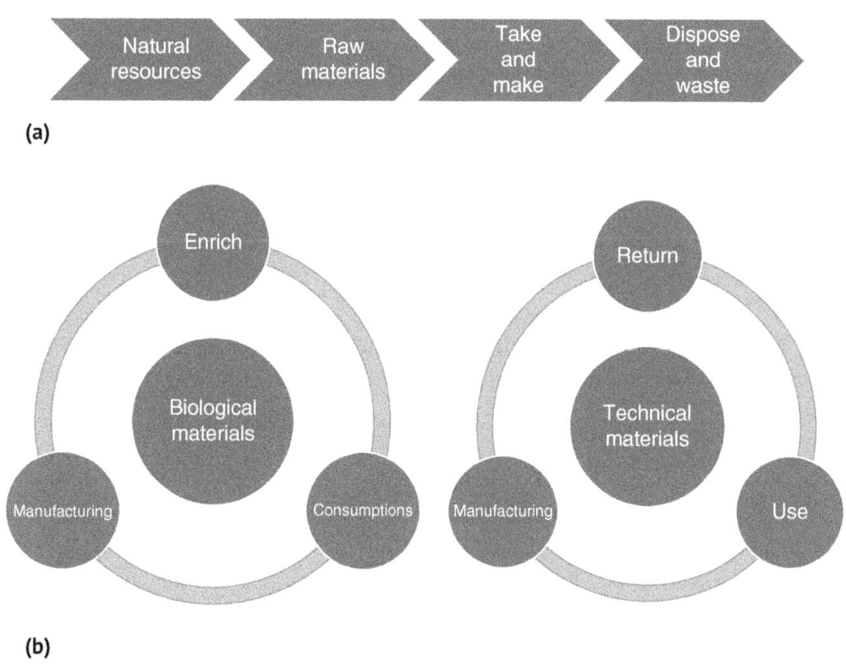

Figure 3.2 Comparison of linear economy and circular economy. (a) Linear economy: energy from finite sources. (b) Circular economy: energy from renewable sources.

3.2.1 Restore Environment

Producing goods using the linear economic model can be costly, and resource limitations contribute to water, air, and soil pollution, harming sensitive ecosystems. By promoting the use of biodegradable materials, the circular economy seeks to help restore the environment by altering manufacturing processes to become more sustainable [13]. Converting goods or products from waste materials significantly diverts waste from landfills and lowers material costs.

3.2.2 Recycling Industry

The circular economy is projected to be worth US$4.5 trillion globally by 2030, partly driven by the recycling industry acting as a powerful economic engine. In 2018, the industry generated over US$110 billion in economic activity, supported 534,000 jobs, and contributed US$13.2 billion in tax revenues to governments nationwide [12]. The circular economy has potential to boost GDP growth and job creation [14]. At the individual level, the circular economy can also save money. For example, instead of purchasing new clothing, repairing old items or taking them to a Repair Cafe can not only save money but also decrease your carbon footprint!

3.2.3 Social Responsibility

Circular economy business models are designed to keep products and materials in use for as long as possible to maximize their value. To meet the increasing consumer demand for more sustainable and eco-friendly products, businesses must adopt new approaches to generate revenue while aligning with consumer values. For instance, utilizing post-consumer waste as a resource not only saves money due to its readily available supply but also attracts a growing number of environmentally conscious consumers.

3.2.4 Reduces Waste

By aiming to minimize waste, the circular economy essentially "closes the loop" of traditional linear economies by reusing waste, thereby reducing the volume of trash sent to landfills. Through practices like upcycling, extending material lifespans, minimizing value loss, and circulating materials, the circular economy can redirect a substantial amount of waste back into productive use within society.

3.2.5 Renewable Energy

As a fundamental element of the circular economy, renewable energy sources simultaneously lessen the exploitation of natural resources and reduce pollution from burning fossil fuels. In industrial societies, renewables offer a dependable and potentially infinite power supply, enhancing energy security and lowering costs associated with finite energy sources. As a core component of the circular economy, renewable energy minimizes GHG emissions throughout the supply chain, addressing critical global issues such as climate change and air pollution.

Table 3.1 Comparison of linear economy and circular economy.

Feature	Linear economy	Circular economy
Approach	Take–make–waste	Reduce–reuse–recycle
Resource use	Resources are extracted, used, and disposed of as waste	Resources are kept in use for as long as possible, extracting the maximum value from them while in use, then recovering and regenerating products and materials at the end of their service life
Waste generation	High levels of waste are generated	Waste generation is minimized
Pollution	Pollution is generated throughout the lifecycle of products	Pollution is minimized through the use of cleaner technologies and processes
Economic model	Based on the sale of goods	Based on the provision of services
Sustainability perspective	Short-term gains	Long-term resilience

Overall, the circular economy provides significant advantages for agriculture, households, and industries. Most importantly, it addresses major global challenges like biodiversity loss, resource scarcity, global warming, and pollution through its carbon-free aspects and its focus on restoring ecosystems [14].

Table 3.1 illustrates the comparison between linear and circular economy from different features.

3.3 Principles of the Circular Economy

The core idea of a circular economy centers on closing resource loops, minimizing waste, and maximizing the value obtained from resources [14]. In the context of sustainable agriculture, this involves designing farming systems that efficiently use resources and minimize environmental impact [9] (Figure 3.3).

3.3.1 Designing for Efficiency

Applying circular economy principles promotes efficient farming practices, leading to less waste generation and optimized resource utilization.

3.3.2 Resource Regeneration

The concept of the circular economy emphasizes promoting regeneration rates for resources, allowing for their effective renewal. In agricultural recycling, this involves returning nutrients from agro-waste to farming systems through methods like composting or biochar production.

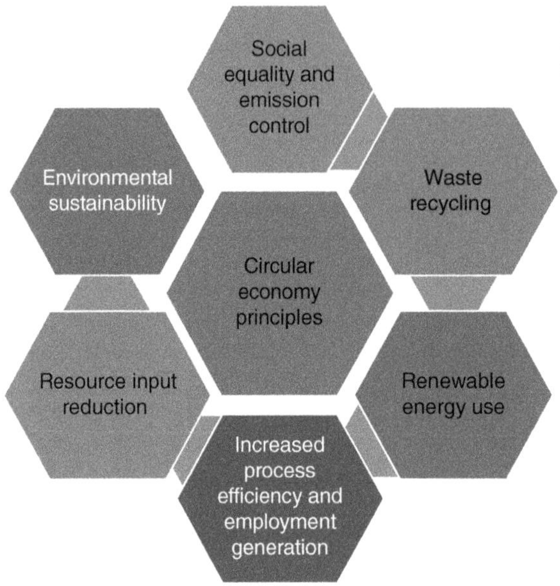

Figure 3.3 The principle of circular economy. Adapted from [12].

3.3.3 Closing the Loop

A key principle of the circular economy is viewing the output of one activity as the input for the next, forming a closed loop that enhances the development cycle. This involves developing integrated systems where waste from one process becomes a resource for another, such as using crop residues for bioenergy production. The emphasis is on closing the loop in processes, maximizing value in products and goods.

3.3.4 Promoting Renewable Energy

Unlike the linear economy, which relies on finite energy sources, the circular economy advocates for utilizing agro-waste for bioenergy generation. This approach reduces dependence on fossil fuels and helps mitigate GHG emissions.

3.3.5 Systems Thinking

The circular economy concept contributes to societal development by encouraging people to adopt a new mindset. It promotes holistic approaches to managing agro-waste, considering environmental, economic, and social consequences.

3.4 Role of Agro-waste in the Circular Economy – Examples

Managing agricultural waste helps reduce environmental problems, promoting the transformation of waste into useful byproducts and creating sources of revenue [14]. Implementing efficient methods for sustainable agricultural development involves treating and processing waste to enhance the bio-economic potential. Improper handling of agricultural waste can negatively impact food production. Agricultural waste can contaminate water sources (including reservoirs, groundwater, and surface water), contribute to soil erosion and sedimentation, and introduce livestock and organic contaminants. Agricultural waste can also obstruct waterways, potentially leading to floods and loss of life and property [15]. Burning agricultural waste or improper dumping causes pollution that harms both the environment and human health [16]. Efficient management of agricultural waste, particularly through recycling, offers the most promising and effective solutions to these problems [17].

Agricultural residues are a source for producing bioethanol through fermentation. Achieving high ethanol yield requires removing lignin and hemicelluloses, which improves the accessibility of cellulosic material to hydrolytic enzymes. Before fermentation, lignin and hemicellulose are separated from lignocellulosic biomass, which is then fermented to produce bioethanol [18]. Various methods for utilizing agricultural waste to enhance the circular economy are explained with examples in Figures 3.4 and 3.5.

3.4.1 Biochar Production

One example of circular economy application involves biochar production (Figure 3.6). Biochar is a type of charcoal, sometimes modified, suggested for use in organic applications, such as in soil. Converting agricultural residues into biochar is commonly done to enhance soil fertility and sequester carbon in agriculture [20, 21]. The use of biochar supports the promotion of biodegradable materials; reusing these materials within a circular economy framework aims to restore the environment by modifying business

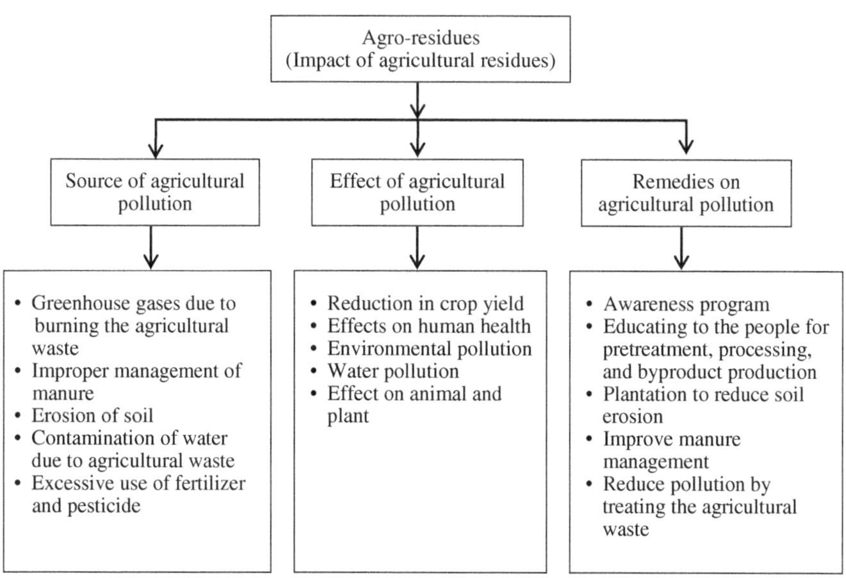

Figure 3.4 Impact of agricultural residues. Adapted from [19].

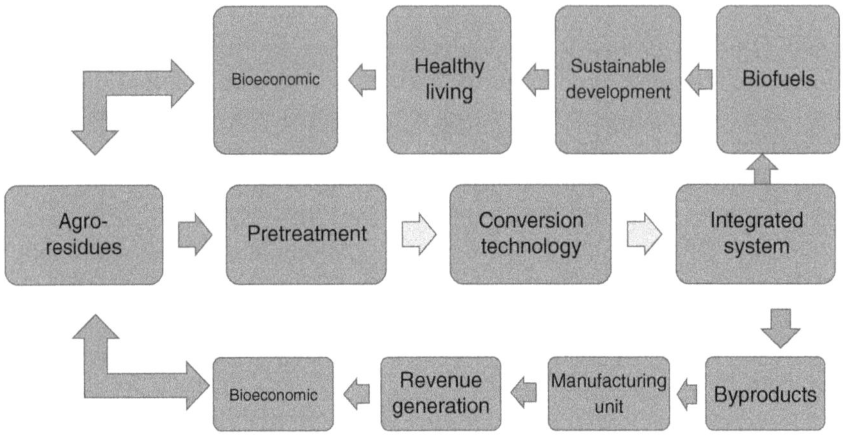

Figure 3.5 Role model of use of agro-waste in circular economy. Adapted from [19].

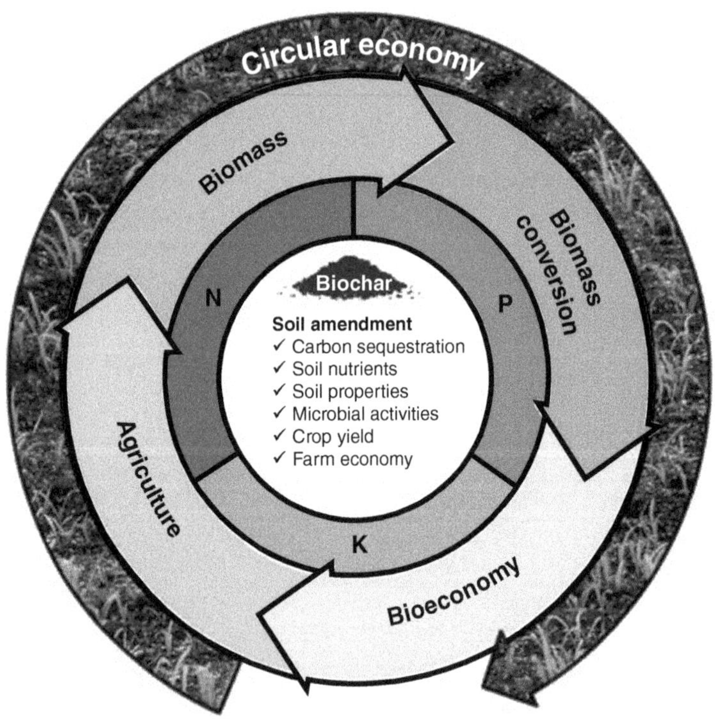

Figure 3.6 Circular economy in biochar production. [22]/Elsevier.

manufacturing processes to be more sustainable. Biochar is created when biomass undergoes high-temperature, limited-oxygen processes, such as pyrolysis or gasification. Biochar is considered a valuable element for regenerative, circular economies because this sustainable material is produced from organic waste, effectively transforming waste into value.

3.4.2 Anaerobic Digestion

Biological processing of organic substances within a digester in the absence of oxygen is called anaerobic digestion. This process involves four main steps: (i) hydrolysis, (ii) acidogenesis, (iii) acetogenesis, and (iv) methanogenesis. The hydrolysis phase breaks down carbohydrates, proteins, and lipids into simpler monomeric forms like sugars, amino acids, and fatty acids, facilitated by hydrolase enzymes within the anaerobic digester.

Biogas production is a potential source of energy that reduces reliance on fossil fuels. Organic waste, such as animal manure, can be converted into biogas for energy generation. Biogas plants process manure into biomethane, a useful energy form. This process not only reduces waste from cow dung but also helps decrease the use of fossil fuels for energy.

3.4.3 Circular Agriculture Models

Processing agricultural waste supports the bio-economic model and the development of sustainable byproducts [23]. Valuable waste streams like carbohydrate and protein waste from agricultural residues can be used for biodiesel production and other value-added products. Converting a large proportion of waste into usable, economically valuable material contributes to increasing agricultural crop productivity, saving energy, promoting environmental sustainability, and reducing dependence on non-renewable energy. Rural communities could establish manufacturing units to process agricultural residues into different environmentally friendly and valuable products, which would lead to improved living standards [24].

Agricultural residues contain numerous nutrients that can be transformed into organic fertilizer, reducing the need for commercial fertilizers. Agricultural waste is a valuable resource that can be used for energy generation through appropriate conversion processes (Figures 3.7 and 3.8). Adopting processes that convert agricultural waste into useful products, such as producing biofuels through recycling,

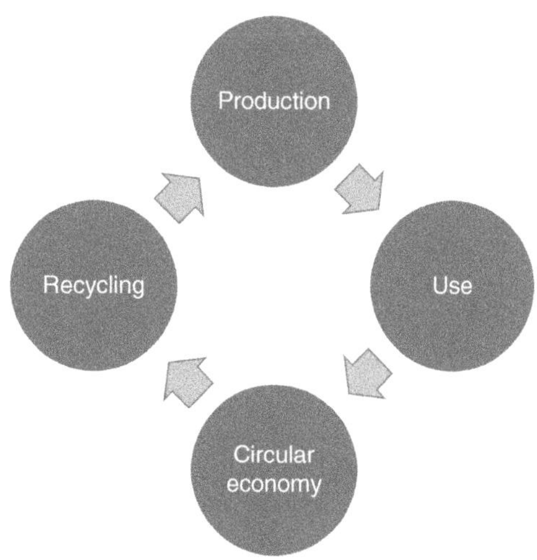

Figure 3.7 Model of circular economy. Adapted from [9].

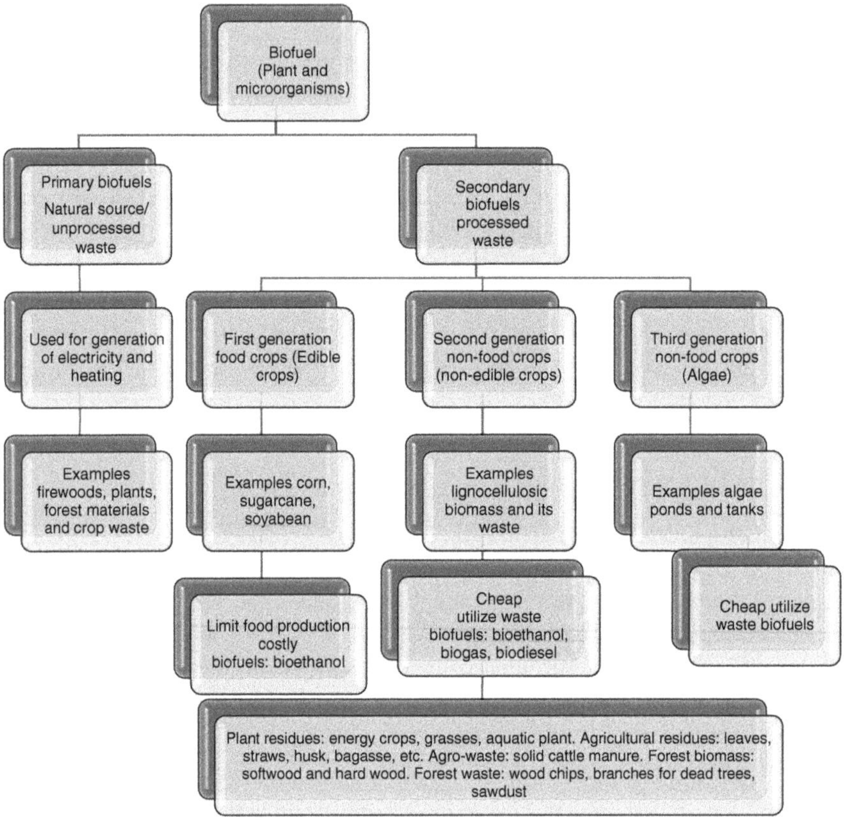

Figure 3.8 Classification of agricultural waste and residues. Adapted from [26].

can minimize this waste [13, 25]. A combined approach where crop residues and organic wastes are recycled together within farms can enhance soil health and crop productivity. Organic waste, particularly cow manure, is effectively used as organic fertilizer for crops instead of synthetic fertilizers, which can pollute the soil and waterways. Cow manure provides high nutrient content for the soil, essentially serving the same function as solid fertilizers but without the associated pollution risks.

3.5 Circular Economy Challenges

The circular economy plays a vital role in resource management, promoting environmental sustainability and reducing environmental impacts through waste management and byproduct conversion that supports financial stability. Some challenges faced by the circular economy include logistical issues, innovation requirements, and policy decisions. Addressing these challenges is essential for future success [27].

3.5.1 Logistical Challenges

Recycling agro-waste to produce biomass, biogas, or biochar involves a closed-loop system with regeneration. This activity is carried out at regeneration centers where agro-waste must be transported. Agro-waste is often bulky, making storage and transport difficult.

3.5.2 Technological Innovation

Successfully implementing the circular economy for waste conversion and energy recovery is possible through the use of modern technology for processing and converting waste [16]. A key challenge is developing cost-effective technologies for waste conversion and energy recovery.

3.5.3 Policy and Economic Incentives

Effective implementation of the circular economy relies heavily on well-designed policies and their enforcement. Encouraging regulatory frameworks and financial incentives are needed to promote circular economy practices in agriculture. Managing agro-waste is crucial for achieving sustainable agricultural practices and mitigating environmental impacts. By applying circular economy principles like resource efficiency, regeneration, and closed-loop systems, agro-waste can be changed from a burden into a valuable resource. Effective implementation requires collaboration among policymakers, researchers, and farmers to create innovative solutions that improve agricultural productivity while conserving natural resources.

3.6 Agro-waste

Agro-waste, the byproduct of agricultural activities, comprises a wide variety of materials generated during crop cultivation, processing, and consumption. Research explores the classification, sources, and generation patterns of agro-waste, highlighting its significance in sustainable agriculture and environmental management. Agro-waste, also referred to as agricultural residue or biomass residue, includes materials left or discarded after farming activities such as harvesting, processing, or consumption. The composition and quantity of these materials vary significantly based on factors like crop type, farming methods, and processing techniques. These materials are increasingly recognized for their potential to be converted into valuable products like biofuels, fertilizers, and biodegradable plastics, thereby promoting a circular economy approach in agriculture [28].

3.6.1 Classification of Agro-waste

Agro-waste can be categorized based on its origin and composition as follows [29]:

1. **Crop Residues:** These are materials left after crop cultivation and harvesting, such as stalks, stems, leaves, and husks, remaining in the field. They are considered effective measures against soil erosion.

2. **Processing Waste:** This refers to byproducts generated during the processing of agricultural produce, such as bran, husks, and shells [30]. Treatment and conditioning are methods used to alter raw material characteristics, making waste suitable for final processing.
3. **Animal Waste:** Generally categorized into solid, slurry, and wastewater types [23]. Solid wastes are often treated through drying or composting. Dried wastes can serve as fertilizer or fuel for energy generation through combustion. This category includes manure and bedding from animal husbandry.
4. **Food-processing Waste:** This is often a complex waste stream. In vegetable and fruit processing, it involves removing oil, starch, sugar, and juice, leaving residues like peels, seeds, and bran. In animal processing, wastes include blood, bone, and neural tissues. This category encompasses waste from food-processing industries, such as peels, pulps, and pomace.

3.6.2 Sources and Generation Patterns

The generation of agro-waste follows specific patterns influenced by farming practices and regional factors. Agro-waste generation and management play a significant role in the circular economy. A large portion of agricultural waste comes from crop cultivation residues, which, when processed, can enhance crop yield [23]. Biofuel production from agro-waste is a potential energy source that reduces the use of fossil fuels. Organic waste, such as animal manure, can be converted into biogas for energy. Biogas plants process manure into biomethane, a useful energy form that reduces waste from cow dung and decreases reliance on fossil fuels for energy [31].

1. **Field Operations:** This refers to activities and tools that support and optimize farming processes and improve efficiency. During crop cultivation, residues like straw, stalks, and leaves are left in fields after harvest. The quantity and composition of these residues vary depending on crop type, yield, and farming practices.
2. **Processing Industries:** These industries, operating either continuously or in batches of indistinguishable materials, generate significant amounts of waste materials like peels, shells, and pulp residues [6]. These wastes are often rich in organic matter and nutrients but require proper handling to avoid environmental contamination.
3. **Livestock Farming:** This sector faces numerous challenges, including food security, job and income creation, animal welfare, and environmental protection. Animal waste, such as manure and bedding materials, is a major component of agro-waste. Effective management of animal waste is essential for mitigating environmental impacts like nutrient runoff and GHG emissions.
4. **Consumer Waste:** This includes residues, materials, and products that have lost their consumer properties. At the consumer level, food waste is a substantial part of agro-waste, comprising edible plant parts discarded due to spoilage, overproduction, or aesthetic reasons.

Agro-waste is a valuable resource that, if managed effectively, can contribute to sustainable agricultural practices, environmental stewardship, and economic development. By examining its classification, sources, and generation patterns, this chapter highlights the importance of integrating agro-waste management into broader strategies for environmental sustainability and circular economy principles in agriculture [32].

3.7 Agro-waste Management: Current Practices

Systematic and organized utilization of agricultural waste from farms, including discarded food items, involves converting them into byproducts for preservation and improving the quality of air, water, soil, and plants in the most sustainable ways [23]. This approach focuses on managing agricultural waste effectively. Biochemical and thermochemical treatments are applied to produce value-added products, addressing significant challenges in agricultural sectors. The sector faces problems such as land availability and crop quality, lack of power sources and machinery, the economic condition of farmers, payment systems between farmers and customers involving intermediaries, adverse effects from geographical and climatic conditions, and insufficient technology. Despite these difficulties, farmers today manage crop residues from cultivation, processing, and consumption, creating opportunities for sustainable development, mitigating environmental impacts, and harnessing waste's potential as a valuable resource.

Farmers protect the ecosystem by controlling the use of harmful pesticides and reducing the risk of contaminating natural resources. Composting provides nutrient-rich fertilizers, supporting water quality protection and sustainable food production. Generating income from bioenergy derived from waste treatment contributes to a more sustainable and environmentally protective approach.

3.7.1 Current Practices in Agro-waste Management

Practices for managing agro-waste vary widely by region, influenced by factors like agricultural methods, infrastructure, and economic conditions. Common approaches include:

1. **Field Incorporation:** Managing agro-waste in the field involves controlling storage, nutrient loss, and soil erosion. A common practice is leaving crop residues like straw and stalks in fields to decompose naturally. This enriches soil organic matter but can potentially lead to nutrient loss and soil erosion.
2. **Burning:** Cultivation sometimes involves clearing fields by setting fires. This method can be quick and economical for clearing weeds and waste, but it is unsustainable due to atmospheric pollution, contributing to air quality issues, GHG emissions, and reduced soil fertility.
3. **Composting:** This is a highly adaptable method for managing biodegradable solid wastes, where agricultural wastes are composted [33]. It offers benefits for plants and animals and reduces methane emissions. Converting organic waste into compost provides organic fertilizer that improves soil health and crop productivity [2].

4. **Biogas Production:** Biogas is a renewable fuel produced from organic matter, such as food or animal waste, through anaerobic digestion, a process where microorganisms break down material without oxygen [32]. The resulting biogas is used for domestic purposes like cooking, heating, and power generation.

3.7.2 Traditional Disposal Methods

Waste generation is a long-standing issue, evolving with human living patterns in terms of quantity and quality of waste produced. Initially, waste was biodegradable, but later forms included toxic substances [7, 34]. The most common and widespread waste management method was open dumping, which is still practiced widely as it is easy and inexpensive. Agro-waste has historically been managed through traditional disposal methods [33].

1. **Open Burning:** Identified as a widely adopted method of waste treatment globally in low- to middle-income countries, open burning of waste contributes to GHG emissions. This practice is prevalent due to inadequate waste disposal infrastructure. Direct burning of crop residues causes environmental pollution through emissions of particulate matter, GHGs, and toxic pollutants.
2. **Dumping and Landfilling:** Open dumping, where waste is deposited at specific sites without environmental protection, is common in many countries. Landfilling involves creating open pits in low-lying areas to dispose of collected waste, which is then covered with soil. Disposing of agro-waste in open dumps or landfills can lead to methane emissions from decomposition and potential leaching of pollutants into soil and groundwater.
3. **Dumping in Water Bodies:** Disposing of waste into water bodies can introduce toxic substances, causing oxygen depletion and harming aquatic life [24]. Dumping waste into rivers, lakes, or seas leads to water pollution, nutrient enrichment, and damage to aquatic ecosystems.
4. **Uncontrolled Decomposition:** This applies to waste types not classified as controlled, special, or hazardous waste, including specific mining and agricultural wastes. Allowing crop residues to decompose naturally in fields without management can deplete soil nutrients and contribute to GHG emissions.

3.7.3 Environmental Impacts of Agro-waste

The primary environmental problems caused by agro-waste are air, water, and soil contamination, largely due to poor waste management, ranging from absent collection systems to ineffective disposal. Contamination of drinking water can cause infections and disease transmission from open, unsanitary landfills. Improper agro-waste management has the following significant environmental consequences:

1. **Air Pollution:** Air pollution results from the open burning of agro-waste, which releases particulate matter, carbon monoxide, nitrogen oxides, and volatile organic compounds, contributing to air pollution and respiratory issues [28]. Developing clean and innovative methods is necessary to combat air pollution.

2. **GHG Emissions:** Carbon dioxide is the primary GHG emitted from burning fossil fuels like coal, oil, and natural gas for electricity generation [31]. However, the decomposition of organic waste in landfills and open fields releases methane, a powerful GHG with a higher global warming potential than carbon dioxide.
3. **Soil Degradation:** Disposing of agro-waste in pits or fields without proper management leads to soil nutrient depletion, reduced fertility, and increased erosion risks. Issues related to water pollutants and infection can also arise from burning and uncontrolled decomposition.
4. **Water Pollution:** Soil pollution is a major contributor to water pollution. This occurs when agro-waste disposed in soil runs off from fields due to mismanagement.
5. **Loss of Biodiversity:** The overall effect of improper agro-waste handling results in soil, water, and air pollution, and GHG emissions. This can lead to habitat destruction due to unsustainable agricultural practices, including poor agro-waste management, threatening biodiversity and ecosystem services.

Properly planned agro-waste management is necessary to minimize environmental impacts and utilize agricultural residues more beneficially. Sustainable methods such as composting, biogas production, and biochar utilization can help agricultural communities improve soil health, reduce GHG emissions, and promote circular economy principles. Supportive policies involving innovation, development, and stakeholder awareness are essential for achieving sustainable agro-waste management practices globally.

3.7.4 Circular Economy Framework

The circular economy improves resource efficiency and adds value by adopting a sustainability approach that minimizes waste through closed-loop systems. The concept of a circular economy aims to reduce waste while maximizing the use of sustainable practices and social benefits. The main principles of the circular economy include (i) designing products to be durable, repairable, and reusable to extend their lifespan; (ii) utilizing renewable resources; (iii) maximizing resource efficiency by reducing waste generation, efficiently using materials, and recycling; (iv) creating closed-loop systems where products, components, and materials are reused and remanufactured; and (v) encouraging innovation across sectors to develop new business models, technologies, and practices that support the circular economy.

3.7.5 Overview of Existing National and International Policies

3.7.5.1 The European Union Waste Framework Directive and Agro-waste in the Circular Economy

The European Union's (EU) legislative approach to waste management is fundamentally guided by the Waste Framework Directive (WFD) (Directive 2008/98/EC), which has been instrumental in shaping waste policy across its member states since its initial adoption and subsequent revisions, notably with the amendments under the Circular Economy Package in 2018 (Directive 2018/851). The WFD establishes overarching principles for waste management, with its cornerstone being the "waste hierarchy." This hierarchy prioritizes waste prevention, followed by preparing for reuse, then recycling, other recovery operations (including energy recovery), and lastly, disposal (such as landfilling), as the least preferred option. The primary objective is to minimize the adverse impacts of waste generation and management on human health and the environment, contribute to the sustainable use of resources, and transition the EU toward a circular economy. The Directive's scope is broad, covering most waste streams, including agricultural waste, although specific rules for certain types of agricultural waste are detailed in other, more specific EU regulations. The explicit inclusion and emphasis on the waste hierarchy within the WFD provide a legally binding framework that mandates member states to move away from linear "take–make–dispose" models toward more circular practices, including for agro-waste streams. The 2018 amendments specifically reinforced the circular economy objectives, setting ambitious targets for recycling and waste reduction across various sectors, compelling member states to integrate circular principles into their national waste management strategies. This legislative evolution signifies a fundamental shift in the EU's environmental policy, positioning waste not merely as a problem to be disposed of, but as a potential resource to be managed strategically within a closed-loop system. For agro-waste, this means a move beyond simple dumping or burning toward valorization through reuse, recycling (e.g., composting, anaerobic digestion), and recovery.

Under the WFD framework and related EU policies, specific measures address agro-waste, particularly focusing on its potential for bioenergy production and material recovery. While the WFD sets the general principles, the Common Agricultural Policy (CAP) and directives related to renewable energy and waste treatment provide more detailed regulations and incentives. Policies encourage the sustainable management of agricultural practices, which inherently includes agro-waste. For example, regulations related to nitrates and soil health influence how manure is managed, promoting its use as fertilizer rather than allowing untreated runoff. Furthermore, the EU's Renewable Energy Directive (Directive 2009/28/EC, revised as Directive 2018/2001/EU) actively promotes bioenergy production to meet renewable energy targets. Agricultural residues and manure are explicitly listed as valuable feedstocks for bioenergy (e.g., biogas via anaerobic digestion) [35]. The policy provides incentives, such as feed-in tariffs or green certificates, to encourage farmers and energy producers to utilize these waste streams for energy generation. Regulations on composting and digestate quality ensure that residues from anaerobic digestion of agro-waste can be safely returned to the soil as nutrient-rich fertilizers, closing nutrient loops within the agricultural system. Policies also support research and innovation in valorization technologies for diverse agro-waste types, including lignocellulosic residues for biofuels or bio-based materials. The aim is to facilitate the conversion of low-value or problematic agro-waste into higher-value products or energy, thereby creating economic opportunities within the rural sector and reducing reliance on synthetic inputs. This integrated policy approach, spanning waste management, agriculture, and energy, is crucial for promoting circularity in the agro-food system. The regulatory framework actively guides the transition from traditional, often environmentally detrimental, agro-waste disposal practices toward more sustainable and economically beneficial pathways aligned with circular economy objectives.

The impact of the EU WFD and associated policies on the management and utilization of agro-waste has been multifaceted, driving significant changes in practices and fostering a growing bioeconomy. Environmentally, the push toward higher levels of the waste hierarchy has led to a reduction in environmentally harmful practices such as uncontrolled burning and landfilling of agro-waste. By encouraging composting and anaerobic digestion, the policies contribute to reducing GHG emissions (particularly methane from manure storage), preventing water and soil pollution from nutrient runoff, and improving soil health through the application of organic fertilizers derived from treated waste. The focus on bioenergy from agro-waste directly contributes to the EU's renewable energy targets and helps mitigate climate change by displacing fossil fuels. Economically, the policies create new markets and revenue streams for farmers and rural businesses involved in collecting, processing, and valorizing agro-waste. Investments in anaerobic digestion plants, composting facilities, and other bio-based industries are stimulated, creating jobs and fostering rural development. The increased availability of organic fertilizers reduces farmer dependence on costly synthetic fertilizers, improving economic resilience in the agricultural sector. Socially, better agro-waste management improves the living conditions in rural areas by reducing odors, pest issues, and health risks associated with improper disposal. However, challenges remain. Implementing these policies effectively requires significant investment in infrastructure, technical expertise, and behavioral change among farmers. Logistics for collecting and transporting diffuse agro-waste can be complex. Ensuring the quality and safety of derived products (like compost and digestate) requires robust monitoring and certification systems. Despite these challenges, the WFD and the broader Circular Economy Package have provided a powerful legislative impetus for transforming agro-waste management in the EU from a disposal problem into a key component of a sustainable, circular bioeconomy, demonstrating that regulatory frameworks can be effective drivers of environmental and economic change when coupled with targeted policies and incentives. The continuous evolution of the WFD reflects the EU's commitment to adapting its regulatory tools to support emerging technologies and evolving circular economy models, including those relevant to agricultural residues.

3.7.5.2 US Environmental Protection Agency Regulations and Agro-waste

In the United States, the regulation of waste management falls primarily under the purview of the Environmental Protection Agency (EPA), governed by key federal statutes such as the Resource Conservation and Recovery Act (RCRA) of 1976 [36]. RCRA is the principal federal law in the United States, governing the disposal of solid and hazardous waste. Its main goals are to protect human health and the environment from the potential hazards of waste disposal, conserve energy and natural resources, reduce the amount of waste generated, and clean up waste that may have spilled, leaked, or been improperly disposed of. While RCRA broadly covers solid waste, including some agricultural waste, many routine agricultural wastes, such as manure and crop residues returned to the soil, are exempt from certain RCRA regulations under specific conditions. However, the EPA's authority and regulatory framework extend beyond RCRA, encompassing other environmental laws like the Clean Water Act (CWA) and the Clean Air Act (CAA), which indirectly or directly impact agro-waste management, particularly regarding pollution control. For example, the CWA regulates discharges from Confined Animal Feeding Operations (CAFOs), influencing how large volumes of manure are managed to prevent water pollution. The CAA regulates air emissions, including those from agricultural activities like burning or large-scale composting facilities, addressing concerns like particulate matter and volatile organic compounds. This multi-faceted regulatory landscape, while perhaps less explicitly branded under a single "circular economy" directive compared to the EU, nonetheless provides a framework for managing agro-waste in ways that align with resource conservation and environmental protection, foundational elements of circularity. The EPA's role involves setting standards, issuing permits, monitoring compliance, and enforcing regulations to ensure that waste management practices, including those in agriculture, minimize environmental harm.

Specific EPA-related policies and programs significantly influence the utilization of agricultural residues and manure, particularly for bioenergy production. A prime example is the Renewable Fuel Standard (RFS), established under the Energy Policy Act of 2005 and expanded by the Energy Independence and Security Act of 2007 [37]. The RFS mandates that a certain volume of renewable fuel must replace or supplement petroleum-based gasoline or diesel transportation fuel. This policy creates a market for biofuels derived from various feedstocks, including agricultural residues, animal fats, and cellulosic biomass from energy crops. The program categorizes biofuels based on their lifecycle GHG emissions, providing greater incentives for advanced biofuels and cellulosic biofuels, which often utilize agro-waste feedstocks, thereby promoting the use of sustainable biomass sources. The RFS provides economic incentives in the form of Renewable Identification Numbers (RINs), which are credits generated for each gallon of renewable fuel produced. These RINs can be traded, providing revenue streams that make bioenergy production from agro-waste economically viable. The EPA also administers programs related to composting and anaerobic digestion, promoting their adoption through guidance documents, research initiatives, and funding opportunities. While not always regulatory mandates for *all* agro-waste, these programs encourage the recovery of nutrients and energy from manure and residues, aligning with resource recovery objectives. Furthermore, research supported by the EPA and other federal agencies explores innovative ways to convert different types of agro-waste into valuable products, such as biochar for soil amendment and carbon sequestration, or biochemicals, broadening the potential for agro-waste valorization beyond traditional uses and supporting a move toward a more circular agricultural system.

The impact of EPA regulations and related federal programs on agro-waste management in the United States has driven changes toward more environmentally conscious practices and spurred growth in the bioenergy sector. Environmentally, regulations under the CWA and CAA have led to improved management practices in large agricultural operations, reducing direct pollution from manure and agricultural burning. The RFS, by promoting biofuels from agro-waste, contributes to reducing GHG emissions in the transportation sector compared to fossil fuels, although the full lifecycle impacts can be complex and depend on feedstock sourcing and production methods. Policies supporting composting and anaerobic digestion encourage nutrient cycling and methane capture, further reducing environmental impacts. Economically, the RFS and bioenergy incentives have stimulated investment in biomass-processing facilities and created new markets for agricultural residues, providing supplemental income for farmers. This has fostered innovation in harvesting and utilizing residues that were previously considered waste. The development of industries around agro-waste valorization creates jobs in rural areas. However, challenges persist. The RFS program has faced implementation challenges and debates over feedstock sustainability. Logistics and infrastructure for collecting and transporting diffuse agricultural residues across vast areas remain a significant barrier. Economic viability can still be marginal without policy support, subject to fluctuations in energy and commodity markets. While RCRA provides baseline environmental protection, its exemptions for many routine agricultural wastes mean that significant volumes of agro-waste may not be subject to the full regulatory scrutiny applied to other waste streams. Despite these limitations, the combination of environmental regulations and bioenergy policies under the EPA and other agencies provides a framework that encourages the sustainable use of agro-waste, pushing practices toward resource recovery and aligning, in principle, with circular economy goals within the agricultural sector.

3.7.5.3 India's National Policy on Biofuels and Agro-waste

India's approach to leveraging its vast agricultural resources for energy and sustainability is significantly shaped by its National Policy on Biofuels, first introduced in 2009 and updated in 2018. This policy is a cornerstone of India's strategy to reduce its reliance on fossil fuel imports, enhance energy security, mitigate climate change, and boost rural incomes. The policy categorizes biofuels into "Basic Biofuels" (first generation) like bioethanol and biodiesel produced from food crops (e.g., sugarcane, edible oils) and "Advanced Biofuels" (second generation [2G]) produced from non-food feedstocks, including agricultural residues, solid waste, and third-generation biofuels derived from algae. The 2018 policy specifically placed a strong emphasis on promoting Advanced Biofuels using lignocellulosic biomass and other agro-waste, recognizing the vast untapped potential of these resources in the country. The policy aims to increase the blending of ethanol in petrol (with a target of 20% blending by 2025–2026) and biodiesel in diesel, while also promoting other biofuels like compressed biogas (CBG). The policy provides a clear framework and targets for expanding biofuel production capacity and feedstock utilization, positioning agro-waste as a critical resource for this expansion [38].

The National Policy on Biofuels includes several specific measures designed to incentivize the production and utilization of biofuels from agro-waste. It promotes the establishment of commercial-scale plants for producing advanced biofuels from non-food biomass, offering viability gap funding, tax incentives, and subsidies to de-risk investments in these technologies. The policy encourages farmers to supply agricultural residues by providing market mechanisms and potentially additional income streams. It explicitly supports the conversion of various types of agro-waste, including rice straw, wheat straw, cotton stalks, bagasse, and corn stover, into biofuels. For instance, it promotes the setting up of 2G ethanol biorefineries that use lignocellulosic agro residues. It also supports the production of CBG from agricultural residues, manure, and other organic wastes, providing price assurance and purchase guarantees for CBG. Furthermore, the policy encourages research and development (R&D) in efficient technologies for biomass collection, logistics, and conversion. It promotes the valorization of other agro-waste streams beyond liquid biofuels, such as using press mud from sugarcane crushing for biogas or producing biochar from crop residues for soil health. The policy creates an enabling ecosystem by facilitating approvals, land allocation, and infrastructure development necessary for setting up biofuel projects utilizing agro-waste. The focus on domestic feedstocks, particularly residues, aligns with the circular economy principle of utilizing waste as a resource, reducing both import dependency and the environmental problems associated with residue burning or improper disposal.

The impact of India's National Policy on Biofuels, particularly its focus on agro-waste, is evident in the growing investments and projects in the biofuel sector, with significant environmental and economic implications. Environmentally, the policy aims to reduce air pollution caused by the widespread practice of burning agricultural residues (like paddy straw) by providing an economic incentive for their collection and use as biofuel feedstocks. This diversion of waste also helps mitigate GHG emissions from burning and reduces the burden on landfills. Promoting CBG from manure helps capture methane, a potent GHG. The use of biofuels in transportation contributes to lower emissions compared to fossil fuels, supporting India's climate goals. Economically, the policy creates new value chains for agricultural waste, providing additional income opportunities for farmers and generating employment in rural areas through biomass collection, processing, and plant operations. It stimulates domestic industry for manufacturing biofuel plant equipment and promotes technological innovation. The policy also contributes to energy security by reducing costly oil imports. Socially, reducing agricultural burning improves air quality, benefiting public health, particularly in regions prone to stubble burning. However, implementation faces challenges, including ensuring efficient biomass supply chains, managing the spatial distribution of residues, ensuring consistent feedstock quality, and overcoming technological hurdles and the high capital costs associated with advanced biofuel technologies. The economic viability of projects often still relies heavily on government support and feedstock pricing. Despite these challenges, the policy represents a significant step in recognizing agro-waste as a valuable national resource and actively promoting its utilization within a framework that supports energy independence, environmental protection, and rural economic development, strongly aligning with circular economy principles for the agricultural sector.

3.8 Challenges in Implementing the Circular Economy for Agro-waste

Transitioning from a linear model of production and consumption to a circular economy, especially within the complex and distributed agro-food system, presents a formidable set of challenges. While the theoretical benefits in terms of resource conservation, waste reduction, and environmental protection are widely acknowledged, the practical implementation faces significant hurdles that can be broadly categorized into economic constraints and technological limitations [39]. A thorough analysis of these challenges is crucial for identifying effective strategies and policy interventions needed to accelerate the adoption of circular economy principles for valorizing agricultural waste streams.

3.8.1 Economic Challenges

The economic viability and competitiveness of circular economy initiatives for agro-waste are influenced by several interconnected factors, often rooted in the established economic structures of the linear system.

1. **Cost of Technology and Infrastructure:** A primary economic barrier is the substantial capital expenditure (CapEx) and operational expenditure (OpEx) associated with developing and implementing the necessary technologies and infrastructure for circular processes. Unlike traditional linear approaches like dumping or burning agro-waste, which often have minimal direct costs (though significant externalized environmental costs), circular solutions such as anaerobic digestion, advanced composting, biorefineries for biochemical production, or specialized recycling facilities require considerable investment in equipment, construction, and skilled labor for operation and maintenance. The cost can vary significantly based on the scale of the operation, the specific type of agro-waste being processed, and the required level of technological sophistication to produce market-ready outputs. For instance, setting up a large-scale facility to convert diverse lignocellulosic residues into advanced biofuels or bio-based materials involves complex chemical or biological processing steps, demanding high-cost reactors, separation units, and process control systems. Furthermore, the distributed nature of agricultural production means that collecting and aggregating sufficient volumes of specific agro-waste streams to feed centralized processing facilities necessitates developing extensive and efficient logistics infrastructure, including collection networks, storage facilities, and transportation fleets. The cost of this collection logistics can be particularly high for low-density, bulky residues like straw or scattered manure from grazing animals, representing a significant portion of the overall OpEx and making these feedstocks economically less attractive compared to more concentrated industrial or urban waste streams. These high initial and ongoing costs can deter potential investors and limit the ability of farmers or small rural enterprises to adopt such technologies without external financial support.
2. **Market Demand and Pricing Volatility:** Products derived from agro-waste within a circular economy framework, such as biogas, bio-based fertilizers (compost, digestate), biochar, or biochemicals, often face challenges in competing with established products from linear value chains, particularly those based on fossil fuels or energy-intensive production methods. The market for these circular products can be nascent, subject to fluctuating demand, and vulnerable to the price volatility of competing conventional products. For example,

the price of organic fertilizers derived from manure or compost can be influenced by the global price of synthetic fertilizers, which are often cheaper due to economies of scale and established distribution networks. Similarly, the economic viability of biogas or biofuels is directly tied to the fluctuating prices of natural gas and petroleum fuels. If fossil fuel prices are low, the economic incentive to produce and use bioenergy decreases significantly, potentially rendering investments in agro-waste to energy facilities unprofitable. Creating stable, high-value markets for novel bio-based materials (e.g., bioplastics from agricultural byproducts) requires demonstrating performance equivalence or superiority to conventional materials, overcoming consumer resistance, and establishing entirely new supply chains and distribution channels. The uncertainty in market demand and the volatility of prices for outputs from agro-waste valorization processes create significant financial risks that impede long-term investment and sustainable operation, making it difficult for circular business models to compete consistently with the more stable (though environmentally costly) economics of the linear economy.

3. **Competing Land Use and Opportunity Costs:** Implementing circular economy practices for agro-waste can introduce competition for resources, particularly land and biomass, leading to opportunity costs. Establishing facilities for processing large volumes of agro-waste, such as anaerobic digesters, composting plants, or biorefineries, requires land, which can be a limited and valuable resource in agricultural areas, potentially competing with land needed for food production, housing, or other economic activities. The allocation of land for waste-processing facilities must be carefully considered in spatial planning, balancing the environmental and economic benefits of waste valorization against other societal needs. Furthermore, utilizing certain agro residues (like straw or corn stover) for energy or material production means diverting them from alternative on-farm uses. These residues can serve valuable functions, such as being returned to the soil to improve organic matter content, prevent erosion, and enhance soil health, or being used as animal bedding. The decision to sell or process these residues for circular products involves an opportunity cost for the farmer – the value lost by not using the residue for its traditional purpose. Evaluating this opportunity cost is complex and depends on factors like soil type, climate, alternative fertilizer costs, and bedding material availability. If the economic return from selling or processing the residue is not sufficiently high to compensate for the loss of its on-farm value (including the long-term benefits to soil health), farmers may be unwilling to participate in the circular economy feedstock supply chain, limiting the availability of materials for valorization.
4. **Lack of Financial Incentives and Supportive Policies:** A significant barrier is the insufficient or inconsistent provision of financial incentives and supportive policies to level the playing field between circular economy initiatives and established linear practices. The linear economy often benefits from mature market structures, economies of scale, and sometimes direct or indirect subsidies (e.g., for fossil fuels, synthetic fertilizers), while externalizing many of its environmental costs. Transitioning to a circular model requires overcoming these embedded advantages. A lack of consistent government support in the

form of subsidies, tax breaks, low-interest loans, grants for R&D, and mandates for recycled content or bio-based products makes it difficult for nascent circular businesses to become competitive. Policy uncertainty and frequent changes in incentive schemes can also deter long-term investment. Furthermore, the regulatory framework may not always adequately account for the environmental benefits of circular approaches (e.g., carbon sequestration from biochar, reduced pollution from treated manure) in a way that translates into economic value. Without policies that internalize the environmental costs of linear practices (e.g., carbon pricing, pollution taxes) or actively promote the demand for circular products through procurement mandates or standards, the economic case for investing in agro-waste valorization technologies remains weak compared to conventional, less sustainable options. The absence of robust, long-term, and well-designed financial and policy support mechanisms is a critical economic challenge that hinders the widespread adoption and scaling of circular economy models for agro-waste.

Collectively, these economic challenges create a complex landscape for the implementation of circular economy principles for agro-waste. High initial costs, market uncertainties, competition for resources, and an uneven playing field due to policy gaps make it difficult for circular initiatives to demonstrate clear economic advantages over established linear practices. Overcoming these barriers requires targeted financial support, the development of stable markets, careful resource planning, and comprehensive policies that recognize and value the environmental and social benefits of circularity.

3.8.2 Technological Limitations

Beyond economic factors, the technical aspects of processing and utilizing diverse agro-waste streams within a circular framework also present significant limitations that need to be addressed through continued research, development, and innovation.

1. **Complexity and Technical Expertise:** Agro-waste streams are inherently diverse in their physical and chemical composition, varying significantly depending on the crop type, farming practices, climate, season, and post-harvest handling. For example, moisture content can range dramatically between dry straw and wet manure. Lignocellulosic biomass (like stalks and husks) requires different processing techniques than protein-rich processing byproducts. The presence of contaminants (e.g., soil, plastics, pesticides, heavy metals from animal feed) further adds to the complexity. This variability and complexity necessitate sophisticated processing technologies capable of handling diverse feedstocks or require extensive pretreatment steps, which can be costly and energy-intensive. Furthermore, operating and maintaining advanced agro-waste valorization technologies (e.g., biochemical reactors, pyrolysis units, advanced separation systems) requires specialized technical expertise and skilled labor, which may not be readily available in rural areas where much of the agro-waste is generated. This lack of local technical capacity can increase operational risks,

maintenance costs, and dependence on external service providers, posing a significant limitation to widespread decentralized implementation.

2. **Scale-up and Integration with Existing Systems:** A major technological hurdle lies in effectively scaling up promising laboratory-scale or pilot-scale technologies for agro-waste valorization to industrial volumes that can handle the vast quantities of waste generated annually. Scaling up involves overcoming engineering challenges related to reactor design, process efficiency, material handling, and product purification at a much larger throughput. Many innovative processes that work well in controlled laboratory environments become significantly more complex, expensive, and less efficient when scaled to the capacities needed for economic viability. Moreover, integrating new agro-waste-processing facilities seamlessly into existing agricultural production systems and broader industrial value chains presents additional technical challenges. This includes developing efficient and cost-effective methods for collecting, storing, and pretreating diffuse agricultural residues at scale, transporting feedstocks to processing sites, and distributing the resulting products (e.g., digestate, biochar, pellets) back into relevant markets or agricultural applications. The technical coordination required to link these different steps and ensure reliable operation across the entire value chain is complex and necessitates robust process design and management systems.
3. **Energy and Resource Intensity:** While the goal of valorizing agro-waste is often to recover energy or resources, some processing technologies themselves can be energy-intensive and may require significant inputs of water or chemicals. For example, drying high-moisture biomass like manure or wet crop residues before thermal processing (e.g., pyrolysis, gasification) requires substantial energy input, potentially reducing the net energy gain or requiring external energy sources. Biochemical conversion processes might require significant volumes of water and specialized enzymes or microorganisms. Although efforts are made to optimize processes for minimal energy and resource consumption and ideally utilize energy generated from the waste itself (e.g., using biogas to power digester operations), achieving a positive net energy and resource balance can be technically challenging for certain feedstocks or technologies. A thorough lifecycle assessment is required to ensure that the energy and resource inputs of the valorization process do not outweigh the benefits gained from utilizing the waste, a technical assessment that is not always straightforward due to methodological complexities.
4. **Technology Readiness and Innovation Gap:** There exists a gap between the current state of technology readiness for various agro-waste valorization processes and the level required for widespread, reliable, and economically competitive commercial deployment. While some technologies like basic composting or anaerobic digestion are relatively mature (though still facing optimization challenges for different feedstocks), others like advanced cellulosic biofuel production, sophisticated biochar production with specific properties, or extraction of high-value biochemicals from residues are still at lower Technology Readiness Levels (TRLs), requiring further research, pilot testing, and demonstration before they can be confidently deployed at scale. The

"valley of death" between promising research results and commercialization is a significant technological limitation. Overcoming this requires sustained investment in R&D, dedicated pilot and demonstration projects, and collaboration between research institutions and industry. Furthermore, the rapid pace of technological innovation in bioengineering, material science, and process engineering means that continuous research is needed to develop next-generation technologies that are more efficient, cost-effective, and capable of handling an even wider range of complex agro-waste streams, ensuring the long-term sustainability and evolution of circular practices in agriculture.

Technological limitations pose significant barriers to the efficient and widespread implementation of circular economy models for agro-waste. The inherent complexity and variability of the feedstocks, coupled with challenges in scaling up and integrating diverse processing technologies into existing agricultural systems, require substantial technical expertise and ongoing innovation. Addressing these limitations through targeted research, development, and demonstration projects is critical for developing robust, reliable, and economically viable technologies capable of transforming agricultural waste from a challenge into a valuable resource within a truly circular economy.

3.9 Conclusion

The comprehensive analysis presented in this study definitively establishes that waste, and particularly agro-waste, is far more than a disposal problem; it is a crucial and undervalued resource. Shifting from the prevailing linear economic model – characterized by a "take-make-dispose" approach that strains finite resources and generates massive environmental burdens – toward a circular economy is imperative for building a sustainable future. This study has underscored the fundamental role that managing waste, especially agro-waste, plays in enabling this critical transition.

The linear economy's inherent flaws have become increasingly apparent. Its relentless consumption of raw materials leads to resource depletion and dependency, while the resulting mountains of waste contribute significantly to environmental degradation. Air, water, and soil pollution, exacerbated by improper disposal methods like open burning and unsanitary landfills, directly harm ecosystems and human health. GHG emissions from burning waste and decomposing organic matter contribute to climate change and global warming. The pressure on natural systems leads to habitat destruction and biodiversity loss. Economically, the linear model is susceptible to volatile resource prices and limits opportunities for value creation at the end of a product's life. This unsustainable path necessitates a fundamental change in how we produce, consume, and manage materials.

The circular economy offers a compelling alternative. Its core philosophy revolves around keeping products and materials in continuous use for as long as possible, effectively "closing the loop" that is left open in the linear model. This is achieved through strategies like designing for durability, promoting repair and reuse, and facilitating high-quality recycling and remanufacturing. By minimizing waste generation and maximizing the value extracted from resources throughout their lifecycle, the circular economy not only reduces environmental harm but also unlocks new economic potential.

Agro-waste serves as a powerful case study for the practical application of circular economy principles. As the study detailed, agricultural activities generate a diverse array of waste materials – from crop residues like straw and husks left in fields to animal manure and byproducts from food-processing industries. The sheer volume of this waste globally presents a significant challenge if mishandled, contributing to pollution, nutrient loss from soils, and the spread of contaminants. However, this same volume represents a vast, often untapped, resource pool. Agro-waste is rich in organic matter and nutrients, making it suitable for soil amendments. It also contains biomass with considerable potential for energy generation.

The chapter explored various methods for valorizing agro-waste within a circular framework. Composting, for instance, transforms organic waste into nutrient-rich fertilizer, directly returning valuable elements to the soil and reducing the need for energy-intensive synthetic fertilizers. Anaerobic digestion converts organic waste, particularly animal manure, into biogas, a renewable energy source that can be used for heating, cooking, or electricity generation, simultaneously addressing waste disposal and providing clean energy. Biochar production, derived from the thermochemical conversion of biomass, offers benefits for soil health, carbon sequestration, and waste reduction. Even practices like field incorporation of crop residues, when managed properly, contribute to soil organic matter. These examples demonstrate how agro-waste can be reintegrated into agricultural systems or used to create new value chains, embodying the circular principle of regeneration.

Implementing these circular practices yields multifaceted benefits. Environmentally, it leads to significant reductions in GHG emissions compared to burning or landfilling. It decreases pollution of air, water, and soil. By cycling nutrients back into the soil, it reduces the environmental impact associated with synthetic fertilizer production and runoff. By finding uses for waste, it lessens the burden on landfills and helps preserve biodiversity by reducing the need to extract virgin resources. Economically, the circular management of agro-waste fosters the development of new industries focused on waste-processing and resource recovery. This creates jobs, stimulates innovation, and provides farmers and rural communities with new income streams. Utilizing agro-waste for energy production enhances energy security and reduces reliance on fossil fuels. The economic viability of these approaches is improving with technological advancements, though costs remain a factor.

The transition to a circular economy, particularly concerning agro-waste, is not automatic. It requires deliberate effort and supportive frameworks. The study highlighted the crucial roles of both policies and technological innovation. Governments need to establish regulatory frameworks and incentives that encourage circular practices, such as mandates for waste separation, subsidies for bio-based products, and support for infrastructure development. Examples from different regions, like the EU's WFD, the US EPA regulations promoting biofuels, and national policies in India and China, illustrate diverse approaches to integrating waste management and circular economy goals.

Technological innovation is equally vital. Developing cost-effective and scalable technologies for processing diverse types of agro-waste into valuable products and energy is a key challenge. Improving efficiency in processes like anaerobic digestion, biochar production, and biofuel conversion is essential for making these alternatives economically competitive with traditional linear approaches. R&D are needed to overcome technical limitations and bridge the gap in technology readiness for widespread adoption.

Despite the clear potential and the growing momentum, significant challenges remain on the path to a fully circular system for agro-waste. Logistical challenges are prominent; collecting and transporting bulky, distributed agro-waste to processing facilities can be difficult and costly. The initial investment required for new technologies and infrastructure can be substantial. Market demand and pricing volatility for recovered resources and bio-based products can create economic uncertainties. Furthermore, effectively coordinating action across different sectors – agriculture, waste management, energy, manufacturing, and policy-making – requires robust collaboration.

In conclusion, the study strongly supports the argument that integrating agro-waste management into broader circular economy strategies is essential for environmental sustainability, economic growth, and societal well-being. While challenges exist, the potential benefits – reducing pollution, mitigating climate change, conserving resources, creating jobs, and enhancing energy security – are profound. Realizing these transformative potential demands sustained R&D to improve technologies, strategic investments in infrastructure and capacity building, supportive and coherent policies that align economic incentives with circular goals, and collaborative efforts among governments, industries, researchers, and communities. By working together to embrace the principles of the circular economy and effectively manage waste as a resource, particularly agro-waste, we can collectively move toward a more resilient, equitable, and sustainable future for all.

References

1 Wasiak, A.L. Effect of biofuel production on sustainability of agriculture. *Procedia Eng*. 2017. https://doi.org/10.1016/j.proeng.2017.03.192.

2 Azadbakht, M., Safieddin Ardebili, S.M., and Rahmani, M. A study on biodiesel production using agricultural wastes and animal fats. *Biomass Convers. Biorefin*. 2021. https://doi.org/10.1007/s13399-021-01393-1.

3 Toldrá-Reig, F., Mora, L., and Toldrá, F. Trends in biodiesel production from animal fat waste. *Appl. Sci*. vol. 10, no. 10, pp. 3644, May. 2020. https://doi.org/10.3390/app10103644.

4 Rocha-Meneses, L., Bergamo, T.F., and Kikas, T. Potential of cereal-based agricultural residues available for bioenergy production. *Data Br*. 2019. https://doi.org/10.1016/j.dib.2019.103829.

5 Onu, P. and Mbohwa, C. (Editors) Front matter. In: *Agricultural Waste Diversity and Sustainability Issues*. Academic Press Inc, 2021, p. iii. https://doi.org/10.1016/B978-0-323-85402-3.01001-9.

6 Nunes, L.J.R., De Oliveira Matias, J.C., and Da Silva Catalão, J.P. Physical pretreatment of biomass. In: *Torrefaction of Biomass for Energy Applications*. Academic Press Inc, 2018. https://doi.org/10.1016/b978-0-12-809462-4.00002-x.

7 Tiwari, R., Agrawal, P.S., Belkhode, P.N., Ruatpuia, J.V.L., and Rokhum, S.L. Hazardous effects of waste transformer oil and its prevention: a review. *Next Sustain*. 2024. https://doi.org/10.1016/j.nxsust.2024.100026.

8 Ruatpuia, J.V.L., Halder, G., Shi, D., Halder, S., and Rokhum, S.L. Comparative life cycle cost analysis of bio-valorized magnetite nanocatalyst for biodiesel production: modeling, optimization, kinetics and thermodynamic study. *Bioresour. Technol*. 2024. https://doi.org/10.1016/j.biortech.2023.130160.

9 Velenturf, A.P.M. and Purnell, P. Principles for a sustainable circular economy. *Sustain. Prod. Consum*. vol. 27, pp. 1437–1457, Jul. 2021. https://doi.org/10.1016/j.spc.2021.02.018.

10 Das, A. and Rokhum, S.L. Green and economical synthesis of glycerol carbonate via microwave-assisted glycerol transesterification over calcined Mangifera indica peel ash derived catalyst. *Sci. Talks* 2023. https://doi.org/10.1016/j.sctalk.2023.100266.

11 Bessette, A.P., Teymouri, A., Martin, M.J., Stuart, B.J., Resurreccion, E.P., and Kumar, S. Life cycle impacts and techno-economic implications of flash hydrolysis in algae processing. *ACS Sustain. Chem. Eng*. 2018. https://doi.org/10.1021/acssuschemeng.7b03912.

12 Brouwer, M.T., Thoden van Velzen, E.U., Augustinus, A., Soethoudt, H., De Meester, S., and Ragaert, K. Predictive model for the Dutch post-consumer plastic packaging recycling system and implications for the circular economy. *Waste Manag*. vol. 71, pp. 62–85, Jan. 2018. https://doi.org/10.1016/j.wasman.2017.10.034.

13 Pathak, G., Rajkumari, K., and Rokhum, L. Wealth from waste: *M. acuminata* peel waste-derived magnetic nanoparticles as a solid catalyst for the Henry reaction. *Nanoscale Adv*. 2019. https://doi.org/10.1039/c8na00321a.

14 Dhanya, B.S., Mishra, A., Chandel, A.K., and Verma, M.L. Development of sustainable approaches for converting the organic waste to bioenergy. *Sci. Total Environ*. vol. 723, pp. 138109, Jun. 2020. https://doi.org/10.1016/j.scitotenv.2020.138109.

15 Yıldız, Ü.Y., Keçili, R., and Hussain, C.M. Green and sustainable chemistry. In: *Green Imprinted Materials*. Elsevier, 2024, pp. 3–25. https://doi.org/10.1016/B978-0-443-15431-7.00014-3.

16 Chew, K.W., Chia, S.R., Chia, W.Y., Cheah, W.Y., Munawaroh, H.S.H., and Ong, W.J. Abatement of hazardous materials and biomass waste via pyrolysis and co-pyrolysis for environmental sustainability and circular economy. *Environ. Pollut*. 2021. https://doi.org/10.1016/j.envpol.2021.116836.

17 Nanda, S., Azargohar, R., Dalai, A.K., and Kozinski, J.A. An assessment on the sustainability of lignocellulosic biomass for biorefining. *Renew. Sustain. Energy Rev*. vol. 50, pp. 925–941, Oct. 2015. https://doi.org/10.1016/j.rser.2015.05.058.

18 Savon, D.Y., Kolotyri, K.P., and Romanov, A.V. Improving the ecological efficiency of the processing industry of agricultural security on the basis of economic

instruments. *Russ. J. Ind. Econ.* 2019. https://doi.org/10.17073/2072-1633-2019-2-305-315.

19 Bhatia, R.K., Ramadoss, G., Jain, A.K., Dhiman, R.K., Bhatia, S.K., and Bhatt, A.K. Conversion of waste biomass into gaseous fuel: present status and challenges in India. *Bioenergy Res.* 2020. https://doi.org/10.1007/s12155-020-10137-4.

20 Ao, S., Prasad Gouda, S., Selvaraj, M., Boddula, R., Al-Qahtani, N., Mohan, S., et al. Transesterification of Jatropha curcas oil to biodiesel using highly porous sulfonated biochar catalyst: optimization and characterization dataset. *Data Br.* 2024. https://doi.org/10.1016/j.dib.2024.110096.

21 Agrawal, P. and Rokhum, S.L. Supported metal/metal oxide catalysts in biodiesel production: an overview. In: *Biodiesel Production: Feedstocks, Catalysts, and Technologies*. Wiley, 2022. https://doi.org/10.1002/9781119771364.ch7.

22 Singh, E. Circular economy based environmental management. *Process Saf. Environ. Prot.* vol. 163, pp. 585–600, 2022.

23 Onu, P., Mbohwa, C., and Pradhan, A. Machine learning: a comprehensive exploration of fault detection and diagnosis in smart grids. In: *2023 International Conference on Electrical, Computer and Energy Technologies (ICECET)*. IEEE, 2023. https://doi.org/10.1109/ICECET58911.2023.10389596.

24 Adejumo, I.O. and Adebiyi, O.A. Agricultural solid wastes: causes, effects, and effective management. In: *Strategies of Sustainable Solid Waste Management.* IntechOpen, 2021. https://doi.org/10.5772/intechopen.93601.

25 Agrawal, P.S., Belkhode, P.N., and Rokhum, S.L. Agro-waste-produced biofuels. In: *Sustainability in Biofuel Production Technology*. Wiley-Blackwell, 2022. https://doi.org/10.1002/9781119888864.ch7.

26 Menyuka, N.N., Sibanda, M., and Bob, U. Perceptions of the challenges and opportunities of utilising organic waste through urban agriculture in the durban south basin. *Int. J. Environ. Res. Public Health*.2020. https://doi.org/10.3390/ijerph17041158.

27 Zhai, Z., Martínez, J.F., Beltran, V., and Martínez, N.L. Decision support systems for agriculture 4.0: survey and challenges. *Comput. Electron. Agric.* 2020. https://doi.org/10.1016/j.compag.2020.105256.

28 Abad-Segura, E., Cortés-García, F.J., and Belmonte-Ureña, L.J. The sustainable approach to corporate social responsibility: a global analysis and future trends. *Sustainability* vol. 11, no. 19, pp. 5382, Sep. 2019. https://doi.org/10.3390/su11195382.

29 Andini, A., Bonnet, S., Rousset, P., and Hasanudin, U. Impact of open burning of crop residues on air pollution and climate change in Indonesia. *Curr. Sci.* vol. 115, no. 12, p. 2259, Dec. 2018. https://doi.org/10.18520/cs/v115/i12/2259-2266.

30 Prabakar, D., Suvetha, K.S., Manimudi, V.T., Mathimani, T., Kumar, G., Rene, E.R., et al. Pretreatment technologies for industrial effluents: critical review on bioenergy production and environmental concerns. *J. Environ. Manage.* vol. 218, pp. 165–180, Jul. 2018. https://doi.org/10.1016/j.jenvman.2018.03.136.

31 Bationo, A. and Mokwunye, A.U. Role of manures and crop residue in alleviating soil fertility constraints to crop production: with special reference to the Sahelian and Sudanian zones of West Africa. *Fertil. Res.* vol. 29, no. 1, pp. 117–125, Jul. 1991. https://doi.org/10.1007/BF01048993.

32 Sarno, M. and Iuliano, M. Biodiesel production from waste cooking oil. *Green Process. Synth.* vol. 8, no. 1, pp. 828–836, Jan. 2019. https://doi.org/10.1515/gps-2019-0053.

33 Singh, D., Sharma, D., Soni, S.L., Sharma, S., Sharma, P.K., and Jhalani, A. A review on feedstocks, production processes, and yield for different generations of biodiesel. *Fuel* 2020. https://doi.org/10.1016/j.fuel.2019.116553.

34 Tiwari, R., Agrawal, P.S., Belkhode, P.N., Brijpuriya, D., Devase, A., and Samarth, P. Utilization of waste transformer oil for the synthesis of biodiesel and glyptal. 2024. https://doi.org/10.1007/978-981-99-4186-5_7.

35 N.S.N and M.B., Jean-Franc. Critical review of sustainability energy. *Renew. Sustain. Energy Rev.* vol. 18, pp. 595–606, 2013.

36 Erickson, B. *U.S. Environmental Protection Agency*, no. Suppl. 1999.

37 Fred Sissine. *Summary of the Energy Independence and Security Act*. Energy Policy Resources, Science, and Industry Division, 2007.

38 Bio Energy National Policy on Biofuels. Minister of Petroleumand Natural Gas, GoI, 2018.

39 Padmanabhan, S., Giridharan, K., Stalin, B., Kumaran, S., Kavimani, V., Nagaprasad, N., et al. Energy recovery of waste plastics into diesel fuel with ethanol and ethoxy ethyl acetate additives on circular economy strategy. *Sci. Rep.* 2022. https://doi.org/10.1038/s41598-022-09148-2.

4

Agro-waste Management

4.1 Introduction

Effective waste management plays a crucial role in reducing the dependency on waste landfilling, potentially to the extent where waste volume becomes insufficient for dedicated electricity generation facilities. City municipalities are increasingly investigating alternative methods to convert waste into energy in an environmentally responsible manner. Traditional waste disposal in landfills often results in unsightly accumulations and the emission of malodorous compounds that negatively impact local environments and commuter experience.

The implementation of recycling processes, as discussed in the literature, can result in negligible emission levels, thereby positioning it as an environmentally benign technology. The residual slurry remaining after energy generation can be processed into organic compost and marketed to fertilizer producers. Plastic waste streams are typically directed toward companies involved in road construction. Municipal corporations are actively developing decentralized models for community-based small- and medium-scale waste-to-energy (WtE) power plants. These initiatives aim to address challenges in proper waste management while simultaneously contributing to resolving potential energy supply deficits [1].

Agriculture, a foundational pillar of human civilization, inevitably generates vast quantities of organic waste collectively termed agro-waste. This encompasses a diverse range of materials, including crop residues (such as straw, stalks, and leaves), byproducts from fruit and vegetable processing (like peels, seeds, and pulps), livestock manure, and other agricultural production wastes. The sheer global volume of agro-waste constitutes a formidable challenge. Improper disposal methods, such as open burning and uncontrolled landfilling, lead to significant detrimental consequences. These include environmental pollution, manifested through the release of greenhouse gases (methane [CH_4], carbon dioxide [CO_2], nitrous oxide), various air pollutants (particulate matter and volatile organic compounds [VOCs]), and the contamination of water bodies by leachate. Resource depletion is another major concern, as valuable organic matter and essential nutrients that could enrich soil fertility are lost, increasing the reliance on synthetic fertilizers. Furthermore,

Agro-Waste Management and Valorization, First Edition. Pratibha S. Agrawal, et al.

public health is jeopardized by the potential spread of pathogens, proliferation of vector-borne diseases, and the generation of odor nuisances associated with decomposing organic waste. Addressing these multifaceted challenges necessitates a fundamental paradigm shift toward sustainable agro-waste management strategies that prioritize the recovery of valuable resources and minimize environmental impact. Chemical engineering, with its core expertise spanning process design, reaction kinetics, separation technologies, and thermodynamics, is indispensable in the development and optimization of these advanced management strategies.

The production of fuel pellets or briquettes from municipal solid waste (MSW), often referred to as Refuse-derived Fuel (RDF), Process Engineered Fuel (PEF), or Solid Recovered Fuel (SRF), involves a defined sequence of operations. These steps include waste segregation, crushing, blending materials with varying calorific values, and solidifying the mixture into a dense form. These fuel pellets serve as a viable alternative to conventional fossil fuels like coal and wood for boiler firing and electricity generation in both domestic and industrial settings. Pelletization technology represents a pathway for renewable energy production, analogous to biomass, wind, solar, and geothermal energy sources. Notably, RDF generally exhibits more favorable emission characteristics compared to coal, resulting in lower emissions of pollutants such as nitrogen oxides (NOx), sulfur oxides (SOx), carbon monoxide (CO), and CO_2. A typical RDF production line comprises several unit operations arranged in series to remove undesirable components and achieve desired material characteristics. These operations include screening, shredding, size reduction, classification, and the separation of materials like metals, glass, or wet organic matter, followed by drying and densification. The specific sequence of these operations can be tailored based on the composition of the raw MSW and the required quality of the RDF product.

The quality of the resultant fuel pellets can be adjusted to meet the specific requirements of the user or market. High-quality RDF is characterized by a higher heating value and lower moisture and ash content. The inherent quality of RDF is often sufficient to warrant its consideration as a preferred fuel for co-firing alongside coal or for dedicated combustion in boilers originally designed for coal utilization.

4.2 Assessment of RDF and SRF

RDF pellets offer notable advantages in terms of storage and long-distance transportation and can serve as a cost-effective substitute for coal. Given that the pelletization process involves significant MSW sorting, it provides an enhanced opportunity to extract environmentally detrimental materials from the incoming waste stream prior to combustion. However, it is important to note that the pelletization process itself is energy-intensive and less suitable for processing wet MSW, particularly during rainy seasons. A critical concern is the potential presence of toxic or hazardous substances within RDF pellets, which renders them unsafe for open burning or residential use. Despite potentially having a favorable calorific value and low chlorine content, the precise composition of RDF can be uncertain for users due to the lack of consistent testing and evaluation according to appropriate standardized methods.

SRF is specifically produced from non-hazardous waste. A fundamental requirement for SRF is that the producer must precisely specify and classify its characteristics, including net calorific value (NCV), chlorine content, and mercury content. This rigorous classification positions SRF as a potentially significant fuel source for future energy needs [2, 3].

4.3 MSW and RDF/SRF Legislation

The ultimate disposal of MSW continues to pose a challenge in numerous countries. Legislative frameworks concerning waste management and the promotion of energy recovery are compelling member states to prioritize the development and implementation of waste recovery techniques before resorting to landfilling. The revised Waste Framework Directive establishes the overarching legislative framework governing the collection, transport, recovery, and disposal of waste within the European Union. This Directive also mandates member states to implement appropriate measures aimed at encouraging, firstly, the prevention or reduction of waste generation and, secondly, the recovery of waste through recycling, reuse, or other processes designed to extract secondary raw materials or utilize waste as an energy source. Consequently, the implementation of this new legislation holds substantial potential for reducing reliance on landfilling, thereby decreasing the overall environmental footprint by mitigating the direct adverse impacts of land disposal and capitalizing on the benefits derived from energy recovery from waste [2, 4].

The processing of MSW yields several valuable outputs. Green fuel (RDF) derived from MSW involves drying, crushing, screening, and compacting dry MSW into brick-like forms that can be used as a substitute for conventional fossil fuels in boilers. Recycled plastic from MSW is produced by segregating plastic from mixed MSW, sorting it into various grades using infrared sorting technologies, and converting it into ingots or granules. This is typically a dry process that minimizes pollution and contamination. These recycled plastic ingots or granules reduce the demand for virgin plastic and are utilized in the manufacturing of products such as trash bags, agricultural pipes, moldings, and ropes. Residual sand from MSW refers to the material remaining after processing, which is further treated in current operating plants to extract any residual recyclables like sand. This ensures that the material ultimately sent to landfills constitutes no more than 20% of the original volume. This processed sand can be used as a supplementary material to dredged sand in construction applications, as visually represented in Figure 4.1, which illustrates a typical MSW disposal site.

4.4 Type of Solid Waste

The term MSW is generally understood to encompass all waste generated within a community's residential, commercial, and institutional sectors, typically excluding waste generated by municipal services' own operations, wastewater treatment plants, and large-scale industrial or agricultural processes. The classification of waste types is depicted in Figure 4.2, illustrating the broad categories of solid waste, construction waste, and special waste, each with subcategories. Table 4.1 provides

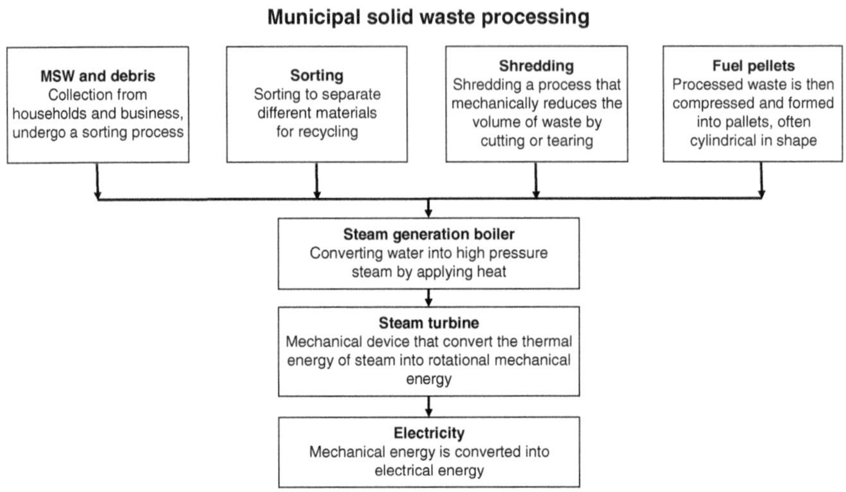

Figure 4.1 Classification of MSW processing.

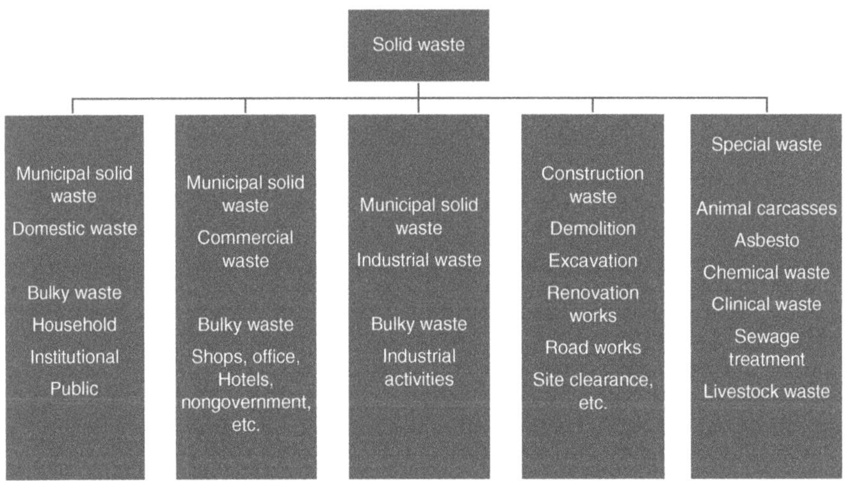

Figure 4.2 Classification of solid waste.

Table 4.1 Solid waste categories based on source.

Sources	Examples
Residential	Single family homes, duplexes, town houses, apartments
Commercial	Office buildings, shopping malls, warehouses, hotels, restaurants
Institutional	Schools, medical facilities, prisons
Industrial	Packaging of components, office wastes, lunchroom and restroom wastes (but not industrial process wastes)

a breakdown of solid waste categories based on their source, offering examples for residential (single-family homes, duplexes, townhouses, apartments), commercial (office buildings, shopping malls, warehouses, hotels, restaurants), institutional (schools, medical facilities, prisons), and industrial sources (packaging components, office/lunchroom/restroom wastes, but generally not core industrial process wastes).

4.5 Pelletization and Incineration

Incineration is a waste disposal methodology in which solid organic wastes are subjected to high-temperature combustion to convert them into residual ash and gaseous products. This technique is applicable for disposing of residues from both solid waste management and wastewater treatment processes. Incineration significantly reduces the volume of solid waste, typically achieving a reduction to 20–30% of the original volume. Along with other high-temperature waste treatment systems, it is sometimes broadly classified as "thermal treatment." Incinerators facilitate the conversion of waste materials into heat, gaseous products, steam, and ash. Incineration is practiced at both small-scale, localized levels and large-scale industrial facilities. It is employed for the disposal of solid, liquid, and gaseous wastes and is recognized as a practical approach for managing certain hazardous waste materials, such as biological and medical waste. However, incineration remains a controversial method of waste disposal primarily due to concerns regarding the emission of gaseous pollutants.

Incineration is particularly common in countries like Japan, where land availability is limited, as these facilities require significantly less space compared to landfills. WtE or energy-from-waste (EfW) are broader terms encompassing facilities that combust waste in furnaces or boilers to generate heat, steam, or electricity. Achieving complete combustion in an incinerator is not always guaranteed, leading to concerns about pollutants released in gaseous emissions from incinerator stacks. Specific attention has been directed toward the potential formation of persistent organic compounds, such as dioxins, furans, and polycyclic aromatic hydrocarbons (PAHs), which can be generated during combustion and pose significant environmental and health risks [5, 6].

The combustible fractions of the waste are separated for the production of RDF. The fine sand material resulting from processing can be utilized as a soil amendment. Inert materials such as large stones can be used as filling material for low-lying areas. RDF serves as an effective substitute for coal, with combustion emissions often considered superior to those from coal burning, exhibiting lower NOx and SO_2 levels. RDF has demonstrated a proven track record as a coal substitute. Instantaneous reduction in MSW volume is primarily achievable through incineration. Energy recovery from garbage can be accomplished either through RDF utilization or direct electricity generation. RDF incineration is generally considered more economically viable than direct garbage incineration and has been reported to result in no dioxin or furan emissions [7, 8]. Table 4.2 presents typical characteristics of RDF, providing data on its proximate analysis (moisture: 7.2%,

Table 4.2 Refuse-derived fuel characteristics.

Proximate analysis (%)	Ultimate analysis (%)
Moisture: 7.2	Carbon: 36.70
Volatile matter: 64.6	Hydrogen: 5.30
Ash content: 21.9	Nitrogen: 1.21
Fixed carbon: 6.3	Sulfur: 0.32

volatile matter: 64.6%, ash content: 21.9%, fixed carbon: 6.3%) and ultimate analysis (carbon: 36.70%, hydrogen: 5.30%, nitrogen: 1.21%, sulfur: 0.32%).

4.6 Case Study: Energy Recovery Potential

Reducing dependence on fossil fuels can be achieved by utilizing solid waste as an energy source, which consequently contributes to a significant reduction in both pollution and greenhouse gas emissions [9]. Municipal authorities often face a dual challenge: the need for proper MSW disposal in compliance with environmental regulations and the financial burden of purchasing electricity from utility providers, especially the high cost associated with powering public infrastructure like street lighting [10]. The potential for power generation from solid waste can be estimated through a practical example.

- Assuming a quantity of dry solid waste generated (W) of 5000 kg.
- The heat released during combustion typically falls within the range of 800–1000 kcal/kg, with an average value of 900 kcal/kg.
- Given a NCV of 900 kcal/kg and a dry solid waste quantity of 5000 kg, the potential energy recovery can be calculated using the conversion factor (1 kcal = 1.16×10^{-3} kWh).
- This calculation yields an energy recovery of 900 kcal/kg $\times$ 5000 kg $\times$ 1.16×10^{-3} kWh/kcal = 5220 kWh.
- Assuming a typical conversion efficiency of 25% from thermal energy to electricity, the net power generation would be 0.25 $\times$ 5220 kWh = 1305 kWh.
- If the street lamps utilized consume 250 W and operate for 12 hours daily, each lamp consumes 250 W $\times$ 12 hours = 3000 Wh = 3 kWh.
- Based on this, the total number of street lamps that could be operated is approximately 1305 kWh/3 kWh/lamp $\approx$ 435 lamps.

This estimation indicates that energy generated from solid waste can power approximately 435 street lamps for 12 hours, thereby helping to minimize the reliance on grid energy to a considerable extent. Implementing a similar approach in developing countries holds the potential to substantially improve their power scenarios [11, 12].

All human activities, spanning domestic, commercial, industrial, healthcare, and agricultural, contribute to the generation of solid waste. The volume and characteristics of this waste are influenced by the specific activity and the level of technological development within a region or country. The substantial increase in the quantity and diversity of waste materials, coupled with their potential adverse effects on the general environment and public health, has underscored the urgent need for adopting scientific methodologies for safe waste disposal. While the primary focus should be on minimizing waste generation and promoting reuse and recycling, technologies for energy recovery from waste can play a vital complementary role in mitigating waste-related problems [13]. These technologies facilitate a substantial reduction in the overall waste quantities requiring final disposal, allowing for the remaining waste to be managed more effectively and safely while simultaneously adhering to pollution control standards.

Waste-derived fuels, specifically RDF and SRF, are produced from MSW. Achieving effective and sustainable MSW management is a critical environmental imperative. The careful selection of appropriate treatment processes is paramount from both economic and environmental perspectives. In this context, SRF production from MSW and similar waste streams should be considered as a subsequent step following extensive source separation, collection, and recycling of valuable materials and composting of the organic fraction. The significant quantities of SRF currently utilized underscore its valuable contribution as a fuel source in various energy-intensive sectors, notably the cement industry and power generation. Waste should fundamentally be perceived as a valuable resource. Developing a formal recycling sector and establishing industries focused on recycling non-biodegradable components recovered from waste can generate employment opportunities, particularly for informal waste collectors, and integrate them into the formal economy. Furthermore, the formulation and implementation of policies incorporating fiscal incentives and quality standards for waste reuse and recycling are essential to encourage producers to manage waste according to established guidelines, thereby alleviating the burden on landfill sites [14].

For large urban centers, establishing zone-wise decentralized composting units is a recommended strategy. This approach involves organizing community participation to ensure the collection and disposal of segregated biodegradable waste from individual households or units into these decentralized composting facilities. WtE plants based on combustion represent highly efficient power generation facilities that utilize MSW as their primary fuel source, offering an alternative to conventional fuels such as coal, oil, or natural gas [15–17]. These plants effectively recover the thermal energy contained within the waste using highly efficient boilers to generate steam. This steam can then be directly supplied to industrial customers or used on-site to drive turbines for electricity production. One notable technology in this domain is the production of fuel pellets from solid waste, which are subsequently utilized in boilers for steam generation and electricity production [18].

4.7 Liquid-waste Management

Liquid waste constitutes a common form of industrial byproduct, frequently containing hazardous chemicals, toxins, and pollutants. Improper disposal of liquid waste leads to the facile contamination of clean water sources, impacting drinking water availability, aquatic ecosystems, and overall environmental health. Such

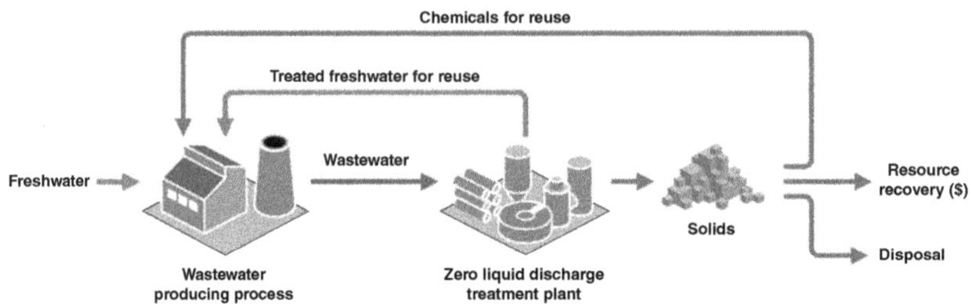

Figure 4.3 Zero liquid discharge wastewater treatment flow.

contamination poses significant risks to human health, disrupts biodiversity, and contributes to long-term environmental degradation [19]. To address these critical issues, a variety of wastewater treatment methods have been developed, broadly categorized into chemical, physical, and biological processes. Chemical treatment focuses on neutralizing harmful substances. Physical methods primarily remove solid contaminants. Biological treatment employs microorganisms to degrade or remove organic pollutants. These comprehensive methods and treatments are designed to ensure that discharged water is cleaner and safer for the environment.

Each wastewater treatment method possesses unique functional characteristics, advantages, and drawbacks. Their effectiveness varies depending on the specific conditions and the type of waste being treated. Consequently, the selection of an appropriate treatment approach is contingent upon factors such as wastewater composition, the nature and concentration of contaminants, operational and material costs, and the anticipated environmental impact [20]. Figure 4.3 provides a schematic illustration of a zero liquid discharge wastewater treatment process flow.

4.8 Physicochemical Treatment

Physicochemical treatments encompass a range of techniques, including coagulation, ozonation, adsorption, ion exchange, chemical oxidation, electrochemical processes, membrane technology, photocatalysis, and advanced oxidation processes (AOPs) [21]. These treatments are highly effective in removing toxic organic and inorganic pollutants from wastewater.

Within physical treatment methods, filtration, sedimentation, separation, centrifugation, and crystallization are common techniques (Table 4.3). Biological treatment, conversely, effectively removes toxic pollutants characterized by high biological oxygen demand (BOD) and chemical oxygen demand (COD). This involves biodegradation processes carried out under aerobic or anaerobic conditions by microorganisms such as fungi, bacteria, and algae, and through the action of enzymes and plants [22]. These biological agents facilitate the breakdown of pollutants, enhancing the safety of wastewater for potential reuse or discharge.

Coagulation is a physicochemical process that effectively separates colloidal particles by using positively charged ions to neutralize the negatively charged ions present in wastewater. Metal salts or polymers are widely used as coagulants. Examples include aluminum sulfate, aluminum chloride, ferric sulfate, ferrous sulfate, ferric

Table 4.3 Types of technique for physicochemical treatment.

Techniques	Advantages	Disadvantages
Coagulation	Satisfactory reduce BOD and COD, cheap technology	Results in a high amount of sludge, depending on pH
Adsorption	Highly reduce BOD and COD, simple method, inexpensive technique	Problem with the utilization of waste, the effectiveness depends on the time
Membranes technology	High efficiency to remove dyes and micro-contaminants, easy to handle and operate	Fouling effects require pressure, expensive equipment, should be precisely controlled
Advanced oxidation process	Fast reactions do not require a large land area and remove heavy metals and hazardous organic components	Complex operating, high cost, need chemical oxidation addition

chloride, melamine-formaldehyde, and tannins. In the dairy industry, $FeCl_3$-$6H_2O$, $Al_2(SO_4)_3 \cdot 6H_2O$, $Ca(OH)_2$, polyacrylamide (PAA), and polyferric sulfate (PFS) are commonly employed coagulants. For instance, aluminum chloride can achieve a 90% removal rate of dyes at a concentration of 40 mg/L. Plant extracts derived from sources like *Plantago ovata, Moringa oleifera*, and *Tamarindus indica* have demonstrated the ability to remove over 90% of turbidity and 60% of color. While these natural options are non-toxic, low-cost, and environmentally friendly, optimizing their performance can be challenging. Adsorption removes pollutants through either chemical bonding or physical interactions [23–25]. This method is capable of reducing COD and BOD by over 90% and is particularly effective for removing various dyes (including acid, basic, and azo dyes) and organic pollutants [26]. Membrane filtration operates as a physical barrier for separating contaminants. There are several types of membrane filtration, including ultrafiltration, nanofiltration, microfiltration, and reverse osmosis. Materials commonly used in membrane filtration include Polyethylene (PE), Polypropylene (PP), Polytetrafluoroethylene (PTFE), cellulose acetate, ceramics, zeolites, and silica. Membrane filtration exhibits high efficacy across various industrial applications. In the textile industry, nanofiltration can remove 57% of COD, 100% of color, and 30% of salinity from wastewater. For the metal finishing industry, a combination of ultrafiltration and reverse osmosis has been shown to eliminate 90–99% of pollutants. In oily wastewater treatment, this same combination can achieve 100% removal of turbidity and oil while reducing COD by 98%. Furthermore, in the paper industry, membrane filtration effectively removes 95.5% of COD and 94.9% of phenol contaminants, underscoring its versatility across different industrial wastewater treatment scenarios [27]. While coagulation is advantageous for its satisfactory reduction of BOD and COD and low cost, it results in a high volume of sludge whose characteristics are pH-dependent. Adsorption, while effective for high reduction of BOD and COD and conceptually simple, can be an expensive technique, and its effectiveness is time-dependent. Membrane technology offers high efficiency for removing dyes [28] and microcontaminants and is easy to handle and operate, but it is susceptible to fouling, requires pressurized systems and expensive equipment, and demands precise control. AOPs facilitate rapid reactions and minimize land area requirements while effectively removing heavy metals and hazardous organic components; however, their operational complexity is high, they incur significant costs, and they necessitate the addition of chemical oxidants.

AOPs employ oxidation mechanisms to degrade organic pollutants into biodegradable compounds [29, 30]. In textile wastewater treatment, a $UV/H_2O_2/TiO_2$ system can degrade 72% of organic compounds while achieving a 57% reduction in COD and a 48% reduction in BOD. For the removal of micropollutants, an Fe(II)/PAA AOP demonstrates a fast reaction rate, particularly effective in a higher pH range. AOPs are highly effective for treating organic pollutants and are capable of breaking down complex chemical structures.

4.9 Physical or Mechanical Treatment

Physical or mechanical treatment serves as a common method for wastewater pretreatment, primarily aimed at removing solid pollutants. Filtration is a widely adopted mechanical treatment technique due to its simplicity, low energy consumption, and cost-effectiveness, functioning by trapping solid pollutants within a porous filter medium. Sedimentation is another straightforward method that leverages gravity to separate suspended solid particles, resulting in clearer wastewater. Crystallization is a well-established technique for separating dissolved substances from liquid waste, where dissolved compounds form solid crystals that are subsequently easier to remove. Crystallization is particularly useful for recovering valuable substances like salt (NaCl), soda ash (Na_2CO_3), ammonia (NH_4OH), and phosphates, as well as for removing dissolved ions such as calcium (Ca^{2+}) and magnesium (Mg^{2+}) [31]. Mechanical treatment plays an important role in enhancing the performance of primary wastewater treatment processes and contributing to a reduction in overall industrial pollution.

4.10 Biological Treatment

Biological wastewater treatment utilizes living organisms, including bacteria, yeast, fungi, and enzymes, to degrade and remove pollutants from wastewater [31]. This method is broadly categorized into two main types: aerobic treatment, which occurs in the presence of oxygen, and anaerobic treatment, which takes place in the absence of oxygen, as illustrated in Figure 4.4. The primary objective of biological treatment is to break down complex organic substances into simpler, harmless compounds. Both aerobic and anaerobic methods are effective in reducing COD and BOD while typically producing less sludge compared to some other treatment methods. However, these biological processes can be expensive and complex to operate. Research efforts are actively focused on developing ways to make these biological treatment processes simpler and more cost-effective [32].

Industrial liquid waste poses significant risks to both the environment and public health, necessitating the implementation of effective treatment and management

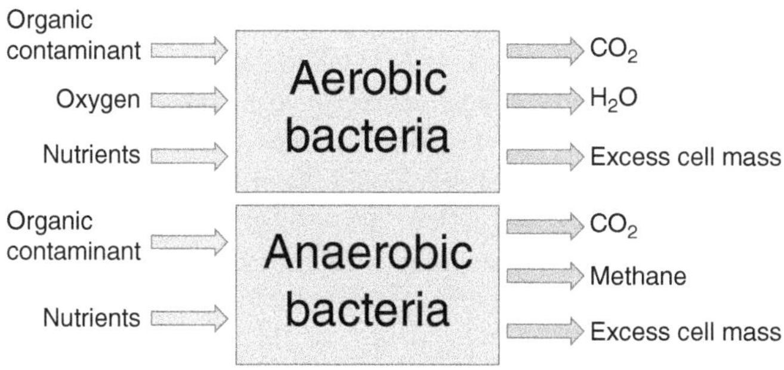

Figure 4.4 Illustration of aerobic and anaerobic processes.

strategies [33]. A variety of methods, including physical, chemical, and biological treatments, have been developed to reduce contaminant levels and mitigate pollution [28]. Each method possesses specific advantages and limitations, making it essential to select the most appropriate approach based on the characteristics of the waste, economic feasibility, and anticipated environmental impact [34]. By employing advanced technologies and adhering to regulatory guidelines, industries can significantly reduce their pollution footprint and contribute to maintaining a cleaner and healthier environment.

4.11 E-waste

The rapid pace of technological innovation leads to the continuous release of new Information and Communication Technology (ICT) products, rendering older devices obsolete. Consequently, the production of electronic waste (e-waste) is increasing exponentially each year. According to available data, the largest contributors to the e-waste stream are large household appliances, accounting for 42.1% of the total volume. This is followed by IT and telecommunications equipment (33.9%), consumer electronics (13.7%), and smaller household appliances (4.6%) [35]. To provide a clearer perspective, other categories collectively contribute smaller proportions: electrical, medical, lighting, and electronic tools account for 2%, while monitoring and control devices, automatic dispensers, sports equipment, and toys make up only 1%. In developing nations, computers, televisions, and mobile phones constitute the majority of the e-waste generated [36].

The initial step in managing e-waste effectively is collection. This involves gathering discarded electronic devices from both households and businesses. Subsequently, the collected e-waste undergoes pretreatment before being transported to dedicated treatment facilities or designated disposal landfills. E-waste can be collected through various mechanisms. Official take-back systems, for instance, provide consumers with the option to return their old devices when purchasing new ones, sometimes in exchange for credits or rewards. In countries with well-established, organized waste management systems, mixed residual waste streams are often collected, which may contain e-waste. For nations lacking properly developed formal take-back or collection systems, reliance is often placed on informal collection methods [37].

Managing e-waste presents a unique set of opportunities alongside significant challenges. E-waste legislation grapples with the inherent duality of electronic materials, which are often both valuable due to recoverable components and toxic due to hazardous substances. While most developed nations have implemented e-waste legislation, a considerable number of developing nations have yet to establish such legal frameworks. In 2014, e-waste laws were in place in 61 nations, covering approximately 44% of the global population. By 2019, this number had increased to 78 countries, regulating e-waste management for an estimated 71% of the world's population [38].

4.12 Environmental and Health Impacts of Waste Mismanagement

The practice of open dumping MSW leads to the anaerobic decomposition of organic matter, resulting in the release of various harmful gases. These include CH_4, a potent greenhouse gas, CO_2, hydrogen sulfide (H_2S), and VOCs. These emissions contribute to climate change and degrade ambient air quality, which in turn can lead to respiratory illnesses and broader environmental degradation.

Waste incineration, particularly the combustion of plastics and e-waste, releases toxic air pollutants. Examples of these pollutants include dioxins and furans, which are highly toxic compounds linked to carcinogenic, reproductive, and immunotoxic effects. Particulate matter (PM2.5 and PM10) consists of fine particles capable of penetrating deep into the lungs, leading to respiratory and cardiovascular diseases. Heavy metals, such as lead, mercury, and cadmium, are toxic elements that can cause damage to the nervous system, kidneys, and other vital organs. Landfills generate greenhouse gas emissions, primarily CH_4, as a result of the anaerobic decomposition of organic waste. These emissions contribute to the generation of foul odors and the deterioration of air quality in nearby communities, and can also increase the risk of fires and explosions, particularly at poorly managed landfill sites [39].

Industrial waste contributes significantly to air pollution, especially when hazardous waste is improperly disposed of by industries. This practice results in the release of SOx, NOx, and VOCs into the atmosphere. These pollutants are known contributors to acid rain, photochemical smog formation, and respiratory diseases in surrounding populations. Mitigation strategies for reducing waste-related air pollution include promoting recycling and composting, which decrease the volume of waste sent to landfills and consequently lower CH_4 emissions. Furthermore, these practices limit air pollution associated with waste burning.

4.13 Disposal Methods of Waste Management

Waste disposal encompasses various methods, including landfilling, incineration, recycling, composting, and WtE technologies. Each of these methods has distinct environmental impacts and holds potential for contributing to sustainable waste management. Landfilling is a common and often cost-effective method, yet if not managed properly, it can lead to leachate contamination and the emission of CH_4. Incineration effectively reduces waste volume and can generate energy, but it necessitates rigorous management of emissions and ash disposal. Recycling transforms waste materials into new products, thereby reducing resource consumption and the volume of waste requiring landfilling; however, it requires established infrastructure and active consumer participation. Composting converts organic waste into a valuable soil amendment, reducing landfill waste and enhancing soil health. WtE processes convert waste into usable energy through techniques such as incineration, gasification, and pyrolysis, addressing energy needs while simultaneously reducing waste volume. Vermicomposting, utilizing earthworms to decompose organic waste into compost, represents a sustainable and environmentally friendly method [40]. Plasma gasification is a high-temperature thermal process that converts waste into a synthesis gas (syngas), which can be subsequently used for energy generation or chemical production. The use of smart waste bins, incorporating wireless sensor networks, IoT devices, and radio frequency identification RFID tags, can optimize waste collection processes and improve overall waste management efficiency.

4.14 Environmental Impacts and Considerations

Waste management practices carry several significant environmental impacts and considerations. Landfill leachate, for instance, is contaminated water that percolates through waste in landfills, posing a serious risk of polluting soil and groundwater resources. CH_4 emissions are a potent greenhouse gas produced by the anaerobic decomposition of organic waste within landfills [41]. Air pollution results from emissions generated during incineration processes and other waste-processing activities. Resource depletion occurs when unsustainable waste management practices lead to the inefficient use and loss of valuable natural resources [42]. Poorly managed waste can pose direct risks to public health through exposure to pathogens and toxins. E-waste, if improperly disposed of, can result in both environmental contamination and health hazards [43].

4.15 Sustainable Waste Management

Sustainable waste management is fundamentally about prioritizing practices that minimize the quantity of waste requiring disposal. Key strategies include the principles of reduce, reuse, and recycle [44]. The waste hierarchy provides a structured approach to waste management that prioritizes actions in a specific order: prevention first, followed by reuse, recycling, and finally disposal. Integrated waste management systems involve implementing comprehensive approaches that combine various waste management techniques synergistically. The concept of a circular economy centers on designing systems that inherently minimize waste generation and maximize the utilization of resources throughout their lifecycle. Raising public awareness and education about sustainable waste management practices and encouraging responsible waste disposal behaviors are crucial components of

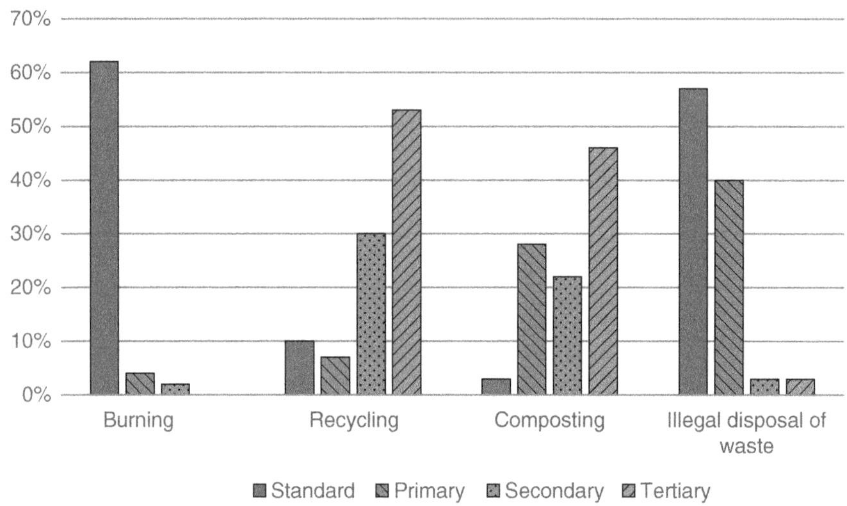

Figure 4.5 Waste management strategy.

an effective system [45]. Technological innovations, such as the development and implementation of smart waste bins and advanced WtE systems, represent key advancements in modern waste management. Figure 4.5 visually presents a survey of waste management strategy adoption, comparing the percentage of respondents engaging in burning, recycling, composting, and illegal disposal across different classification "Standards" (Standard, Primary, Secondary, Tertiary).

4.16 Biological Conversion Techniques

The increasing global demand for sustainable waste management solutions and resource recovery pathways has spurred significant development in various biological conversion techniques. These methods leverage the metabolic activities of microorganisms to break down organic matter present in agro-waste, yielding valuable products including compost, biogas, and bioelectricity [46, 47]. The two most widely adopted biological conversion methods are composting and anaerobic digestion (AD). This section delves into the technical intricacies of these processes and highlights the essential role of chemical engineering in optimizing their performance and designing efficient systems.

4.16.1 Composting: Processes, Chemical Engineering Aspects, and Applications

Composting is an aerobic biological process where microorganisms decompose organic waste in the presence of oxygen, leading to the formation of a stable, humus-like material rich in nutrients and organic matter [48]. This process involves a complex cascade of biochemical reactions facilitated by diverse communities of bacteria, fungi, and other microorganisms.

From a chemical engineering perspective, process optimization is paramount in composting. Chemical engineers play a critical role by meticulously controlling key parameters such as temperature, moisture content, aeration rate, and the carbon-to-nitrogen (C/N) ratio [19]. Modeling the composting process using fundamental principles of reaction kinetics and mass transfer equations can provide valuable insights for identifying optimal operating conditions and consequently improving process efficiency. A thorough understanding of the various factors that influence the composting process is indispensable for consistently achieving a high-quality, stable end product. Different types of composting reactors and systems have been developed to suit varying scales and feedstock types, including windrow composting, aerated static pile composting, and in-vessel composting systems [49]. Chemical engineers contribute significantly to the design and optimization of these reactor systems to ensure uniform aeration, effective temperature distribution, and precise moisture control throughout the composting matrix. Computational fluid dynamics (CFD) simulations are increasingly utilized to analyze airflow patterns and temperature gradients within the reactor, facilitating improved reactor designs, enhanced process efficiency, and a reduced environmental footprint.

Odor control is a critical aspect of composting operations, as the process can generate unpleasant odors stemming from the release of VOCs and ammonia. Chemical engineers are actively involved in developing and implementing odor control strategies, such as biofiltration systems, activated carbon adsorption, and chemical scrubbing techniques, to effectively mitigate odor emissions from composting facilities. By gaining a deeper understanding of the underlying mechanisms responsible for odor generation, chemical engineers can design and implement more effective and environmentally sound odor control solutions.

4.16.2 AD: Processes, Chemical Engineering Aspects, and Applications

AD is a biological process in which microorganisms decompose organic matter in the complete absence of oxygen [48]. The primary products of AD are biogas, which is a gaseous mixture predominantly composed of CH_4 and CO_2, and digestate, a nutrient-rich solid or semi-solid residue. Biogas is recognized as a valuable renewable energy source, while digestate can often be utilized as a fertilizer.

Chemical engineering plays a vital role in the optimization of AD processes. Key aspects include understanding microbial kinetics and designing efficient bioreactors. AD is a complex process that relies on the coordinated metabolic activity of several distinct microbial populations working synergistically to convert organic matter into biogas. Chemical engineers apply principles of microbial kinetics and bioreactor design to optimize the performance of the AD process (Figure 4.6). This involves selecting appropriate microbial consortia, designing bioreactors that facilitate efficient substrate utilization and gas production, and precisely controlling operating parameters such as temperature, pH, and mixing intensity [50, 51].

Process optimization aims to enhance the efficiency of AD by fine-tuning parameters such as the organic loading rate, hydraulic retention time, and substrate composition. Chemical engineers employ mathematical modeling and simulation

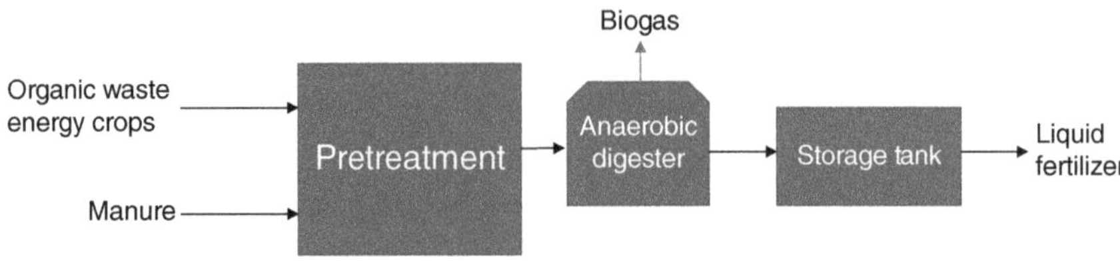

Figure 4.6 Biogas production.

techniques to predict the performance of AD systems under varying operating conditions and to identify optimal operational strategies. Advanced process control strategies, such as model predictive control (MPC), can be implemented to further optimize AD processes and minimize energy consumption [52].

Biogas upgrading is often a necessary step, as raw biogas typically contains 40–70% CH_4 and 30–60% CO_2, along with trace amounts of other gases including H_2S and ammonia (NH_3) [53]. To utilize biogas as a transportation fuel or inject it into the natural gas grid, it must be upgraded to remove CO_2 and other impurities. Chemical engineers develop and optimize various biogas upgrading technologies such as pressure swing adsorption (PSA), membrane separation, and chemical absorption. The selection of the most appropriate upgrading technology is influenced by factors including the composition of the biogas, required purity levels, and cost-effectiveness. Digestate management is another crucial consideration. The digestate produced by AD can serve as a valuable fertilizer [54], but it may still contain residual organic matter and nutrients that, if not managed properly, can contribute to water pollution. Chemical engineers develop and optimize technologies for digestate treatment, such as solid–liquid separation techniques, nutrient recovery processes (e.g., struvite precipitation), and composting, to minimize environmental impacts and maximize resource recovery from the digestate stream.

4.16.3 Emerging Biological Conversion Technologies

Microbial Fuel Cells (MFCs) represent an emerging technology that utilizes microorganisms to directly convert the chemical energy stored in agro-waste into electrical energy. Chemical engineers are actively engaged in developing MFCs specifically tailored for agro-waste treatment and energy generation, focusing on optimizing electrode materials, reactor designs, and the composition and activity of the microbial community. Current challenges in MFC technology include relatively low power output, limitations in scaling up systems for practical applications, and high capital costs.

Bioelectrochemical Systems (BES) are a broader class of technologies that integrate biological and electrochemical processes to enhance agro-waste treatment and generate valuable products [34]. Examples include microbial electrolysis cells (MECs), which can be used for hydrogen production, and microbial electrosynthesis (MES), which facilitates CO_2 fixation [55]. Chemical engineers are contributing to the development of new materials, the design of efficient reactor configurations, and the optimization of microbial communities to improve the overall performance of BES.

Enzymatic hydrolysis employs enzymes to break down cellulose and hemicellulose components in agro-waste into fermentable sugars [53]. These sugars can subsequently be converted into biofuels or other valuable bioproducts through fermentation. Chemical engineers are developing and optimizing enzymatic hydrolysis processes, with research focusing on enzyme production, enzyme immobilization techniques, and process optimization to enhance efficiency and yield. Enzymatic hydrolysis holds significant promise for increasing the efficiency of downstream bioconversion processes and reducing overall energy consumption.

Biological conversion techniques, including established methods like composting and AD, along with emerging technologies, offer substantial potential for sustainable waste management and resource recovery. Chemical engineering plays a pivotal role in optimizing and designing these biological systems through the precise control of process parameters, the development of advanced reactor designs, and the integration of novel control strategies [15, 56]. The continued advancement of emerging biological conversion technologies, such as MFCs and BES, presents exciting opportunities for achieving increased efficiency, further reducing the environmental footprint of waste treatment, and producing valuable products from agro-waste. Through sustained research and innovation, the field of biological conversion techniques is poised for significant growth and is expected to have a profound impact in the coming years.

4.17 Thermochemical Conversion Techniques

Agro-waste, encompassing a wide variety of materials such as agricultural residues, forestry byproducts, and food-processing wastes, constitutes a significant and largely underutilized biomass resource that can be harnessed through various thermochemical conversion techniques. These methods employ heat and chemical reactions to transform agro-waste into valuable products, including biofuels, chemicals, and energy carriers [57]. This review specifically focuses on three primary thermochemical conversion methods – pyrolysis, gasification, and combustion – and examines the critical role of chemical engineering in optimizing these processes for efficient agro-waste valorization, while also outlining associated challenges and future research directions.

4.17.1 Pyrolysis: Unlocking the Potential of Bio-oil, Biochar, and Syngas

Pyrolysis, defined as the thermal decomposition of organic matter in an oxygen-deficient or anaerobic environment, holds immense promise for agro-waste valorization by producing three primary product streams: bio-oil, biochar, and syngas [58]. The relative proportions and specific properties of these products are highly dependent on the precise pyrolysis conditions employed, including temperature, heating rate, residence time, and the specific reactor configuration. Chemical engineers are essential in the design, optimization, and scaling-up of pyrolysis processes, addressing key scientific and technological hurdles to achieve efficient and economically viable agro-waste conversion.

Chemical engineering principles are central to the optimization of pyrolysis. Reactor design is paramount for achieving high conversion rates, controlling product selectivity, and ensuring efficient heat transfer within the system [59]. A variety of reactor configurations have been developed, each possessing distinct advantages and limitations.

Fixed-bed reactors, for instance, operate by maintaining the agro-waste feedstock stationary within the reactor while heat is supplied externally [59]. While relatively simple in design and operation, fixed-bed reactors are limited by factors such as poor heat transfer, non-uniform temperature distribution throughout the bed, and the potential for channeling effects, which can reduce contact efficiency. Advanced research in this area focuses on improving bed designs, such as utilizing packed beds with enhanced heat transfer media, and developing reactive fixed beds that incorporate catalysts to promote specific product formation pathways. CFD modeling plays a vital role in optimizing the bed geometry and heat distribution profiles to minimize temperature gradients and improve overall reactor performance. More recent advancements also explore staged pyrolysis within fixed beds, where different temperature zones are implemented sequentially along the reactor length to decompose the agro-waste in a controlled manner, leading to tailored product yields.

Fluidized-bed reactors offer superior heat transfer characteristics and better temperature uniformity compared to fixed beds by suspending the agro-waste particles in a fluidizing gas, typically nitrogen or recycled syngas [59]. These reactors are well-suited for continuous operation and can effectively handle a wide range of particle sizes. However, they present challenges such as particle attrition, the elutriation of fine particles from the bed, and potential back-mixing phenomena, which can negatively impact product yields. Chemical engineers are actively engaged in optimizing fluidized-bed hydrodynamics through advanced CFD simulations, focusing on parameters such as fluidization velocity, distributor plate design, and particle size distribution to minimize these issues and enhance overall performance. Novel designs incorporating internal baffles or draft tubes are also being explored to promote improved particle mixing and further reduce back-mixing.

Ablative reactors rely on the principle of rapid heating of the agro-waste surface, which causes rapid volatilization and simultaneously minimizes secondary cracking reactions that could lead to the formation of char [59]. These reactor systems often employ mechanical mechanisms to ensure continuous and efficient contact between the agro-waste material and a heated surface. Challenges associated with ablative reactors include maintaining consistent contact, preventing the buildup of char on the heated surface, and the complexities of scaling up the mechanical components. Research efforts are focused on developing novel ablative reactor designs with enhanced heat transfer capabilities, such as rotating cone reactors or vortex reactors, and exploring the use of advanced materials for the heated surface to minimize char adhesion.

Vacuum pyrolysis reactors operate under reduced pressure conditions, which lowers the boiling points of the volatile pyrolysis products, thereby facilitating their rapid removal from the reactor and minimizing secondary reactions [59]. This approach typically results in higher bio-oil yields with improved quality. However, vacuum pyrolysis systems are generally more complex and energy-intensive. Chemical engineers focus on optimizing vacuum levels and reactor designs to minimize energy consumption while maximizing product yield and quality. Integrated systems combining vacuum pyrolysis with in situ vapor upgrading strategies are also gaining significant attention.

Process optimization is crucial as the yield and composition of pyrolysis products are significantly influenced by key process parameters such as temperature, heating rate, and residence time [50]. Understanding the complex interplay between these parameters is critical for optimizing the pyrolysis process to target specific applications and desired products.

Temperature plays a dominant role in dictating the extent of thermal decomposition and the selectivity toward different products. Lower temperatures (e.g., 400–500 °C) typically favor the production of biochar, while intermediate temperatures (e.g., 500–600 °C) tend to maximize bio-oil yield, and higher temperatures (>700 °C) promote the formation of syngas [60]. Chemical engineers employ sophisticated kinetic modeling techniques and conduct extensive experimental studies to map the product distribution as a function of temperature, enabling precise control over the pyrolysis process to obtain desired product streams. Advanced analytical techniques, such as temperature-programmed pyrolysis (TPP) coupled with mass spectrometry, provide valuable insights into the reaction pathways and the thermal stability characteristics of different agro-waste components.

Heating rate significantly influences the distribution of pyrolysis products [60]. Rapid heating rates (characteristic of fast pyrolysis) favor the production of bio-oil by minimizing secondary cracking reactions, while slow heating rates (typical of slow pyrolysis) promote the formation of biochar. Chemical engineers utilize advanced heat transfer models to optimize the heating rate profiles within the reactor, ensuring uniform and rapid heating of the agro-waste particles, which is particularly critical in fast pyrolysis applications. Novel heating methods, such as microwave-assisted pyrolysis, are also being explored as a means to achieve rapid and uniform heating throughout the biomass feedstock.

Residence time of the volatile products within the reactor directly affects the extent of secondary cracking and polymerization reactions [60]. Short residence times are generally desirable for maximizing bio-oil yield by quickly removing volatile products before they can undergo degradation reactions. Conversely, longer residence times can favor the formation of syngas. Chemical engineers optimize the reactor design and operating conditions to precisely control the residence time of volatile products, aiming to balance the need for complete initial decomposition with the minimization of undesirable secondary reactions. CFD simulations are employed to model the flow patterns within the reactor and predict the residence time distribution, allowing for accurate control over the final product composition.

Catalysis plays a pivotal role in selectively enhancing the formation of specific pyrolysis products [61]. Chemical engineers are actively involved in the development and evaluation of catalysts specifically for pyrolysis applications, focusing on key parameters such as catalyst activity, selectivity, and stability under pyrolysis conditions.

In situ catalysis involves incorporating the catalyst directly into the pyrolysis reactor, where it interacts concurrently with the volatile products as they are formed [61]. This approach can be particularly effective for upgrading bio-oil by reducing its oxygen content and improving its fuel properties. Zeolites, such as ZSM-5 and Y-zeolite, are commonly employed as in situ catalysts due to their high surface area, inherent acidity, and shape selectivity [61]. Chemical engineers modify zeolite catalysts by incorporating metal oxides or altering their pore structure to tailor their activity and selectivity toward specific products. For example, the addition of metal oxides such as nickel (Ni) or molybdenum (Mo) to zeolites can enhance their ability to deoxygenate bio-oil components through hydrodeoxygenation reactions.

Ex situ catalysis involves passing the volatile products generated in the pyrolysis reactor over a separate catalyst bed [61]. This approach offers greater control over the reaction conditions and the catalyst properties. Supported metal catalysts, such as platinum (Pt), palladium (Pd), or ruthenium (Ru) dispersed on supports like alumina or silica, are commonly used for ex situ catalytic cracking and reforming of bio-oil. Chemical engineers focus on optimizing the catalyst composition, support material characteristics, and reaction conditions to maximize the conversion of bio-oil into valuable chemicals and fuels. The development of bifunctional catalysts possessing both acidic and metallic sites is also gaining attention, as these catalysts can promote a wider range of reactions, leading to improved product selectivity [32].

Bio-oil upgrading is a necessary process because bio-oil, being a complex mixture of oxygenated organic compounds, typically requires further processing to improve its fuel properties, such as heating value, stability, and corrosivity [32]. Chemical engineers are continuously developing and refining bio-oil upgrading technologies to effectively address these challenges.

Hydrotreating (HDO) involves reacting bio-oil with hydrogen over a catalyst to remove oxygen atoms through hydrodeoxygenation reactions [32]. HDO can significantly improve the stability and heating value of bio-oil, making it more suitable for use as a transportation fuel. However, HDO processes typically require high pressures and temperatures, which can lead to catalyst deactivation due to coke formation. Chemical engineers are developing novel HDO catalysts with improved activity and stability, as well as exploring alternative HDO processes that can operate at lower temperatures and pressures. The use of non-sulfided catalysts and the addition of promoters to the catalyst surface are also being investigated to enhance the catalyst's resistance to coke formation.

Catalytic cracking involves using a catalyst to break down the larger molecules present in bio-oil into smaller, more volatile molecules [32]. This process can improve the volatility and fluidity of bio-oil, making it easier to transport and utilize in engines. Zeolites are commonly employed as catalytic cracking catalysts due to their high surface area and inherent acidity. Chemical engineers are optimizing the zeolite structure and composition to tailor their activity and selectivity toward desired cracking products. The use of hierarchical zeolites, which possess both micropores and mesopores, is also being explored to improve the accessibility of the catalyst surface to larger bio-oil molecules and enhance their conversion.

Esterification involves reacting the carboxylic acids present in bio-oil with alcohols to form esters, which have improved fuel properties [61]. This process effectively reduces the acidity and corrosivity of bio-oil, making it more compatible with existing infrastructure and downstream processing. Acid catalysts, such as sulfuric acid (H_2SO_4) or Amberlyst resins, are commonly used for esterification. Chemical engineers are optimizing the reaction conditions and catalyst properties to maximize the ester yield while minimizing the formation of undesirable byproducts [62, 63]. The use of solid acid catalysts, such as sulfated zirconia, is also gaining attention due to their ease of separation from the product mixture.

Emulsification involves creating a stable mixture of bio-oil with conventional fuels, such as diesel, allowing it to be used in existing engines [61]. Surfactants are typically added to the mixture to stabilize the emulsion and prevent phase separation. Chemical engineers are focused on identifying suitable surfactants and optimizing the emulsion formulation to achieve stable and homogeneous mixtures with acceptable combustion properties. The use of nano-additives is also being explored as a means to further improve the stability and performance of bio-oil emulsions.

4.17.2 Alkaline Hydrolysis: Extracting Lignin and Enhancing Cellulose Digestibility

Alkaline hydrolysis, also known as alkali pretreatment or alkaline pulping, utilizes strong alkaline solutions, such as sodium hydroxide (NaOH), potassium hydroxide (KOH), or ammonia (NH_3), to disrupt the complex lignin structure within agro-waste, thereby improving the digestibility of cellulose and facilitating the extraction of lignin-derived products [31]. This process is particularly effective for delignification because alkali solutions solubilize lignin, significantly increasing the accessibility of cellulose to subsequent enzymatic hydrolysis or fermentation processes.

The reaction mechanisms and kinetics of alkaline hydrolysis of lignin involve a series of complex reactions. Saponification involves the alkaline hydrolysis of ester bonds present within the lignin structure, leading to the breakdown of its cross-linked network. Ether bond cleavage primarily targets the β-O-4 linkages, the most predominant type of ether bond in lignin, which are cleaved by the alkali, resulting in a reduction of the lignin's molecular weight [37]. Phenolic hydroxyl ionization involves the ionization of the phenolic hydroxyl groups in lignin by the alkali, increasing the lignin's solubility in water.

The rate of delignification is influenced by several key factors, including the type and concentration of the alkali used, temperature, reaction time, the liquid-to-solid ratio of the mixture, and the specific composition of the agro-waste feedstock [31]. The kinetics of alkaline hydrolysis can be modeled using various approaches, considering factors such as the diffusion of alkali into the lignocellulosic matrix and the inherent heterogeneity of the lignin structure.

Operational parameters and optimization of alkaline hydrolysis require careful consideration to maximize efficiency and yield [54, 64]. Alkali concentration is a critical parameter; higher alkali concentrations generally enhance delignification, but excessively high concentrations can lead to undesirable cellulose degradation and the formation of inhibitory byproducts [65]. The optimal alkali concentration is thus dependent on the type of agro-waste and the desired degree of delignification.

Temperature significantly affects the reaction rate; elevated temperatures accelerate delignification but can also increase the rate of cellulose degradation. Conversely, lower temperatures may necessitate longer reaction times, potentially impacting process economics. Reaction time should be optimized to be sufficient for achieving significant delignification without causing excessive degradation of the cellulose component. Liquid-to-solid ratio influences the contact between the alkali solution and the lignocellulosic material; a higher liquid-to-solid ratio generally promotes better contact and enhances delignification effectiveness. Pretreatment methods, such as mechanical treatments like milling or chipping, can increase the surface area of the agro-waste material, thereby improving the accessibility of lignin to the alkali and enhancing the efficiency of subsequent alkaline hydrolysis.

Engineering challenges and technological advancements in alkaline hydrolysis are areas of ongoing research. Efficient alkali recovery is crucial for the economic viability and environmental sustainability of the process. Technologies for alkali recovery include evaporation, causticization, and membrane separation [66]. Lignin utilization involves valorizing the extracted lignin, which is a valuable byproduct. Lignin can potentially be utilized as a fuel source, a building block for producing various materials, or a source of valuable aromatic compounds. Lignin valorization requires developing efficient and cost-effective processes for its purification and subsequent conversion into value-added products [66]. Cellulose degradation during alkaline hydrolysis is a challenge as it can reduce the yield of fermentable sugars. Process optimization and the use of protective additives can help minimize cellulose degradation [65]. Wastewater treatment is required for the effluent from alkaline hydrolysis, which typically contains high levels of organic matter and residual alkali, necessitating effective treatment technologies such as AD and chemical oxidation.

Technological advancements in alkaline hydrolysis include the development of organosolv pretreatment, which uses organic solvents (e.g., ethanol, acetone) in combination with alkali to extract lignin. This approach often results in a higher quality lignin fraction compared to traditional alkaline hydrolysis methods. Ammonia fiber expansion (AFEX) is another pretreatment technique that involves treating agro-waste with ammonia under high pressure and temperature, followed by rapid decompression. This process effectively disrupts the lignocellulosic structure and significantly improves cellulose digestibility [31]. Deep eutectic solvents (DESs) represent a new class of solvents that are being explored as sustainable alternatives to traditional alkali solutions for delignification. DESs are typically composed of two or more components that form a eutectic mixture with a significantly lower melting point than the individual components, offering tunable properties [31].

4.17.3 Transesterification: Transforming Agro-waste-derived Oils and Fats into Biodiesel

Transesterification is a fundamental chemical process employed to convert triglycerides, which are the main components of vegetable oils and animal fats, into fatty acid methyl esters (FAMEs), commonly recognized as biodiesel [67]. Agro-waste, particularly lipid-rich residues such as oilseed residues and animal fats, can serve as a sustainable and renewable feedstock for biodiesel production via transesterification. The process involves reacting the triglycerides with an alcohol, typically methanol or ethanol, in the presence of a catalyst, which can be either an acid or a base. This reaction yields FAMEs (biodiesel) and glycerol as a primary byproduct.

The reaction mechanisms and kinetics of transesterification proceed through a series of reversible steps [68]. The reaction is initiated by a nucleophilic attack of the alkoxide ion, which is formed by the reaction of the alcohol with the catalyst, on the carbonyl carbon of the triglyceride molecule [60]. This initial attack leads to the formation of a tetrahedral intermediate. Subsequently, these tetrahedral intermediates collapse, resulting in an ester exchange and the release of a FAME and a diglyceride molecule. This sequence of steps is repeated two more times, progressively converting the diglyceride to a monoglyceride, and finally to glycerol, with the release of 1 M of FAME in each step [69–71].

The overall rate of the transesterification reaction is influenced by several key factors, including the type and concentration of the catalyst, the molar ratio of alcohol to oil, the reaction temperature, the reaction time, the mixing intensity, and the presence of water and free fatty acids (FFAs) in the feedstock [59]. The kinetics of transesterification can be accurately modeled using various approaches that account for factors such as the reversibility of the reaction, the mass transfer limitations within the reaction mixture, and the inhibitory influence of impurities.

Optimizing the operational parameters of transesterification is crucial for achieving high biodiesel yield and purity. Careful control of the following parameters is necessary: catalyst type and concentration are critical; both acid and base catalysts can be used for transesterification. Base catalysts, such as NaOH and KOH, are more commonly employed due to their higher catalytic activity and lower cost [64]. However, base catalysts are highly sensitive to the presence of water and FFAs in the feedstock, which can lead to saponification reactions and a reduction in biodiesel yield. Acid catalysts, such as H_2SO_4 and hydrochloric acid (HCl), are less sensitive to water and FFAs but typically require higher reaction temperatures and longer reaction times [36]. Enzyme catalysts, specifically lipases, offer the advantage of high selectivity and can catalyze the reaction under mild conditions, but they are generally more expensive and susceptible to inhibition by various components in the feedstock or products.

The alcohol-to-oil molar ratio is typically maintained in excess of the stoichiometric requirement to drive the equilibrium toward the formation of FAMEs. However, an excessive amount of alcohol can complicate the separation of FAMEs from glycerol and increase overall production costs [64]. The optimal alcohol-to-oil molar ratio is dependent on the specific type of catalyst used and the quality of the feedstock.

Temperature significantly impacts the reaction rate; elevated temperatures generally increase the rate but can also lead to the formation of undesirable byproducts and increased evaporation of alcohol [72]. The optimal reaction temperature is dependent on the type of catalyst and the stability of the feedstock components, typically ranging near the boiling point of the alcohol used [59]. Reaction time should be sufficient to ensure a high conversion of triglycerides to FAMEs without promoting excessive byproduct formation. Mixing intensity is essential to ensure good contact between the reactants (oil, alcohol, and catalyst) and to overcome mass transfer limitations, which can otherwise limit the reaction rate.

Engineering challenges and technological advancements in transesterification are areas of active research and development. Feedstock variability is a significant challenge, as agro-waste-derived oils and fats can exhibit considerable variability in their composition and quality depending on the source and processing, which can affect the performance of the transesterification process. Pretreatment methods and blending strategies can help mitigate the impact of feedstock variability. Efficient catalyst separation and recovery are crucial for the economic viability and environmental sustainability of continuous or large-scale processes. Catalyst recovery technologies include washing, filtration, and adsorption [66]. Glycerol utilization is important, as glycerol is a significant byproduct of transesterification. It can be potentially utilized as a fuel source, a chemical feedstock for producing other chemicals, or a building block for various materials. Valorizing glycerol requires developing efficient and cost-effective processes for its purification and conversion [66]. Biodiesel purification is necessary because the crude biodiesel product typically contains impurities such as residual alcohol, glycerol, catalyst residues, and soaps, which can adversely affect its fuel properties and compliance with quality standards. Biodiesel purification techniques include washing, distillation, and adsorption.

Technological advancements in transesterification include the development of heterogeneous catalysis, utilizing solid acid and base catalysts. These offer advantages over homogeneous catalysts, including ease of separation from the product stream, potential for recyclability, and reduced reactor corrosion. Enzymatic transesterification employs lipases as catalysts, which can catalyze the reaction under mild conditions with high selectivity, potentially reducing energy requirements and byproduct formation. Supercritical transesterification utilizes alcohols in their supercritical state, where they can act as both a solvent and a reactant, facilitating the transesterification reaction without the need for a separate catalyst. Reactive distillation is an integrated process that combines the transesterification reaction and the separation of products within a single unit operation, potentially reducing capital costs and energy consumption. Microreactors offer improved mixing, enhanced heat transfer efficiency, and improved mass transfer characteristics due to their small scale, leading to potentially higher reaction rates and efficiency.

Chemical conversion techniques collectively provide a powerful toolkit for valorizing agro-waste and producing a range of valuable products, thereby contributing significantly to the establishment of a circular economy and promoting sustainable resource management. Acid hydrolysis, alkaline hydrolysis, and transesterification represent distinct yet complementary pathways for transforming agro-waste into fermentable sugars, lignin-derived compounds, and biodiesel, respectively [73]. While each of these techniques faces specific engineering challenges, ongoing research and technological advancements are continuously improving their efficiency, reducing costs, and enhancing their environmental sustainability for agro-waste valorization. Future research efforts should strategically focus on developing robust and long-lasting catalysts, optimizing process parameters across different technologies, integrating conversion technologies with downstream processing steps, and exploring novel valorization pathways to fully unlock the inherent potential of agro-waste as a valuable resource [73–75].

4.18 Techno-economic Analysis and Life Cycle Assessment: Evaluating Sustainability

Techno-economic Analysis (TEA) and Life Cycle Assessment (LCA) are indispensable tools for comprehensively evaluating the feasibility, economic viability, and environmental sustainability of various agro-waste management technologies.

4.18.1 TEA: The Bottom Line

TEA is an analytical framework used to evaluate the economic performance of a technology or process by systematically considering various factors, including capital costs, operating costs, and potential revenues [34]. Key aspects analyzed within a TEA include capital costs, which represent the initial investment required to construct and commission the technology or facility. This encompasses costs for equipment procurement, civil construction, and land acquisition. Operating costs encompass the ongoing expenses incurred during the operation of the technology. These costs typically include feedstock procurement, energy consumption, chemical inputs, labor costs, and maintenance expenses. Revenues are the income streams generated from the sale of primary products and any valuable byproducts derived from the process. Discounted Cash Flow (DCF) Analysis is a widely used technique within TEA to evaluate the profitability of a project over its entire operational lifespan, taking into account the time value of money. Key financial metrics calculated using DCF analysis include Net Present Value (NPV), Internal Rate of Return (IRR), and the payback period. Sensitivity analysis is employed to assess the robustness of the economic performance to variations in key input parameters, such as feedstock price, product price, and capital costs [34]. This analysis helps identify the parameters that have the most significant impact on profitability and quantifies the associated risks.

TEA serves as a valuable tool for identifying the most cost-effective agro-waste management options and guiding informed investment decisions. It is crucial to incorporate several external factors into the TEA framework to capture the full economic landscape. Government subsidies and incentives, often provided by governments to promote renewable energy and sustainable waste management technologies, can significantly enhance the economic viability of agro-waste projects. Carbon markets allow entities to buy and sell carbon credits, providing an additional potential revenue stream for agro-waste management technologies that contribute to reducing greenhouse gas emissions. Social benefits, while not directly financial, represent positive externalities such as job creation, improved air quality, and reduced landfill volume, which should be considered to provide a holistic view of the project's value. Sensitivity to market fluctuations is a crucial consideration; assessing the sensitivity of economic viability to changes in feedstock prices, energy costs, and market values for products helps in understanding the inherent risks and opportunities associated with a specific technology [76, 77].

4.18.2 LCA: Environmental Footprint

LCA is a standardized methodology used to evaluate the potential environmental impacts associated with a technology or product throughout its entire life cycle, from raw material extraction ("cradle") to end-of-life disposal or recovery ("grave") [78]. This assessment considers various environmental aspects, including energy consumption, greenhouse gas emissions, water pollution, and land use [79]. Key stages and components of an LCA typically include goal and scope definition, which involves clearly defining the purpose of the LCA and establishing the boundaries of the system being analyzed. Inventory analysis entails quantifying all relevant inputs (e.g., raw materials, energy, chemicals) and outputs (e.g., emissions to air, water, and soil; final products; waste streams) associated with the defined system [80]. Impact assessment involves evaluating the potential environmental impacts of the quantified inputs and outputs. This includes assessing impacts across various categories such as climate change potential, air quality effects, water quality degradation, and impacts on human health [81]. Interpretation involves analyzing the results of the inventory and impact assessment phases, drawing conclusions, and formulating recommendations based on the findings of the LCA.

LCA is a powerful tool for identifying the most environmentally sustainable agro-waste management options and informing policy decisions aimed at promoting environmentally sound practices. Chemical engineers frequently utilize TEA and LCA in conjunction to optimize agro-waste management processes, focusing on simultaneously minimizing costs and environmental impacts. They are also instrumental in the development of novel methodologies and tools specifically tailored for conducting TEA and LCA studies on agro-waste management technologies.

4.19 Emerging Technologies and Future Trends: Shaping the Future of Agro-waste Management

The field of agro-waste management is undergoing rapid evolution, driven by the emergence of new technologies and innovative approaches that hold significant promise for enhancing efficiency, improving sustainability, and increasing the economic viability of agro-waste valorization [82].

Nanomaterials are finding increasing application in agro-waste management across various domains. Examples include their use in catalyst development (e.g., enhancing enzyme immobilization for biological processes), the fabrication of biosensors (e.g., for real-time monitoring of fermentation parameters), and advanced wastewater treatment applications (e.g., for the removal of persistent pollutants from agro-waste-derived effluents).

Artificial Intelligence (AI) and Machine Learning (ML) techniques are being applied to optimize agro-waste management processes [83]. AI and ML algorithms can be trained on historical and real-time operational data to predict process performance, identify optimal operating conditions, and enable proactive control of process parameters. For example, ML algorithms can be employed to predict the yield of bioethanol from enzymatic hydrolysis of specific agro-waste feedstocks, facilitating real-time adjustments to maximize product output.

Integrated biorefineries represent a future trend that involves combining multiple agro-waste conversion technologies within a single facility to produce a diverse portfolio of products from a single agro-waste feedstock [72]. This integrated approach offers several advantages over single-product biorefineries, including increased overall resource utilization efficiency, minimized waste generation through synergistic process integration, and improved economic viability due to the production of multiple revenue streams. Integrated biorefineries can be designed to produce a range of valuable outputs, including biofuels, biochemicals, bio-based materials, and fertilizers.

The future trajectory of agro-waste management will undoubtedly be shaped by continued innovation in these emerging technologies, leading to the development of increasingly efficient, sustainable, and economically viable solutions. Furthermore, advances in genetic engineering and synthetic biology offer significant potential for developing microorganisms with enhanced capabilities specifically tailored for agro-waste conversion processes. The development of closed-loop systems that facilitate the efficient recycling of water and nutrients within agro-waste-processing facilities will also be critical for achieving long-term sustainability. Agro-waste management is inherently a complex and multifaceted field that requires a multidisciplinary approach, drawing expertise from various scientific and engineering disciplines. Chemical engineers play a pivotal role in developing and optimizing technologies across the entire agro-waste valorization chain, including pretreatment methods, product upgrading processes, TEA, and LCA. The ongoing development and successful implementation of innovative technologies, such as nanomaterials, AI, and integrated biorefineries, will be essential for achieving sustainable agro-waste valorization and contributing effectively to the transition toward a more circular economy [54]. Future research should prioritize addressing key remaining challenges in this field, such as reducing the cost of enzyme production, improving the efficiency of product upgrading technologies, and developing more accurate and comprehensive models for TEA and LCA. By effectively addressing these challenges, the full potential of agro-waste as a valuable, renewable resource can be unlocked, contributing significantly to a more sustainable future.

4.20 Conclusion

Agro-waste represents a significant environmental and economic challenge globally, but it simultaneously serves as a valuable and abundant resource with immense potential for the production of fuels, chemicals, materials, and fertilizers [9, 84]. Chemical engineering plays a critical and indispensable role in the development, optimization, and implementation of advanced agro-waste management technologies, thereby contributing significantly to the realization of a circular economy and promoting sustainable agricultural practices [85]. By rigorously applying chemical engineering principles to process design, reaction kinetics analysis, separation techniques, and thermodynamic evaluations, it is technically feasible to transform agro-waste from an environmental liability into a valuable resource stream. Continued research and development efforts focused on emerging technologies, such as nanomaterials, AI/ML, and integrated biorefineries, will further enhance the efficiency, sustainability, and economic viability of agro-waste management processes. The successful implementation of technological innovation, coupled with supportive policy frameworks and active community engagement, will be essential for fully realizing the substantial potential of agro-waste as a resource [86]. By proactively addressing the inherent challenges associated with agro-waste and strategically leveraging the opportunities for resource recovery, it is possible to contribute to the development of a more sustainable and resilient agricultural sector, foster a circular economy, and effectively mitigate the adverse environmental impacts of waste disposal. This requires a concerted and collaborative effort involving researchers across various disciplines, engineers, policymakers, and the agricultural community to jointly develop and implement innovative agro-waste management solutions that yield substantial benefits for both the environment and society.

References

1 Laskar, I.B., Deshmukhya, T., Biswas, A., Paul, B., Changmai, B., Gupta, R., et al. Utilization of biowaste-derived catalysts for biodiesel production: process optimization using response surface methodology and particle swarm optimization method. *Energy Adv.* vol. 1, no. 5, pp. 287–302, May 2022. https://doi.org/10.1039/D2YA00011C.

2 Ragazzi, M. and Rada, E.C. RDF/SRF evolution and MSW bio-drying. *WIT Trans. Ecol. Environ.* vol. 163, pp. 199–208, Jun. 2012. https://doi.org/10.2495/WM120191.

3 Bessi, C., Lombardi, L., Meoni, R., Canovai, A., and Corti, A. Solid recovered fuel: an experiment on classification and potential applications. *Waste Manag.* vol. 47, pp. 184–194, Jan. 2016. https://doi.org/10.1016/J.WASMAN.2015.08.012.

4 Nunes, L.J.R., De Oliveira Matias, J.C., and Da Silva Catalão, J.P. Physical pretreatment of biomass. In: *Torrefaction of Biomass for Energy Applications.* Academic Press Inc., 2018. https://doi.org/10.1016/b978-0-12-809462-4.00002-x.

5 Sujauddin, M., Huda, S.M.S., and Hoque, A.T.M.R. Household solid waste characteristics and management in Chittagong, Bangladesh. *Waste Manag.* vol. 28, no. 9, pp. 1688–1695, Jan. 2008. https://doi.org/10.1016/j.wasman.2007.06.013.

6 Hamad, T.A., Agll, A.A., Hamad, Y.M., and Sheffield, J.W. Solid waste as renewable source of energy: current and future possibility in Libya. *Case Stud. Therm. Eng.* vol. 4, pp. 144–152, Nov. 2014. https://doi.org/10.1016/j.csite.2014.09.004.

7 Ramakrishna G.V. Electricity generation from municipal solid waste - CDM perspective. In: *Selco International Limited, Hyderabad.* 2013.

8 Sarno, M. and Iuliano, M. Biodiesel production from waste cooking oil. *Green Process. Synth.* vol. 8, no. 1, pp. 828–836, Jan. 2019. https://doi.org/10.1515/gps-2019-0053.

9 Shah, M.P. Industrial wastewater treatment: a challenging task in the industrial waste management. *Adv. Recycl. Waste Manag.* vol. 01, no. 03, 2017. https://doi.org/10.4172/2475-7675.1000115.
10 Gutierrez, V.J. Municipal solid waste as a source of energy. In: *Proceedings of the IEEE Power Engineering Society*. 2000.
11 Psomopoulos, C.S., Bourka, A., and Themelis, N.J. Waste-to-energy: a review of the status and benefits in USA. *Waste Manag.* vol. 29, no. 5, pp. 1718–1724, May 2009. https://doi.org/10.1016/j.wasman.2008.11.020.
12 Rousta, K., Bolton, K., Lundin, M., and Dahlén, L. Quantitative assessment of distance to collection point and improved sorting information on source separation of household waste. *Waste Manag.* vol. 40, pp. 22–30, Jun. 2015. https://doi.org/10.1016/j.wasman.2015.03.005.
13 Vijay, R., Jain, P., Sharma, N., Bhattacharyya, J.K., Vaidya, A.N., and Sohony, R.A. GIS based solid waste management information system for Nagpur, India. *J. Env. Sci. Eng.* vol. 55, no. 1, pp. 114–119, 2013.
14 Shariar K.F. and Bustam, H.A. Waste to energy: a new dimension in generating electricity in Bangladesh. *Int. J. Eng. Technol.* vol. 4, no. 4, pp. 480–483, 2012.
15 Agrawal, P.S., Belkhode, P.N., and Rokhum, S.L. Microalgae –biofuel production trends. In: *Sustainability in Biofuel Production Technology*. Wiley-Blackwell, 2022. https://doi.org/10.1002/9781119888864.ch6.
16 Agrawal, P.S., Belkhode, P.N., and Rokhum, S.L. Biofuel production technologies – an overview. In: *Sustainability in Biofuel Production Technology*. Wiley-Blackwell, 2022. https://doi.org/10.1002/9781119888864.ch4.
17 Agrawal, P.S., Belkhode, P.N., and Rokhum, S.L. Agro-waste-produced biofuels. In: *Sustainability in Biofuel Production Technology*. Wiley-Blackwell, 2022. https://doi.org/10.1002/9781119888864.ch7.
18 Timotius, E. and Sukmarani, G. Industrial waste treatment management: a review. *J. Sci. Appl. Eng.* vol. 4, no. 1, pp. 1–8, May 2021. https://doi.org/10.31328/jsae.v4i1.2400.
19 Iloms, E., Ololade, O.O., Ogola, H.J.O., and Selvarajan, R. Investigating industrial effluent impact on municipal wastewater treatment plant in Vaal, South Africa. *Int. J. Environ. Res. Public Health*. vol. 17, no. 3, p. 1096, Feb. 2020. https://doi.org/10.3390/ijerph17031096.
20 Zajda, M. and Aleksander-Kwaterczak, U. Wastewater treatment methods for effluents from the confectionery industry – an overview. *J. Ecol. Eng.* vol. 20, no. 9, pp. 293–304, Oct. 2019. https://doi.org/10.12911/22998993/112557.
21 Badrus, Z. Potential of natural flocculant in coagulation-flocculation wastewater treatment process. *E3S Web Conf.* vol. 73, p. 05006, Dec. 2018. https://doi.org/10.1051/e3sconf/20187305006.
22 Loloei, M., Alidadi, H., Nekonam, G., and Kor, Y. Study of the coagulation process in wastewater treatment of dairy industries. *Int. J. Environ. Health Eng.* vol. 3, no. 1, p. 12, 2014. https://doi.org/10.4103/2277-9183.132684.
23 Deshpande, B.D., Agrawal, P.S., Yenkie, M.K.N., and Dhoble, S.J. Wastewater purification using nano-scale techniques. In: *Water Pollution Sources and Purification: Challenges and Scope*. Bentham Science Publishers, 2022. https://doi.org/10.2174/9789815050684122010009.

24 Deshpande, B.D., Agrawal, P.S., and Yenkie, M.K.N. Nanoparticles aided AOP for degradation of p-nitro benzoic acid. *Mater. Today: Proc.* 2019. https://doi.org/10.1016/j.matpr.2020.02.924.

25 Ashtaputrey, S.D. and Agrawal, P.S. Acoustic cavitation oxidation of ionic liquids: a brief review. *Water Environ. J.* vol. 38, no. 3, pp. 359–372, Aug. 2024. https://doi.org/10.1111/wej.12924.

26 Nayl, A.E.A., Elkhashab, R.A., El Malah, T., Yakout, S.M., El-Khateeb, M.A., Ali, M.M., et al. Adsorption studies on the removal of COD and BOD from treated sewage using activated carbon prepared from date palm waste. *Environ. Sci. Pollut. Res.* vol. 24, no. 28, pp. 22284–22293, Oct. 2017. https://doi.org/10.1007/s11356-017-9878-4.

27 Obotey Ezugbe, E. and Rathilal, S. Membrane technologies in wastewater treatment: a review. *Membranes (Basel).* vol. 10, no. 5, p. 89, Apr. 2020. https://doi.org/10.3390/membranes10050089.

28 Chowdhury, A.P., Anantharaju, K.S., Keshavamurthy, K., and Rokhum, S.L. Recent advances in efficient photocatalytic degradation approaches for azo dyes. *J. Chem.* 2023. https://doi.org/10.1155/2023/9780955.

29 Ashtaputrey, S.D. and Agrawal, P.S. Fenton and photo-assisted advanced oxidative degradation of ionic liquids: a review. *Environ. Sci. Pollut. Res.* 2023. https://doi.org/10.1007/s11356-023-29777-y.

30 Deshpande, B.D., Agrawal, P.S., Yenkie, M.K.N., and Dhoble, S.J. Nanoparticle-aided AOP for treatment of benzoic acid. In: *Water Pollution Sources and Purification: Challenges and Scope.* Bentham Science Publishers, 2022. https://doi.org/10.2174/9789815050684122010008.

31 Goswami, L., Manikandan, N.A., Dolman, B., Pakshirajan, K., and Pugazhenthi, G. Biological treatment of wastewater containing a mixture of polycyclic aromatic hydrocarbons using the oleaginous bacterium Rhodococcus opacus. *J. Clean. Prod.* vol. 196, pp. 1282–1291, Sep. 2018, https://doi.org/10.1016/j.jclepro.2018.06.070.

32 Liu, X., Li, C., Yao, Z., Zhang, S., Liu, Z., Liu, L., et al. Zeolite and silica fume as enhancers for cementitious solidification of borate radioactive waste. *Nucl. Eng. Technol.* vol. 57, no. 9, p. 103609, Sep. 2025. https://doi.org/10.1016/j.net.2025.103609.

33 Ab Halim, M.H., Nor Anuar, A., Abdul Jamal, N.S., Azmi, S.I., Ujang, Z., and Bob M.M. Influence of high temperature on the performance of aerobic granular sludge in biological treatment of wastewater. *J. Environ. Manage.* vol. 184, pp. 271–280, Dec. 2016. https://doi.org/10.1016/j.jenvman.2016.09.079.

34 Kumar, S. and Kumar, V. (Editors) *Electronic Waste Management: Policies, Processes, Technologies, and Impact.* 2024.

35 Needhidasan, S., Samuel, M., and Chidambaram, R. Electronic waste – an emerging threat to the environment of urban India. *J. Environ. Heal. Sci. Eng.* vol. 12, no. 1, p. 36, Dec. 2014. https://doi.org/10.1186/2052-336X-12-36.

36 Khanna, R., Mukherjee, P.S., and Park, M. A critical assessment on resource recovery from electronic waste: Impact of mechanical pre-treatment. *J. Clean. Prod.* vol. 268, p. 122319, Sep. 2020. https://doi.org/10.1016/j.jclepro.2020.122319.

37 Masud, M.H., Akram, W., Ahmed, A., Ananno, A.A., Mourshed, M., Hasan, M., et al. Towards the effective E-waste management in Bangladesh: a review. *Environ.*

Sci. Pollut. Res. vol. 26, no. 2, pp. 1250–1276, Jan. 2019. https://doi.org/10.1007/s11356-018-3626-2.
38 Batinic, B., Vaccari, M., Savvilotidou, V., Kousaiti, A., Gidarakos, E., Marinkovic, T., et al. Applied WEEE pre-treatment methods: opportunities to maximizing the recovery of critical metals. *Glob. NEST J.* vol. 20, no. 4, pp. 706–711, Sep. 2018. https://doi.org/10.30955/gnj.002589.
39 Hajam, Y.A., Kumar, R., and Kumar, A. Environmental waste management strategies and vermi transformation for sustainable development. *Environ. Chall.* vol. 13, p. 100747, Dec. 2023. https://doi.org/10.1016/j.envc.2023.100747.
40 Gritsch, L., Breslmayer, G., and Lederer, J. Quantity and quality of paper-based packaging in mixed MSW and separate paper collection – a case study from Vienna, Austria. *Resour. Conserv. Recycl.* vol. 215, p. 108091, Apr. 2025. https://doi.org/10.1016/j.resconrec.2024.108091.
41 Singh, E., Mishra, R., Kumar, A., Shukla, S.K., Lo, S.L., and Kumar, S. Circular economy-based environmental management using biochar: driving towards sustainability. *Process Saf. Environ. Prot.* vol. 163, pp. 585–600, Jul. 2022. https://doi.org/10.1016/J.PSEP.2022.05.056.
42 Onu, P. and Mbohwa, C. (Editors) Front Matter. In: *Agricultural Waste Diversity and Sustainability Issues*. Academic Press Inc., 2021, p. iii. https://doi.org/10.1016/B978-0-323-85402-3.01001-9.
43 Zhai, Z., Martínez, J.F., Beltran, V., and Martínez, N.L. Decision support systems for agriculture 4.0: survey and challenges. *Comput. Electron. Agric.* 2020. https://doi.org/10.1016/j.compag.2020.105256.
44 Sanitarc.si, To promote public and environmental health profession, *Int. J. Sanit. Eng. Res*, 2014.
45 Jain, P., Yadav, S.K., and Priyadarshini, I. (Editors) Waste management for smart cities. In: *Springer Science and Business Media LLC*. Springer, 2025.
46 Tiwari, R., Agrawal, P.S., Belkhode, P.N., Ruatpuia, J.V.L., and Rokhum, S.L. Hazardous effects of waste transformer oil and its prevention: a review. *Next Sustain.* 2024. https://doi.org/10.1016/j.nxsust.2024.100026.
47 Casas-Godoy, L., Barrera-Martínez, I., Ayala-Mendivil, N., Aguilar-Juárez, O., Arellano-García, L., Reyes, A.L., et al. Biofuels. In: *Biobased Products and Industries*. 2020. https://doi.org/10.1016/B978-0-12-818493-6.00004-X.
48 Rao, M.N., Sultana, R., and Kota, S.H. *Solid and Hazardous Waste Management Science and Engineering*. 2016.
49 Evodc, N., Qamar, S.A., Bilal, M., Barceló, D., and Iqbal, H.M.N. Plastic waste and its management strategies for environmental sustainability. *Case Stud. Chem. Environ. Eng.* vol. 4, p. 100142, Dec. 2021. https://doi.org/10.1016/j.cscee.2021.100142.
50 Kim, S.-J., Park, A., Nam, S.-E., Park, Y.-I., and Lee, P.S. Practical designs of membrane contactors and their performances in CO 2 /CH 4 separation. *Chem. Eng. Sci.* vol. 155, pp. 239–247, Nov. 2016. https://doi.org/10.1016/j.ces.2016.08.018.
51 Ribeiro, R.M.P. Bioenergy production from waste: metagenomics and metatranscriptomics analyses of full scale anaerobic digesters. *Univ. do Minho*. 2023.

52 Verma, C. and Dubey S. (Editors) *Biomass Wastes for Sustainable Industrial Applications A Waste-to-Wealth Approach*. CRC Press, 2024.
53 Speight, J.G. *Synthesis Gas*. Wiley, 2020.
54 Okoro, O.V., Sun, Z., and Birch, J. Techno-economic assessment of a scaled-up meat waste biorefinery system: a simulation study. *Materials (Basel)*. vol. 12, no. 7, p. 1030, Mar. 2019. https://doi.org/10.3390/ma12071030.
55 Yang, E., Mohamed, H.O., Park, S.G., Obaid, M., Al-Qaradawi, S.Y., Castaño, P., et al. A review on self-sustainable microbial electrolysis cells for electro-biohydrogen production via coupling with carbon-neutral renewable energy technologies. *Bioresour. Technol*. vol. 320, p. 124363, Jan. 2021. https://doi.org/10.1016/j.biortech.2020.124363.
56 Bhujade, R. and Agrawal, P.S. Algal biocrude as drop-in feedstock for green fuel and green chemicals. In: *The Production of Biodiesel and Related Fuel Additives*. Bentham Science Publishers, 2024, pp. 178–198. [Online]. Available: https://www.benthamdirect.com/content/books/9789815196740.chapter-6.
57 Spurek, G. The classification of solid recovered fuels for the needs of Polish cement industry. *Env*. vol. 87, no. 1, pp. 37–42, 2005.
58 Julie Chandra, C.S., Sasi, S., Bindu Sharmila, T.K., and Varghese, J.R. Modern biomass conversion technologies. In: *Handbook of Biomass*. Singapore: Springer Nature Singapore, 2023, pp. 1–31. https://doi.org/10.1007/978-981-19-6772-6_40-1.
59 Singh, B. and Oraon, R. (Editors) *Advanced Nanocatalysts for Biodiesel Production*. Routledge, 2022.
60 Sharma, N., Sodhi, A.S., and Batra, N. (Editors) *Basic Concepts in Environmental Biotechnology*. CRC Press, 2021.
61 Santillan-Jimenez, E. and Crocker, M. Catalytic deoxygenation of fatty acids and their derivatives to hydrocarbon fuels via decarboxylation/decarbonylation. *J. Chem. Technol. Biotechnol*. vol. 87, no. 8, pp. 1041–1050, Aug. 2012. https://doi.org/10.1002/jctb.3775.
62 Agrawal, A., Karadbhajne, V., Agrawal, P., Bawa, S., Tiwari, R., and Murthy, S.S. Formulation of biogreases derived from Mahua and Karanja Oil. *Res Sq*. 2022. https://doi.org/10.21203/rs.3.rs-1802614/v1.
63 Agrawal, A.J., Karadbhajne, V.Y., Agrawal, P.S., Arekar, P.S., and Chakole, N.P. Synthesis of biolubricants from non edible oils. *Int. Res. J. Eng. Technol*. 2017.
64 Khan, T.F., Ullah, M.W., and Huq, S.M.I. Heavy metal contents of different wastes used for compost. *J. Miner. Mater. Charact. Eng*. vol. 4, no. 3, pp. 241–249, May 2016. https://doi.org/10.4236/jmmce.2016.43022.
65 Anouzla, A. and Souabi, S. (Editors) *Technical Landfills and Waste Management*. Springer Science and Business Media LLC, 2024.
66 Mohanta, Y.K., Mishra, B., and Pudake, R.N. *Nano-microbiology for Sustainable Development*. 2025.
67 Shovon, S.M., Akash, F.A., Rahman, M.A., Rahman, W., Chakraborty, P., Monir, M.U., et al. The potential for sustainable waste management and energy recovery in Bangladesh: a review. *Sustain. Energy Technol. Assessments* vol. 64, p. 103705, Apr. 2024. https://doi.org/10.1016/j.seta.2024.103705.

68 Sirohi, R., Tarafdar, A., de Souza Vandenberghe, L.P., Taherzadeh, M.J., and Pandey, A. (Editors) *Sustainable Technologies for Food Waste Management*. CRC Press, 2025.

69 Das, A., Shi, D., Halder, G., and Lalthazuala Rokhum, S. Microwave-assisted synthesis of glycerol carbonate by transesterification of glycerol using *Mangifera indica* peel calcined ash as catalyst. *Fuel*. 2022. https://doi.org/10.1016/j.fuel.2022.125511.

70 Laskar, I.B., Rokhum, L., Gupta, R., and Chatterjee, S. Zinc oxide supported silver nanoparticles as a heterogeneous catalyst for production of biodiesel from palm oil. *Environ. Prog. Sustain. Energy*. 2019. https://doi.org/10.1002/ep.13369.

71 Agrawal, P. and Rokhum, S.L. Supported metal/metal oxide catalysts in biodiesel production. In: *Biodiesel Production*. John Wiley & Sons, Ltd, 2022, pp. 119–141. https://doi.org/10.1002/9781119771364.ch7.

72 Toldrá-Reig, F., Mora, L., and Toldrá, F. Trends in biodiesel production from animal fat waste *Appl. Sci*. vol. 10, no. 10, pp. 3644, May. 2020. https://doi.org/10.3390/app10103644.

73 Mukhtar Ahmad. *Operation and Control of Renewable Energy Systems*. Wiley, 2017.

74 Oluseun Adejumo I. and Adebukola Adebiyi, O. Agricultural solid wastes: causes, effects, and effective management. In: *Strategies of Sustainable Solid Waste Management*. IntechOpen, 2021. https://doi.org/10.5772/intechopen.93601.

75 Jonathan W. C. Wong, Rao Y. Surampalli, Tian C. Zhang, Rajeshwar D. Tyagi, Ammaiyappan Selvam. Sustainable Solid Waste Management, American Society of Civil Engineers (ASCE). 2016.

76 Yıldız, Ü.Y., Keçili, R., and Hussain, C.M. Green and sustainable chemistry. In: *Green Imprinted Materials*. Elsevier, 2024, pp. 3–25. https://doi.org/10.1016/B978-0-443-15431-7.00014-3.

77 Chew, K.W., Chia, S.R., Chia, W.Y., Cheah, W.Y., Munawaroh, H.S.H., and Ong, W.J. Abatement of hazardous materials and biomass waste via pyrolysis and co-pyrolysis for environmental sustainability and circular economy. *Environ. Pollut*. 2021. https://doi.org/10.1016/j.envpol.2021.116836.

78 Behnia, M., Ghahderijani, M., Kaab, A., and Behnia, M. Evaluation of sustainable energy use in sugarcane production: a holistic model from planting to harvest and life cycle assessment. *Environ. Sustain. Indic*. vol. 26, Jun. 2025. https://doi.org/10.1016/j.indic.2025.100617.

79 Singhania, R.R., Guo, W., de Souza Vendenberghe, L.P., Mannina, G., and Kim, S.H. Bioresource technology for bioenergy, bioproducts & environmental sustainability. *Bioresour. Technol*. vol. 347, Mar. 2022. https://doi.org/10.1016/j.biortech.2022.126736.

80 Scarlat, N., Dallemand, J.F., and Banja, M. Possible impact of 2020 bioenergy targets on European Union land use. A scenario-based assessment from national renewable energy action plans proposals. *Renew. Sustain. Energy Rev*. vol. 18, pp. 595–606, 2013. https://doi.org/10.1016/j.rser.2012.10.040.

81 ur Rehman, K., Schwennen, C., Visscher, C., Plötz, M., Grabowski, N.T., Sultana, M.U., et al. Closing the loop with pretreatment and black soldier fly technology for recycling lignocellulose-rich organic by-products: a progressive review. *Carbohydr.*

Polym. Technol. Appl. vol. 9, Mar. 2025. https://doi.org/10.1016/j.carpta.2024.100630.

82 Dey, T., Bhattacharjee, T., Nag, P., Ritika, Ghati, A., and Kuila, A. Valorization of agro-waste into value added products for sustainable development. *Bioresour. Technol. Rep.* vol. 16, Dec. 2021. https://doi.org/10.1016/j.biteb.2021.100834.

83 Onu, P., Mbohwa, C., and Pradhan, A. Machine learning: a comprehensive exploration of fault detection and diagnosis in smart grids. *Int. Conf. Electr. Comput. Energy Technol. ICECET 2023*, 2023. https://doi.org/10.1109/ICECET58911.2023.10389596.

84 Bakhti, A. and Lashani, E. Chapter 13 Microbial bioremediation for industrial waste valorization. In: *Biotechnological Applications in Industrial Waste Valorization*. Springer, 2025.

85 Johannes Lehmann, S.J. *Biochar for Environmental Management - Science, Technology and Implementation*. Routledge, 2015.

86 Bakindi, A., Wiberg, A.H., Norman, J., Marsh, E., and Allen, S. Hybrid life cycle assessment (H-LCA) for buildings and construction materials: a systematic review and meta-analysis. *Build. Environ.* vol. 272, p. 112630, Mar. 2025. https://doi.org/10.1016/j.buildenv.2025.112630.

5

Waste Biorefinery

5.1 Introduction

Contemporary global society confronts significant interconnected challenges encompassing food, water, and energy security, exacerbated by relentless population growth, escalating greenhouse gas emissions, and pronounced climate change impacts primarily attributed to anthropogenic activities. This confluence of factors widens the gap between economic development aspirations and the imperative of environmental sustainability. In developing nations, while policies are increasingly oriented toward fostering a circular economy and promoting sustainable technologies to ensure consistent feedstock and energy supplies, inadequate infrastructure and limited financial resources frequently render traditional waste disposal complex and environmentally detrimental. Empirical observations in countries such as India, Bangladesh, and Pakistan reveal that effective waste collection rates hover around 60%, leaving substantial quantities of waste improperly managed in public spaces like streets, railway corridors, vacant plots, and low-lying areas [1]. It is widely acknowledged that waste recycling offers substantial environmental and economic benefits. Fundamentally, all forms of waste – whether solid, liquid, or gaseous – contain inherent energy and valuable components that can be recovered and reused. This potential for recuperation and valorization supports a crucial shift toward a low-carbon bioeconomy. The concept of a waste biorefinery signifies a transformative paradigm shift in waste management, moving beyond conventional disposal methodologies to embrace a more sustainable and resource-efficient approach centered on biological and thermochemical conversion processes [2]. Essentially, a waste biorefinery operates as a sophisticated facility designed to process various types of waste materials, converting them into valuable bio-based products [3, 4]. This function parallels that of a petroleum refinery, which processes crude oil into a diverse range of fuels, chemicals, and other useful products. Consequently, the waste biorefinery approach provides a comprehensive solution that simultaneously addresses the production of potentially marketable bio-based products and the critical need for waste remediation. The strategic introduction and widespread adoption of waste biorefineries are pivotal in confronting several pressing issues inherent in

Agro-Waste Management and Valorization, First Edition. Pratibha S. Agrawal, et al.

modern waste management systems. These include reducing the volume of waste sent to landfills, mitigating the associated environmental pollution and greenhouse gas emissions, conserving valuable resources embedded in waste, and contributing to the development of a more circular and sustainable economy. The inherent energy contained within organic waste, if harnessed effectively through biorefining, can displace the need for fossil fuels, leading to a reduction in greenhouse gas emissions and enhancing energy security. Furthermore, the diverse range of products that can be generated from waste biorefineries, such as biofuels, biochemicals, organic acids, enzymes, fertilizers, and biodegradable plastics, offers significant economic opportunities and can integrate into various industrial, agricultural, and pharmaceutical value chains. By closing the loop on waste streams and maximizing resource recovery, these advanced facilities play a crucial role in facilitating the transition toward a truly circular and sustainable economic model. Overall, the conceptual framework and practical implementation of waste biorefineries represent an innovative strategy for waste management that is intrinsically aligned with the principles of sustainability, resource efficiency, and environmental stewardship. By effectively harnessing the latent potential within waste materials and converting them into valuable products through advanced biological, chemical, and thermochemical processes, waste biorefineries offer a highly promising solution to the escalating challenges of modern waste management and are instrumental in constructing a more sustainable future. Some of the advantages of waste biorefineries are as shown in Figure 5.1.

This chapter aims to provide a detailed analysis of the various waste feedstocks suitable for biorefining, delve into the kinetic analysis of biomass conversion processes, explore different waste feedstock conversion pathways, provide insights into water purification-based biorefineries, and examine the economic analysis of waste biorefineries to assess their potential for future deployment.

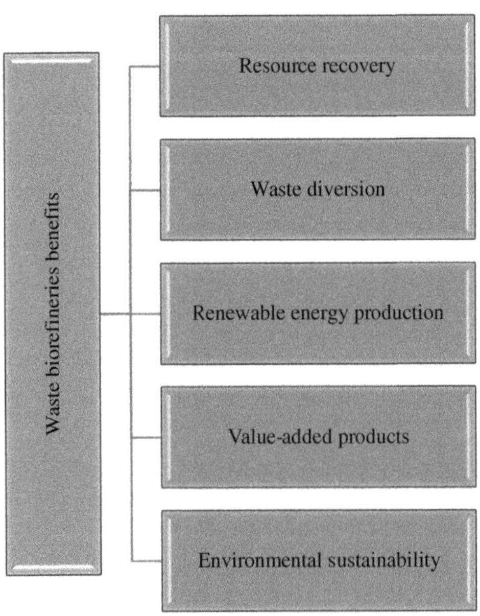

Figure 5.1 Benefits offered by waste biorefineries.

5.2 Waste Feedstock for Biorefinery

In the context of producing bio-based chemicals and biofuels through biorefining, a wide spectrum of waste materials – including solids, liquids, and gases – can be effectively utilized as feedstock. Consequently, the waste feedstock landscape for biorefineries is vast, encompassing a diverse array of organic materials amenable to biological, chemical, or thermochemical conversion processes [5]. These feedstocks originate from numerous sources, each possessing unique characteristics and inherent potential for conversion. However, as highlighted by Gaffey et al. [6], a significant global challenge remains the prevalent practice of disposing of industrial, municipal, and agricultural wastes in landfills or through incineration. These conventional methods not only consume vast tracts of land but also pose serious threats to environmental health, underscoring the urgent need for robust reutilization and recycling strategies for these waste streams [7].

The global biorefinery market is experiencing substantial growth, indicative of increasing investment and interest in waste valorization technologies. Figure 5.2 illustrates this trajectory, estimating the global biorefinery market size at US$135.43 billion in 2022 and projecting its expansion to reach US$328.04 billion by 2032 [8].

This projected growth underscores the significant economic potential inherent in waste biorefining. Common categories of waste feedstocks utilized in biorefineries include:

- **Municipal Solid Waste (MSW):** MSW constitutes a heterogeneous mixture of waste generated from households, commercial establishments, and institutions. Its composition typically includes organic materials such as food waste, yard waste, paper, and cardboard. Biorefineries can process MSW through anaerobic digestion (AD) to produce biogas or through thermochemical routes like pyrolysis and gasification to generate biofuels and biochemicals.

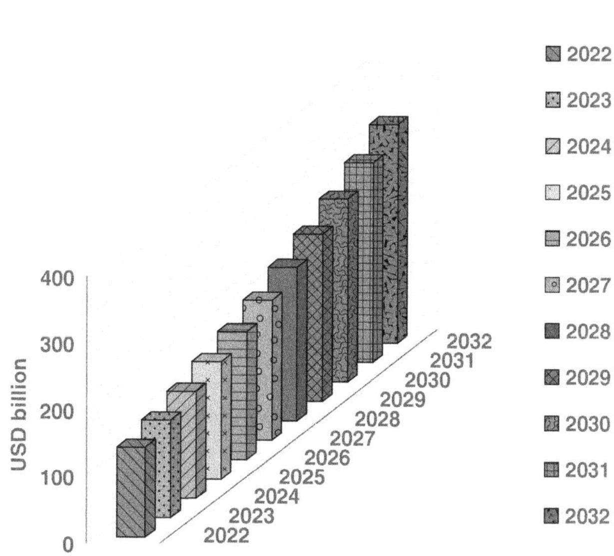

Figure 5.2 The global biorefinery market size from 2022 to 2032. Adapted from [8].

- **Agricultural Residues:** These are abundant byproducts of agricultural activities, encompassing crop residues (e.g., straw, stalks), animal manure, and waste generated during agricultural processing (e.g., fruit pomace, bagasse). These residues are rich in organic matter and serve as valuable feedstocks for biorefineries to produce bioenergy, biofuels [9], and high-value biochemicals.
- **Forestry Residues:** Derived from forestry operations such as logging, thinning, and timber processing, forestry residues comprise woody biomass materials including branches, tops, bark, and sawdust. These residues can be converted into biofuels like wood pellets and bio-oil or utilized for the production of biochemicals and biocomposites within a biorefinery framework.
- **Food-processing Waste:** This category includes organic waste streams generated during the processing and manufacturing of food products, such as fruit and vegetable trimmings, dairy waste, and brewery waste. These waste streams are particularly amenable to conversion into biofuels, enzymes, and other value-added products through enzymatic hydrolysis, fermentation, and other bioprocessing techniques.
- **Wastewater Sludge:** Also known as sewage sludge, this material is a byproduct of wastewater treatment processes, containing concentrated organic matter, nutrients, and pathogens removed from sewage [10]. Wastewater sludge can be processed through AD to produce biogas or subjected to thermochemical conversion to yield biochar, bio-oil, and syngas.
- **Organic Industrial Waste:** Various industrial sectors generate distinct organic waste streams as byproducts of their manufacturing processes. Examples include agro-industrial wastes (e.g., sugarcane bagasse, palm oil mill effluent), waste from the food industry (e.g., brewery spent grains, olive pomace), and pulp and paper mill waste (e.g., black liquor, lignin) [6]. These diverse organic industrial wastes can be efficiently converted into bioenergy, biochemicals, and biopolymers in biorefineries.

Waste feedstocks for biorefineries represent a diverse and abundant resource base that underpins sustainable bioconversion processes. Their effective utilization contributes significantly to the production of renewable energy, biofuels, and bio-based products, while simultaneously mitigating environmental pollution and addressing critical waste disposal challenges. The selection and preparation of appropriate feedstocks are crucial initial steps in the design and operation of any waste biorefinery.

5.3 Kinetic Analysis of Biomass

Kinetic analysis of biomass refers to the systematic study of the rates at which biochemical and thermochemical reactions and processes occur during the conversion of biomass into valuable products, including biofuels, biochemicals, and bioproducts. This analytical approach is fundamental to understanding and optimizing biomass conversion pathways such as enzymatic hydrolysis, fermentation, pyrolysis, and gasification. As emphasized by Dhyani and Bhaskar [11], kinetic analysis plays a critical role in understanding the progression of biomass pyrolysis reactions and determining their rate dependence on specific process parameters. This is often conceptualized as a two-stage process, beginning with experimental studies conducted at various temperatures under isothermal or non-isothermal conditions, followed by the mathematical determination of kinetic parameters based on the experimental data. The primary areas of focus within the kinetic analysis of biomass include:

1. **Reaction Rate Modeling:** Developing accurate kinetic models is essential for describing the rates of complex biomass conversion reactions. These models can be empirical, semi-empirical, or mechanistic, depending on the level of understanding of the underlying reaction mechanisms and the availability of sufficient experimental data [12]. Reaction rate equations are typically formulated based on the influence of factors such as reactant concentrations, temperature, pressure, and the presence and activity of catalysts.
2. **Experimental Kinetics Studies:** Conducting rigorous experimental studies under controlled laboratory conditions is indispensable for obtaining the data required for kinetic analysis. These studies involve precisely measuring reaction rates and determining key kinetic parameters such as rate constants and activation energies. Laboratory-scale reactors, ranging from simple batch reactors to more complex continuous flow systems, are employed, often coupled with analytical techniques like spectroscopy, chromatography, and calorimetry to monitor reaction progress and product formation.
3. **Parameter Estimation:** Once experimental data on reaction rates is obtained, kinetic parameters (e.g., rate constants, activation energies) are estimated by fitting the experimental data to the proposed kinetic models. This process often involves advanced parameter estimation techniques, including nonlinear regression and optimization algorithms, aimed at identifying the set of parameters that best minimize the discrepancy between the experimental observations and the model predictions.
4. **Effect of Operating Conditions:** The rates and selectivities of biomass conversion reactions are profoundly influenced by variations in operating conditions such as temperature, pressure, pH, initial substrate concentration, and the specific type and loading of the catalyst. Kinetic analysis provides the quantitative framework necessary to elucidate the precise effects of these operating parameters on reaction rates and to optimize process conditions to maximize the yield and selectivity of desired products while minimizing byproduct formation [13].

In essence, kinetic analysis of biomass is a critical tool for advancing biomass conversion technologies. It provides the fundamental scientific understanding necessary for the rational design, optimization, and scaling-up of biorefinery processes. By enabling the prediction of reactor performance under various conditions, kinetic models facilitate the sustainable utilization of renewable biomass resources for the production of bio-based fuels, chemicals, and materials. This analytical approach finds broad application across various sectors of industry and society. In Biorefinery Process Development, kinetic analysis directly informs the design and optimization of processes aimed at converting biomass into biofuels, biochemicals, and bioproducts. In Biochemical Reaction Engineering, a deep understanding of the kinetics of biomass conversion reactions is fundamental for designing efficient biocatalytic systems and enzyme-based processes. For thermochemical conversion technologies, kinetic analysis provides crucial insights for the design and operation of processes like pyrolysis, gasification, and liquefaction, which are used to convert biomass into bioenergy and biofuels [14]. Finally, in bioprocess optimization, kinetic models are invaluable for optimizing process parameters and reactor configurations to achieve maximum product yields, minimize the formation of undesirable byproducts, and improve overall process efficiency [15].

5.4 Conversion Processes

Biorefineries function as sophisticated facilities dedicated to transforming biomass into a diverse array of valuable products, encompassing biofuels, biochemicals, and biomaterials. The specific suite of conversion processes deployed in a biorefinery is dictated by the type of biomass feedstock utilized and the desired characteristics of the final products. However, these processes generally involve a series of key steps to break down complex biomass structures and convert them into usable forms.

- **Feedstock Preparation:** This initial stage is critical and involves the collection and preparation of biomass feedstocks. This can include a wide variety of materials such as agricultural residues, forestry waste, MSW, or purpose-grown energy crops [16]. Preparation typically involves physical size reduction methods like shredding, chipping, or grinding to increase the surface area, along with drying or other pre-processing steps designed to enhance the suitability of the biomass for subsequent conversion processes.
- **Pretreatment:** Biomass pretreatment is often a necessary step to overcome the inherent recalcitrance of the biomass structure, particularly for lignocellulosic feedstocks. Pretreatment aims to disrupt the complex matrix of cellulose, hemicellulose, and lignin components, thereby increasing their accessibility to enzymes or catalysts used in subsequent conversion steps [17]. Various pretreatment methods can be employed, including mechanical (e.g., milling), chemical (e.g., acid or alkaline hydrolysis), thermal (e.g., steam explosion), or biological approaches.
- **Hydrolysis:** The process of breaking down complex carbohydrate polymers, such as cellulose and hemicellulose, into simpler monosaccharide sugars like glucose and xylose. This crucial step can be achieved through chemical hydrolysis using acids or bases, or more commonly through enzymatic hydrolysis, where specific enzymes like cellulase and hemicellulase catalyze the breakdown of polysaccharides into fermentable sugars.
- **Fermentation:** A biological process carried out by selected microorganisms, typically bacteria or yeast, which convert the sugars obtained from hydrolysis into biofuels or biochemicals. In the context of biofuel production, sugars derived from biomass are fermented to produce ethanol, butanol, or other types of biofuels. For biochemical production, fermentation pathways can be

engineered to yield a wide range of valuable products, including organic acids, enzymes, and specialty chemicals.

- **Distillation and Separation:** Following the conversion step, the desired products are present in a complex mixture (e.g., fermentation broth or pyrolysis liquid). Effective separation and purification techniques are essential to isolate and concentrate the target compounds from this mixture and remove impurities [18]. Common techniques include distillation, chromatography, filtration, solvent extraction, and membrane separation.
- **Upgrading and Refining:** In some cases, the biofuels or biochemicals produced may require further upgrading or refining to meet specific quality standards or ensure compatibility with existing infrastructure and end-use applications. Upgrading processes, such as esterification (for bio-oil), hydrogenation (for removing oxygen), or catalytic conversion, can be used to improve the properties of the final products.
- **Waste Management:** Throughout the entire biorefining process, various byproducts and residual waste streams are generated. These may include residual lignin (from hydrolysis/fermentation), ash (from thermochemical conversion), wastewater, and other residues. Implementing effective waste management strategies is crucial to minimize the environmental impact of the biorefinery operation and enhance its overall sustainability. Techniques such as AD (for organic residues), composting, or thermal conversion can be applied to treat or convert these waste streams into other valuable products or generate energy.

By strategically integrating these various conversion processes, biorefineries are capable of efficiently transforming diverse waste biomass streams into a spectrum of valuable products [19], thereby generating economic value while simultaneously reducing reliance on fossil fuels and mitigating associated environmental impacts.

5.4.1 Thermochemical Conversions

Thermochemical conversion processes are central to waste biorefining, leveraging elevated temperatures and pressures to transform the chemical energy embedded within different types of biomass into value-added products. These methods involve controlled thermal and chemical reactions, often utilizing catalysts or solvents, to break down complex organic structures. Thermochemical conversions are particularly effective for converting waste biomass into energy carriers, fuels, and chemicals and are integral to biorefinery operations [20]. The thermochemical conversion processes utilized in waste-to-wealth biorefineries include:

1. **Pyrolysis:** The thermal decomposition of organic material in the absence of oxygen, yielding three primary product fractions: bio-oil, syngas, and biochar. Syngas is a gaseous mixture predominantly composed of hydrogen, carbon monoxide, and methane (CH_4). Bio-oil and syngas are promising intermediates that can be further processed into various fuels, including biodiesel, ethanol, or hydrogen, while biochar is a solid residue with potential applications as a soil amendment or for carbon sequestration [21]. Pyrolysis typically

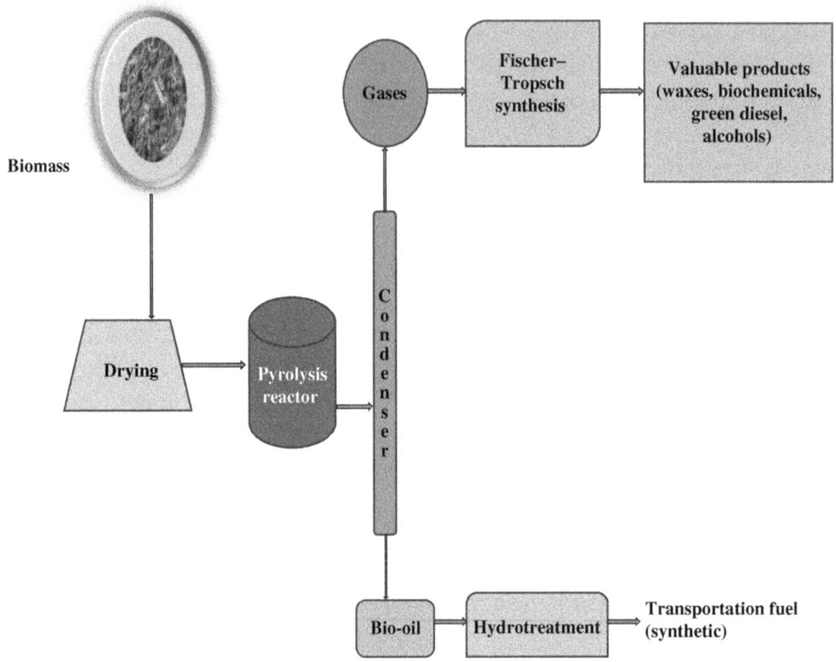

Figure 5.3 Representation of pyrolysis and gasification process.

occurs at temperatures ranging from 300 to 700 °C under anaerobic conditions. Classification of pyrolysis often distinguishes between types based on heating rate, temperature, and vapor residence time, such as slow pyrolysis (conventional), fast pyrolysis, flash pyrolysis, and microwave pyrolysis, alongside torrefaction. Researchers [22] demonstrated the conversion of agricultural and food-processing biomass (wheat straw, brewery grains, spent coffee grounds) into biochar via pyrolysis at 400–500 °C under nitrogen, showing varying carbon content (45–77%) and adsorption properties of the resulting biochar. The general pyrolysis process is illustrated in Figure 5.3, showing biomass undergoing drying and reaction in a pyrolysis reactor to produce gases, which can be condensed into bio-oil, followed by optional hydrotreatment and Fischer–Tropsch synthesis to produce transportation fuel [23].

2. **Gasification:** A thermochemical process that converts biomass into syngas by reacting it at high temperatures (typically 800–1600 °C) with a limited amount of oxidizing agent, such as oxygen, air, steam, or carbon dioxide (CO_2) [24]. The resulting syngas is primarily composed of H_2, CO, CO_2, CH_4, N_2 (if air is used), and H_2O, and can contain various contaminants like tars and particulates. Syngas is a versatile intermediate; it can be directly combusted for heat and power generation or used as a feedstock for the synthesis of liquid fuels and chemicals through processes like Fischer–Tropsch synthesis. According to Korberg et al. [25], biomass is a key renewable energy feedstock, and syngas from gasification can supplement biogas in power plants or be used for CO_2 electro fuels. Various gasification reactor types exist, including fluidized-bed,

Table 5.1 Thermochemical conversion processes and products. Adapted from [25].

Thermochemical conversion processes	Heating rate (°C minutes^{-1})	Residence time (minutes)	Temperature (°C)	Major product	Byproduct	% Biochar yield
Flash pyrolysis	>1000	~0.041	750–1000	Bio-oil	—	37.1
Slow pyrolysis	<10	>60	350–700	Biochar	Bio-oil combustible gas	31–56
Fast pyrolysis	>200	~0.02	400–600	Bio-oil	Combustible gas	10–25
Gasification	~1000	0.21–0.4	800–1600	Syngas	Biochar tars	14.0–26
Hydrothermal carbonization	<1	5.0–240	400–1000	Hydrochar	—	50–78.9

dual fluidized-bed, fixed-bed (updraft and downdraft), supercritical water, and plasma gasification. Table 5.1 summarizes characteristics of different thermochemical processes, including typical operating parameters (heating rate, residence time, temperature), major products, byproducts, and biochar yields. For example, fast pyrolysis (heating rate >200 °C/minutes) at 400–600 °C with ~0.02 minutes residence time yields bio-oil as the major product and combustible gas, with a biochar yield of 10–25%. Gasification (heating rate ~1000 °C/minutes) at 800–1600 °C with 0.21–0.4 minutes residence time yields syngas and biochar tars, with a biochar yield of 14.0–26%.

3. **Hydrothermal Liquefaction (HTL):** HTL is a thermochemical conversion process suitable for wet biomass feedstocks, converting them into a liquid bio-oil (often called bio-crude) at moderate temperatures (200–350 °C) and high pressures (40–220 bar) in the presence of water [26]. Unlike pyrolysis, HTL does not require prior drying of the feedstock, making it energy-efficient for wet materials like algae or sewage sludge. The process involves complex ionic and free radical reactions facilitated by the unique properties of subcritical or supercritical water. The resulting bio-oil is a complex mixture of oxygenated compounds, which can be further upgraded into transportation fuels or specialty chemicals. HTL is considered a highly efficient strategy for wet biomass valorization with lower energy consumption compared to processes requiring drying.
4. **Hydrothermal Carbonization (HTC):** HTC is a thermochemical process that converts biomass into a solid, carbon-rich material called hydrochar (or biochar) by heating it in water at moderate temperatures (180–250 °C) and autogenous pressures [27]. This process typically operates below 200 °C in a sealed vessel. In addition to hydrochar, HTC produces a small amount of liquid (aqueous phase containing dissolved organic compounds) and gas (primarily CO_2). Applied temperature, pH, and the type of biomass significantly affect the characteristics and product distribution. For hydrochar formation, temperatures around 159 °C are often reported as providing high yields (up to ~200 °C yield reported).

Table 5.2 Catalysts utilized in catalytic pyrolysis and gasification processes. [17]/Royal Society of Chemistry.

Catalyst	Reactor	Catalyst to biomass ratio	Temperature (K)	Gas wt.%	Bio-oil wt.%	Char wt.%
H-Beta	Fluidized-bed	0.41	723	49.3	28.7	22.0
CoO	Fixed-bed	0.45	773	21.28	56.49	22.23
NiO	Fixed-bed	0.45	773	27.75	47.44	24.86
ZSM-5	Fixed-bed	0.02	763–813	19.44	38.28	42.29
Co_3O_4	Fixed-bed	0.45	773	33.77	55.46	22.59
ZSM-5	Auger reactor	5.00	773	26.2	50.1	13.8
Al-MCM 41	Fixed-bed	0.45	773	7.46	48.44	37.45

Hydrochar's hydrophobic nature and homogeneous properties facilitate its separation from the liquid phase. Mass yields typically range from 35 to 80%, and hydrochar possesses a higher carbon content than lignite. Hydrochar can be used as a solid fuel, a soil amendment to improve soil properties, or a carbon sequestration agent in agricultural applications.

5. **Catalytic Pyrolysis and Gasification:** The integration of catalysts into pyrolysis and gasification processes can significantly enhance reaction rates, improve process efficiency, and steer selectivity toward desired products [28]. Catalysts can help control the distribution of products, reduce energy requirements, and facilitate the production of high-value chemicals or fuels. The presence of a catalyst notably enhances biomass decomposition and influences the quantity and composition of pyrolysis products. In catalytic pyrolysis, two main configurations are used: catalyst mixing, where catalyst is blended directly with biomass before the reactor, and using a fixed catalytic bed (in situ) through which pyrolytic vapors pass, influencing the char, bio-oil, and gas composition. Table 5.2 provides examples of catalysts used in catalytic pyrolysis and gasification, showing their impact on product yields (gas, bio-oil, char) under specific conditions.

Thermochemical conversion processes, when integrated into biorefineries, offer several advantages, including high energy efficiency, flexibility in handling diverse feedstocks, and the capacity to process a wide range of biomass types and waste streams. By optimizing these processes and their integration, waste biomass can be effectively converted into valuable products, generating economic value and contributing to sustainability goals.

5.4.2 Combined Gasification-fermentation Processes

Combined gasification-fermentation (CGF) processes represent an advanced integrated approach within waste biorefining that synergistically couples biomass gasification with microbial fermentation. This combination aims to convert waste biomass into energy-rich products such as biofuels or biochemicals by leveraging the distinct advantages of both thermochemical and biological conversion pathways to maximize overall energy efficiency and product yields [29]. The initial step in the CGF process involves the gasification of waste biomass feedstock, which can include a variety of materials such as agricultural residues, forestry waste, or MSW [30]. This biomass is converted into a synthesis gas (syngas) primarily composed of carbon monoxide (CO), hydrogen (H_2), and CH_4, along with CO_2, H_2O, and other trace components depending on the gasification agent and conditions. Gasification is typically performed by heating the biomass in a low-oxygen environment, often in the presence of steam or CO_2, to produce a combustible gas mixture.

The crude syngas produced from biomass gasification invariably contains impurities, including tars, particulates, sulfur compounds (e.g., H_2S), and other trace contaminants. These impurities are detrimental to subsequent microbial fermentation processes and must be removed through rigorous syngas cleanup procedures. Syngas cleanup techniques are critical and can involve various methods such as filtration for particulates, scrubbing for acid gases, catalytic conversion for tar reduction, or other purification methods tailored to the specific syngas composition. Ensuring the cleaned syngas meets the purity requirements of the microbial fermentation is paramount for process success.

Once the syngas is cleaned to the required specifications, it serves as the primary feedstock for microbial fermentation. Specific microorganisms, such as acetogenic bacteria (e.g., *Clostridium ljungdahlii, Clostridium autoethanogenum*) or methanogenic archaea, are capable of metabolizing the components of syngas (primarily CO, H_2, and CO_2) to produce various biofuels or biochemicals. For instance, acetogenic bacteria can convert CO and H_2 into acetate, which can then be further utilized or converted into ethanol or other desired products through subsequent microbial fermentation pathways [31]. This biological conversion of syngas occurs under relatively mild temperatures and pressures compared to thermochemical synthesis methods, offering potential cost and energy savings.

After the microbial fermentation stage, the desired biofuels or biochemicals are recovered from the fermentation broth and purified using appropriate separation techniques. These techniques can include distillation (for alcohols), chromatography, filtration, or solvent extraction, depending on the target compounds. Figure 5.4 provides a conceptual flowchart illustrating the key steps in a gasification-fermentation process, showing wet biomass entering gasification, syngas cleaning, and subsequent fermentation of syngas to biofuels and value-added chemicals.

Throughout the CGF process, various byproducts and waste streams are generated, including residual biomass ash from gasification, spent fermentation broth, and wastewater. Effective waste management strategies for these streams are essential to minimize the overall environmental impact of the CGF process and maximize resource recovery within the biorefinery. These streams can often be valorized or utilized within the biorefinery itself for energy generation, nutrient recovery, or recycling.

CGF processes offer several compelling advantages for waste-to-energy conversion and biorefining. Firstly, they enable the efficient utilization of a wide range of waste biomass feedstocks, including those with high moisture content or complex compositions that may be less suitable for other conversion methods [32]. Secondly, by

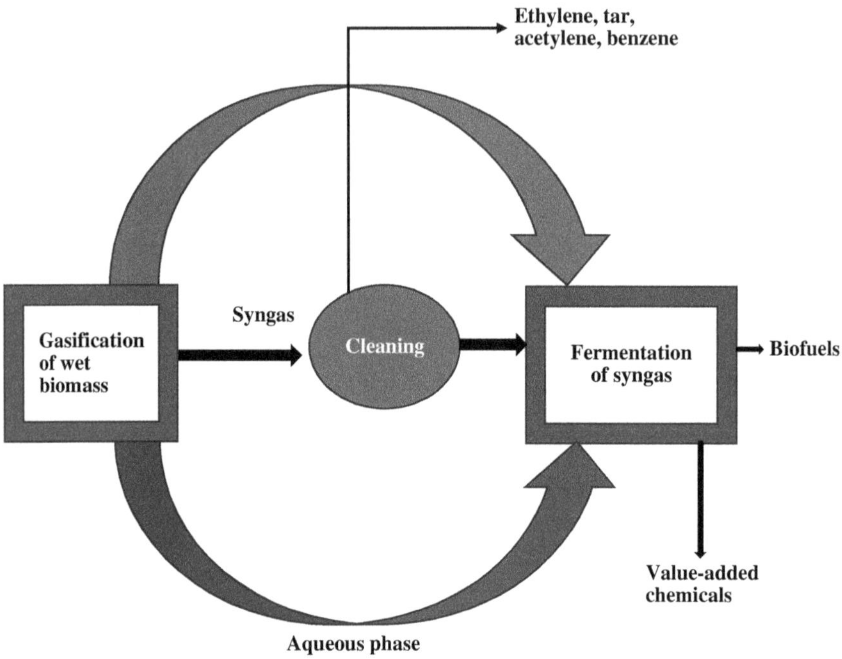

Figure 5.4 Flowchart representing gasification-fermentation process.

integrating both thermochemical gasification and biochemical fermentation, CGF can achieve high overall energy efficiency. Thirdly, CGF can produce a variety of valuable biofuels (e.g., ethanol, butanol) and biochemicals (e.g., organic acids, solvents) from waste materials, contributing to energy security and sustainability goals. Lastly, the modular nature of CGF processes holds potential for the development of decentralized and distributed biorefinery systems, enabling the utilization of locally available biomass resources and reducing transportation costs associated with feedstock logistics. By optimizing the integration of gasification and fermentation processes and addressing challenges such as syngas cleanup and microbial inhibition, CGF biorefineries can effectively convert waste biomass into valuable energy products while simultaneously minimizing environmental impact and enhancing resource efficiency.

5.4.3 Food Waste Biorefinery

Food waste biorefineries represent a highly promising strategy for addressing the significant environmental and economic challenges associated with the disposal of food waste. The escalating volume of food waste globally, generated across the entire food supply chain from production and processing to retail and consumption, has underscored the urgent need for effective management solutions beyond landfilling and incineration. The concept of food waste biorefining emerged in response to the growing recognition of food waste not merely as a waste problem but as a valuable resource rich in organic matter, carbohydrates, proteins, and lipids. Early research and development efforts focused on developing technologies to convert food waste into basic value-added products, such as biogas through AD or compost. However, the potential of food waste extends far beyond these primary applications.

Over the past few decades, significant advancements in biotechnology, process engineering, and integrated waste management have enabled the development of more efficient and cost-effective biorefinery processes specifically tailored for the valorization of food waste. These processes encompass a range of technologies, including biological methods like fermentation and AD, enzymatic hydrolysis to break down complex polymers, and thermochemical conversion technologies [8]. These advancements have transformed the concept of food waste biorefining from a theoretical idea into a practical and increasingly implemented solution. Food waste biorefineries aim to extract a wider spectrum of high-value products. This includes not only biofuels (e.g., biogas, bioethanol, biodiesel) but also biochemicals (e.g., organic acids, enzymes, platform chemicals), bioplastics, and nutrient-rich fertilizers.

Food waste biorefining continues to be a dynamic area of research and development globally. Numerous academic institutions, research organizations, and private companies are actively exploring innovative technologies and optimizing existing processes to efficiently and economically convert diverse food waste streams into high-value products. The feasibility and scalability of food waste valorization technologies have been demonstrated through the establishment of several pilot- and commercial-scale food waste biorefineries in various regions worldwide. These facilities serve as important models for future deployments. Governments and regulatory agencies in many countries are increasingly recognizing the importance of reducing food waste and promoting its valorization. Policy measures such as setting waste diversion targets, providing financial incentives for biorefinery development, and establishing supportive regulatory frameworks are playing a crucial role in encouraging the deployment of food waste biorefining solutions [14].

Continued technological innovation is anticipated to further enhance the efficiency, improve the economics, and strengthen the environmental sustainability of food waste valorization processes. Advances in areas such as enzyme engineering for more efficient hydrolysis, fermentation microbiology for improved product yields and expanded product ranges, and process integration for synergistic operation of different conversion steps are likely to significantly expand the potential of food waste biorefining. Food waste biorefineries play a crucial role in advancing the principles of the circular economy by establishing a closed loop for organic waste streams, thereby reducing the need for landfilling and minimizing associated environmental impacts. Ebikade et al. [15] provided an example of an integrated biorefinery approach for valorizing potato peels, a common food-processing residue. Their study utilized ultrasonic extraction for recovering valuable extractives, followed by optimized hydrolysis and glucose dehydration, which resulted in a 54% yield of HMF (5-hydroxymethylfurfural), a valuable bio-based chemical precursor. The residual lignin from this process was further pyrolyzed to produce biochar with potential for remediating pesticide-contaminated water. Tsegaye et al. [16] highlighted the vital role of food waste biorefineries in producing biofuels, bio-based substances, and platform chemicals, contributing to reducing the environmental

Table 5.3 Value-added products generated from food waste biorefinery. Adapted from [16].

Type of Bioprocess	Raw material	Reactor type/condition	Product
Batch fermentation	Discard juices of carrot	250 ml flask	Bioethanol
Batch fermentation	Calcium alginate	250 ml flask	Bioethanol
Immobilization of oxidase and glucoamylase	Kitchen food waste	Simultaneous scarifications, fermentations, pH 6.3, 328 K	Ethanol
Immobilization of lipase	Waste cooking oil	Hydrolysis and transesterification	Biodiesel
Anaerobic digestion	Food waste	45 L fluidized-bed reactor, 313 K	Biogas (CH_4, 60%)
Ensiling and centrifugation	Orange peel waste	Freezing and thawing	Lactic acid
Hydrolysis and fermentation	Pastry and cake waste	Lab-scale fermentation	Succinic acid
Solvent extraction	Grape stalk	–	Phenols
Pulsed electric field	Blue berries waste	Energy input, 10 kJ/kg	Pectin
Lactic acid fermentation	Brewery waste	500 ml flask	Proteinase
Batch fermentation	Discarded carrot juices	250 ml flask	Single cell protein

burden of waste disposal and providing sustainable resources. These facilities significantly enhance the circular economy by enabling eco-friendly and sustainable production processes, aligning with sustainable development goals. Table 5.3 provides examples of valuable products obtained from food waste feedstocks using various bioprocesses, demonstrating the diverse range of outputs possible from this waste stream. Examples include bioethanol from fruit juices and kitchen waste, biodiesel from waste cooking oil, biogas from AD, and lactic acid from orange peel waste.

Future developments are expected to involve integrating food waste biorefinery systems more broadly across the entire food supply chain to optimize resource recovery and minimize waste generation at every stage. As consumer awareness regarding sustainability issues increases, there is a growing market demand for products derived from food waste, including biofuels, bioplastics, and renewable chemicals. The commercialization of food waste biorefinery products is anticipated to accelerate, driven by both increasing market demand and favorable regulatory environments. Collaboration among stakeholders across the entire food value chain, including food producers, processors, retailers, waste management companies, and biorefinery operators, will be essential for the widespread and successful adoption of food waste biorefinery solutions. Public–private partnerships and industry consortia can play a key role in facilitating knowledge sharing, technology transfer, and investment in the necessary infrastructure. In summary, food waste biorefineries have evolved from a conceptual idea into a practical and effective solution for addressing food waste challenges. With ongoing technological innovation, supportive policies, and growing market demand, food waste biorefineries are well-positioned to play a significant role in the transition toward a more sustainable and circular food system.

5.4.4 Municipal Waste Biorefinery

MSW represents a significant and complex feedstock for biorefineries, often referred to as urban biorefineries or waste-to-resource facilities. The concept of valorizing MSW through biorefining gained traction in the late twentieth century as urban areas faced increasing volumes of waste and sought sustainable management solutions. Early approaches primarily focused on basic waste management practices like sorting, recycling, and composting to divert waste from landfills. While these methods remain foundational, the biorefinery paradigm seeks to extract maximum value by converting the organic fractions of MSW into a wider range of products beyond traditional recycling outputs. MSW is inherently a heterogeneous mixture, varying significantly in composition depending on socio-economic factors, geographic location, and seasonal influences. It typically includes diverse organic materials such as food waste, yard waste, paper, cardboard, plastics, textiles, glass, and metals. This heterogeneity presents a primary challenge for biorefining, necessitating effective pretreatment steps to separate and prepare the organic fraction for subsequent conversion.

In the early twenty-first century, advancements in biotechnology, process engineering, and waste management significantly propelled interest in applying biorefinery concepts to MSW valorization. Researchers began exploring various biological, chemical, and thermal conversion technologies tailored to extract valuable components from the organic fractions. As illustrated in Figure 5.5, municipal waste biorefineries integrate several distinct technological approaches. AD is a widely adopted biological technology specifically well-suited for treating the putrescible organic waste fractions within MSW [20].

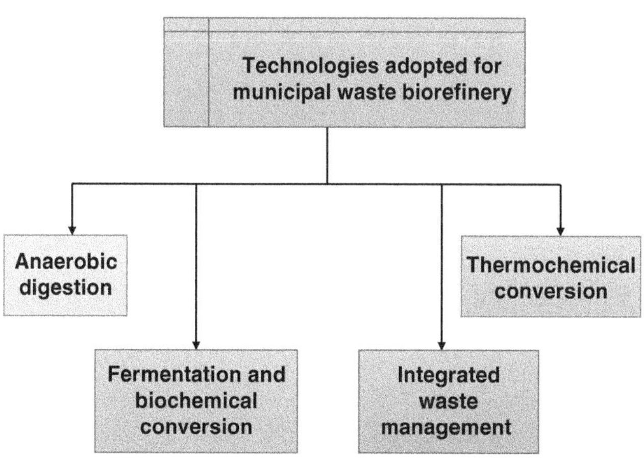

Figure 5.5 Technological advancement in municipal waste biorefinery.

In the AD process, a diverse community of anaerobic microorganisms breaks down organic matter in the absence of oxygen, yielding biogas, a mixture primarily of CH_4 and CO_2, and digestate. The produced biogas is a valuable renewable energy carrier that can be used for heat and power generation, while the nutrient-rich digestate can often be utilized as a soil amendment.

Beyond AD, fermentation processes are employed to convert specific organic components of MSW into biochemicals such as organic acids (e.g., lactic acid, acetic acid), alcohols (e.g., ethanol, butanol), and biopolymers. These processes rely on selected microorganisms metabolizing simple sugars or more complex substrates derived from the waste. The resulting biochemicals can serve as building blocks for various industrial applications.

Thermochemical technologies are also integral to MSW biorefining, particularly for fractions less amenable to biological conversion or for generating energy carriers [7, 33]. Pyrolysis is a process that heats organic matter in the absence of oxygen to produce bio-oil, syngas, and biochar. Gasification converts organic material at high temperatures with a limited amount of oxygen or steam to produce syngas. These thermal products can be used as fuels or further processed into chemicals. Hydrothermal treatment, such as HTC or HTL, offers pathways specifically for wet organic fractions of MSW.

Modern municipal waste biorefineries increasingly operate as integrated systems that combine multiple conversion technologies in series or parallel to maximize resource recovery and minimize residual waste. These systems may include initial steps like advanced mechanical–biological sorting to separate organics from recyclables and inerts, followed by hydrolysis (to break down complex carbohydrates), fermentation or AD (for biological conversion), and potentially thermochemical steps for residual organics or energy recovery. The choice of technologies is often tailored to the specific composition of the MSW stream and the desired product portfolio.

Rajendran et al. [20] highlighted the potential of utilizing MSW for value-added product production through such integrated approaches. While acknowledging that biological processes can be more convenient and economically feasible for certain fractions compared to energy-intensive thermal processes, the integration of both is often necessary for comprehensive valorization. For instance, the H-M-Z-S concept studied by Escamilla-Alvarado et al. [21] demonstrates a multi-stage integrated biorefinery focusing on the organic fraction of MSW (OFMSW). This specific system involves fermentation to produce hydrogen, followed by subsequent steps for enzyme and CH_4 production, and saccharification. As illustrated in Figure 5.6, this H-M-Z-S biorefinery integrates technologies like Proton Exchange Membrane Fuel Cell PEMFC and Combined Cycle Power Plant (CCPP) for converting intermediate products (CH_4 and H_2) into electricity and heat.

The study showed a significant energy recovery (approximately 205 kWh of electricity and 268 MJ of gross heat from CH_4 and H_2) and positive environmental outcomes, including a reduction in landfill dependency.

The complexity of MSW requires smart integration of technologies to optimize energy efficiency and process time. Major challenges in MSW biorefining include managing feedstock variability, removing contaminants (which can inhibit biological processes or produce hazardous emissions during thermal conversion),

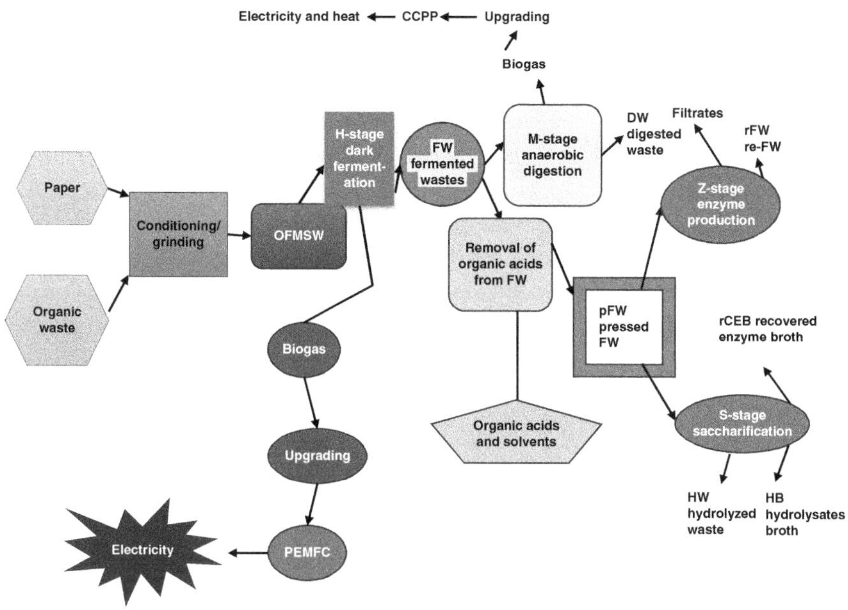

Figure 5.6 A brief representation of H-M-Z-S biorefinery process.

and developing efficient, cost-effective separation and purification technologies for diverse products. Despite these challenges, municipal waste biorefineries are critical for transitioning urban areas toward more sustainable and circular resource management models. They reduce the burden on landfills, create renewable energy and materials, and contribute to local economic development by generating revenue streams from waste. Continued research focuses on optimizing integrated system design, developing robust technologies for heterogeneous feedstocks, and improving the economics of downstream product recovery.

5.4.5 Lignocellulosic Biorefinery

Lignocellulosic biorefinery technology focuses on the conversion of lignocellulosic biomass, a vast and abundant renewable resource, into a range of valuable products. Lignocellulosic biomass primarily consists of the structural components of plants, including agricultural residues (such as crop stalks, straw, and bagasse), forestry residues (like wood chips, bark, and sawdust), and dedicated energy crops (e.g., switchgrass, miscanthus) [23]. Unlike first-generation biofuels, which rely on edible feedstocks like corn grain or sugarcane, lignocellulosic biomass is non-food competing and widely available, making it a sustainable feedstock source for biofuels, biochemicals, and biomaterials. However, valorizing lignocellulosic biomass presents significant challenges due to its complex and rigid structure, primarily composed of cellulose, hemicellulose, and lignin. The recalcitrance of this structure, particularly the strong linkages between lignin and carbohydrates and the crystalline nature of cellulose, hinders access for enzymes and catalysts required for deconstruction. Furthermore, the high oxygen content in lignocellulosic biomass can limit its energy density.

To overcome these challenges, effective pretreatment and fractionation methods are essential to disrupt the lignocellulosic matrix and separate its main components (cellulose, hemicellulose, and lignin) for subsequent conversion. Pretreatment methods aim to increase the accessibility of cellulose and hemicellulose to hydrolysis enzymes or catalysts. Various pretreatment techniques exist, including mechanical (e.g., milling), chemical (e.g., dilute acid, alkaline, ionic liquids, organosolv), thermal (e.g., steam explosion), and biological methods. Fractionation processes aim to separate the biomass into its constituent polymers. As illustrated in Table 5.4, different pretreatment and fractionation methods applied to various lignocellulosic feedstocks (e.g., wheat straw, pinewood, black liquor, walnut shells, mango seed

Table 5.4 Phenolic and monosaccharide monomers formed by the fractionation of lignocellulosic biomass. Adapted from [24].

Feedstocks	Method/conditions	Monosaccharide yield %	Phenolic monomers yield %
Wheat straw	Hydrothermal treatment, 493 K	44%	—
Pinewood	Dimethyl carbonate (green solvent), ethylene glycol as stabilizing agent; H_2SO_4 as catalyst	84.8%	9 wt.%
Black liquor	Hydrothermal treatment that is followed by wet oxidation, 115–221 °C, N_2	<74%	—
Paulownia wood	Microwave treatment followed by hydrothermal, 200–231 °C	78–80%	—
Flax shave	Reductive catalytic fractionation, 3 wt% Ru/C, and ethanol	—	9.6 wt.% lignin
Black liquor solid	Hydrothermal, 298–300 °C	—	86.8%
Pine wood	Reductive catalytic fractionation, 5 wt% Pd/C, and methanol	—	56.4 wt.% lignin
Black locust bark	Reductive catalytic fractionation followed by liquid/liquid extraction, Ru/C, and methanol	—	35.2 wt.% lignin
Birch sawdust	Reductive catalytic fractionation, Co/C, and ethanol	—	34.1 wt.% lignin
Eucalyptus wood chips	Organic solvent followed by acid catalysis, methyl isobutyl ketone : methanol : water (25 : 42 : 33) with 5%H_3PO_4	17.9 wt.%	13.6 wt.%
Wheat straw	Organic solvent followed by acid hydrolysis, 50% *w*/*w* aqueous EtOH	86.0%	84.0%
Corn cob	Mild acid followed by hydrothermal, 310 °C	—	1.28 wt.% lignin
Switchgrass	Ionic liquids, 1-butyl-3-methylimidazolium acetate	<60.2%	<86.5%

shells [MSS]) yield varying amounts of phenolic compounds (derived from lignin) and monosaccharides (derived from cellulose and hemicellulose) [24]. For instance, hydrothermal treatment of wheat straw yielded 44% monosaccharides, while reductive catalytic fractionation of pinewood yielded 56.4 wt.% lignin.

Following effective pretreatment and fractionation, the separated carbohydrate streams (cellulose and hemicellulose hydrolysates containing sugars) can be converted into biofuels or biochemicals through fermentation or other catalytic processes. The lignin fraction, a complex aromatic polymer, can also be valorized into various products, including fuels, chemicals, or materials like adhesives and carbon fibers. Hydro-thermolysis (hydrothermal treatment), using hot compressed water, is a promising method for pretreating lignocellulosic biomass, disrupting its structure and hydrolyzing hemicellulose into oligosaccharides or monosaccharides [34]. This process has been investigated for extracting carbohydrates and lignin from various biomass sources, including pine and agricultural residues like rapeseed flour and walnut shells [35–37]. Monteiro et al. [38] applied hydrothermal processing to MSS to selectively depolymerize and solubilize the xylan (hemicellulose) fraction, optimizing the extraction of xylooligosaccharides (XOS) and studying the kinetics of the process.

Lignocellulosic biorefineries offer significant advantages over first-generation biofuel processes, primarily by utilizing non-food feedstocks, thereby avoiding the food versus fuel debate. They also have the potential for higher greenhouse gas emission reductions and potentially higher product yields per unit of land. However, despite the progress, challenges remain, including the high capital and operating costs associated with pretreatment and enzyme production, the technological complexity of integrating multiple conversion steps, and the need for highly efficient separation and purification of diverse products. Making lignocellulosic biorefineries economically viable on a large scale requires addressing these challenges [39].

Life Cycle Assessment (LCA) is a crucial tool for evaluating the environmental impacts of lignocellulosic biorefineries throughout their entire life cycle, from biomass cultivation or collection to the production and end-use of the final products and management of residues [40]. An LCA involves defining the goal and scope, compiling an inventory of all inputs and outputs (including energy, water, chemicals, emissions, waste), assessing the environmental impacts (e.g., climate change, air/water pollution, land use), interpreting the results, and identifying opportunities for improvement. Conducting a Life Cycle Cost Assessment alongside the LCA is also beneficial to evaluate the economic feasibility and competitiveness [41]. By systematically evaluating both environmental and economic performance, comprehensive assessments guide the development of more sustainable and viable lignocellulosic biorefinery pathways.

5.5 Water-based Biorefinery

The concept of a water-based biorefinery represents an innovative and integrated approach that merges the principles of biorefining with advanced water treatment technologies. This type of facility is designed to purify water while simultaneously recovering valuable resources and producing value-added products from the organic contaminants and waste streams present in the water [42, 43]. Unlike conventional wastewater treatment plants that primarily focus on removing pollutants to meet discharge standards, a water-based biorefinery adopts a holistic approach aimed at maximizing resource recovery from the water matrix and associated sludge. This concept embodies a sustainable strategy for managing water resources and waste concurrently.

At the core of a water-based biorefinery is often a robust biological treatment system. This system utilizes the metabolic activities of diverse microorganisms to effectively degrade organic pollutants present in the water. Common biological treatment processes integrated into water-based biorefineries include activated sludge systems, biological filters, membrane bioreactors, and constructed wetlands. These biological systems are designed to break down complex organic pollutants into simpler, less harmful compounds, thereby purifying the water.

Beyond pollutant removal, a key objective of a water-based biorefinery is resource recovery. This involves extracting valuable components from both the treated water stream and the separated sludge. Energy recovery is a primary focus, often achieved through the AD of organic sludge, which produces biogas (CH_4) that can be used for heat and power generation [44]. Nutrient recovery is another critical aspect, involving the extraction of valuable nutrients such as nitrogen and phosphorus from the water or sludge for potential reuse as fertilizers or in other applications. Furthermore, water-based biorefineries can aim to produce bio-based chemicals or materials from the organic contaminants present in the water, leveraging them as feedstock for fermentation or other conversion processes.

Advanced treatment processes are typically integrated into water-based biorefineries to further purify the water to meet specific quality standards, whether for discharge, reuse (e.g., for irrigation or industrial purposes), or even potable water production. These advanced treatments can include membrane filtration technologies (such as reverse osmosis or ultrafiltration) to remove suspended solids, pathogens, and dissolved contaminants, advanced oxidation processes to degrade recalcitrant organic compounds, or ion exchange for removing specific ions.

The successful implementation of a water-based biorefinery relies heavily on the effective integration and optimization of the various unit processes involved. This includes the biological treatment, resource recovery steps, and advanced treatment processes. Optimization efforts are aimed at maximizing resource recovery (e.g., biogas yield, nutrient extraction efficiency) while simultaneously ensuring efficient water purification and meeting stringent water quality standards [45, 46]. Process modeling, advanced control systems, and real-time monitoring technologies play crucial roles in optimizing the overall performance of the integrated biorefinery system.

Water purification biorefineries offer several compelling advantages compared to traditional water treatment plants. They can lead to reduced overall energy consumption, lower operating costs through energy recovery and product sales, and generate potential revenue streams through the sale of value-added products [47]. Additionally, they contribute significantly to the circular economy by minimizing waste (sludge) and promoting the reuse of water and resources. However, challenges exist, including the scalability of some technologies, the need for supportive regulatory frameworks for resource recovery and reuse, and ensuring public acceptance of products derived from wastewater streams [48]. Addressing these challenges is crucial for realizing the full potential of water purification biorefineries in promoting sustainable water management and resource utilization.

5.6 The Economic Aspects of Waste-to-energy Biorefineries

The economic feasibility of waste-to-energy biorefineries is a critical determinant of their widespread adoption and long-term sustainability. A thorough analysis of both costs and potential revenue streams is essential to evaluate their economic viability [49]. Waste-to-energy biorefineries offer a sustainable solution that simultaneously addresses environmental concerns and contributes to energy security. By converting waste into biomaterials and bioenergy, they can provide alternatives to petroleum-based products and energy sources, thereby contributing to a cleaner and low-carbon environment [2]. The integration of waste valorization into bioprocesses for the production of useful products and metabolites facilitates a transition toward a sustainable circular bioeconomy [50]. These facilities contribute to a low-carbon economy by reducing greenhouse gas emissions and offer significant prospects for a sustainable and greener future [51]. Key economic factors influencing the viability of waste-to-energy biorefineries include:

a. **Capital Investment:** The initial capital investment required to establish a waste-to-energy biorefinery can be substantial, representing a significant upfront cost [52–54]. This includes expenses related to land acquisition, civil engineering and construction of the facility, procurement and installation of complex process equipment, and development of necessary infrastructure (e.g., power connection, feedstock receiving facilities). The magnitude of the capital investment is highly dependent on the scale of the biorefinery, the specific technologies employed (e.g., thermal vs. biological, type of reactors), and the location of the facility.
b. **Operating Costs:** Ongoing expenses incurred during the day-to-day operation of the biorefinery. These costs include feedstock procurement costs (which can vary depending on availability and market dynamics), labor costs for operating and maintaining the facility, energy consumption for various processes, chemical inputs (e.g., for pretreatment, catalysts), maintenance and repair expenses, and costs associated with managing residual waste streams. Optimizing operational efficiency and minimizing these costs are paramount for maintaining the economic viability of the biorefinery [55, 56]. Advanced process control systems, automation, and predictive maintenance strategies can help reduce labor and energy costs over time.
c. **Feedstock Availability and Cost:** The reliable and consistent availability of suitable waste feedstock at a competitive cost is a critical factor influencing the economics of a waste-to-energy biorefinery [57]. Access to a steady supply of

organic waste materials at predictable prices is essential for maintaining consistent production and profitability. Factors influencing feedstock cost include collection and transportation expenses, the need for pre-processing (e.g., sorting, shredding, drying), and market dynamics influenced by competing uses for the waste.

d. **Energy Revenue:** Revenue generated from the sale of energy products, such as electricity fed into the grid, heat supplied to local users, or biogas and biofuels sold as transportation fuels, constitutes a primary income stream for waste-to-energy biorefineries [58]. The profitability of energy production is directly influenced by prevailing energy market prices, the availability of favorable feed-in tariffs or power purchase agreements, government incentives supporting renewable energy, and regulatory frameworks that encourage the use of waste-derived energy.
e. **Byproduct Valorization:** Waste-to-energy biorefineries often produce various byproducts or co-products in addition to primary energy carriers [59]. Examples include digestate from AD, biochar from pyrolysis or HTC, or various biochemicals. Valorizing these byproducts into marketable products can generate additional revenue and significantly enhance the overall economic viability of the biorefinery [60, 61]. The contribution of byproduct valorization to revenue depends on their market demand, potential applications, and the efficiency and cost-effectiveness of recovering and upgrading them.
f. **Economic Incentives and Policies:** Supportive government policies and economic incentives play a crucial role in promoting the deployment of waste-to-energy biorefineries. This can include direct financial support such as subsidies, grants, and tax credits [62]. Favorable regulatory frameworks that mandate waste diversion from landfills, set targets for renewable energy use, and facilitate permitting processes can also reduce financial risk and enhance project profitability [63]. Policies that create or support markets for waste-derived products are also beneficial.
g. **Financial Risk and Return on Investment (ROI):** Waste-to-energy biorefineries, like other large-scale industrial projects, involve financial risks. These risks can stem from volatility in energy and product prices, technological uncertainties associated with novel processes, potential changes in regulatory policies, and challenges in securing project financing. Investors and project developers evaluate the expected ROI based on metrics such as project cash flows, payback period, internal rate of return (IRR), and net present value (NPV) to assess the overall economic viability and attractiveness of the investment.
h. **Economic Viability and Market Opportunities:** The ultimate economic viability of waste-to-energy biorefineries depends on their ability to successfully compete within the existing waste management and energy sectors. This involves competing with traditional waste disposal options (like landfilling) and other energy generation methods (including fossil fuels and other renewables). Identifying and capitalizing on niche markets for specific waste-derived products, developing a diversified product portfolio, and effectively managing revenue streams can enhance the competitiveness and market opportunities for biorefinery projects [64].

Thus, the successful development and operation of waste-to-energy biorefineries require comprehensive economic analysis, meticulous strategic planning, and proactive risk management to ensure profitability, sustainability, and long-term viability in a competitive market environment. Collaboration among diverse stakeholders, the development of innovative and efficient business models, and the presence of supportive policies are essential for unlocking the full economic potential of waste-to-energy biorefineries and contributing to a cleaner, more sustainable future.

5.7 Conclusion

Waste biorefining represents a transformative and sustainable solution for managing waste and harnessing its inherent potential. It offers a compelling pathway to convert diverse biomass resources into a range of valuable products, significantly contributing to a reduction in reliance on finite fossil fuels, mitigating waste accumulation, and minimizing associated environmental impacts. Through continuous technological innovation and process optimization, biorefineries are capable of producing renewable energy, valuable biochemicals, and sustainable bioplastics, thereby fostering the principles of a circular economy. By maximizing resource efficiency and closing material loops, these facilities actively contribute to the development of a cleaner, more sustainable future. Ultimately, the widespread implementation of waste biorefineries is paving the way toward a greener and more resilient global society.

In summation, waste-to-energy biorefineries stand as a promising solution capable of addressing both critical environmental concerns and economic challenges by efficiently converting organic waste streams into renewable energy and valuable co-products. These facilities offer a multitude of benefits, including significant waste diversion from landfills, clean renewable energy generation, enhanced resource recovery, and improved overall environmental protection. However, the economic viability and successful implementation of waste-to-energy biorefineries are contingent upon the careful consideration and management of various factors, including initial capital investment, ongoing operational costs, reliable feedstock availability and cost, potential revenue from energy and product sales, the value derived from byproducts, the availability of supportive incentives and policies, effective financial risk assessment, and understanding prevailing market opportunities. Successfully deploying waste-to-energy biorefineries necessitates rigorous economic analysis, informed strategic planning, and proactive risk management to ensure profitability, long-term sustainability, and viability. The collaborative efforts of researchers across disciplines, engineers, policymakers, and the agricultural community are crucial in developing and implementing innovative waste management solutions that provide substantial benefits for both the environment and society. These collaborations, coupled with the development of innovative business models and the implementation of supportive policies, are essential for unlocking the full economic potential of waste-to-energy biorefineries and contributing effectively to a cleaner, more sustainable future.

References

1 Amulya, K., Dahiya, S., and Venkata Mohan, S. Building a bio-based economy through waste remediation: innovation towards sustainable future. In: *Bioremediation and Bioeconomy*. Elsevier, 2016, pp. 497–521. https://doi.org/10.1016/B978-0-12-802830-8.00019-8.

2 Ahmed, S.F., Kabir, M., Mehjabin, A., Oishi, F.T.Z., Ahmed, S., Mannan, S., et al. Waste biorefinery to produce renewable energy: Bioconversion process and circular bioeconomy. *Energy Rep*. 2023. https://doi.org/10.1016/j.egyr.2023.09.137.

3 Pathak, G., Rajkumari, K., and Rokhum, L. Wealth from waste: *M. acuminata* peel waste-derived magnetic nanoparticles as a solid catalyst for the Henry reaction. *Nanoscale Adv*. 2019. https://doi.org/10.1039/c8na00321a.

4 Tiwari, R., Agrawal, P.S., Belkhode, P.N., Brijpuriya, D., Devase, A., and Samarth, P. Utilization of waste transformer oil for the synthesis of biodiesel and glyptal. In: *Recent Advances in Sustainable Waste Management Practices. SWMP 2023*. Lecture Notes in Civil Engineering, Springer, Singapore, 2024. https://doi.org/10.1007/978-981-99-4186-5_7.

5 Rahimi, Z., Anand, A., and Gautam, S. An overview on thermochemical conversion and potential evaluation of biofuels derived from agricultural wastes. *Energy Nexus*. 2022. https://doi.org/10.1016/j.nexus.2022.100125.

6 Gaffey, J., Rajauria, G., McMahon, H., Ravindran, R., Dominguez, C., Ambye-Jensen, M., et al. Green biorefinery systems for the production of climate-smart sustainable products from grasses, legumes and green crop residues. *Biotechnol. Adv*. 2023. https://doi.org/10.1016/j.biotechadv.2023.108168.

7 Agrawal, P.S., Belkhode, P.N., and Rokhum, S.L. Biofuel production technologies – an overview. In: *Sustainability in Biofuel Production Technology*. John Wiley & Sons, Ltd, 2022, pp. 75–97. https://doi.org/10.1002/9781119888864.ch4.

8 Etzold, H., Röder, L., Oehmichen, K., and Nitzsche, R. Technical design, economic and environmental assessment of a biorefinery concept for the integration of biomethane and hydrogen into the transport sector. *Bioresour. Technol. Rep*. 2023. https://doi.org/10.1016/j.biteb.2023.101476.

9 Agrawal, P.S., Belkhode, P.N., and Rokhum, S.L. Agro-waste-produced biofuels. In: *Sustainability in Biofuel Production Technology*. John Wiley & Sons, 2022. https://doi.org/10.1002/9781119888864.ch7.

10 Hari, S., Balaji Harshan, P., Punniavan, S., Sudharsan, M.S., and Ramadoss, G. Valorization of industrial wastes. In: *Valorization of Wastes for Sustainable Development*. Elsevier, 2023, pp. 371–394. https://doi.org/10.1016/B978-0-323-95417-4.00014-7.

11 Dhyani, V. and Bhaskar, T. Kinetic analysis of biomass pyrolysis. In: *Waste Biorefinery Potential Perspect*. Elsevier, 2018, pp. 39–83. https://doi.org/10.1016/B978-0-444-63992-9.00002-1.

12 Agrawal, P.S., Belkhode, P.N., and Rokhum, S.L. Modelling and simulation to predict the performance of the diesel blends. In: *Sustainability in Biofuel Production Technology*. John Wiley & Sons, 2022, pp. 209–266. https://doi.org/10.1002/9781119888864.ch10.

13 Rahimpour, M.R., Makarem, M.A., and Kiani, P. (Editors) *Hydrogen Production from Renewable Resources and Wastes*. 2024. https://doi.org/10.1201/9781003382270.

14 Weber, C.T., Trierweiler, L.F., and Trierweiler, J.O. Food waste biorefinery advocating circular economy: bioethanol and distilled beverage from sweet potato. *J. Clean. Prod.* 2020. https://doi.org/10.1016/j.jclepro.2020.121788.

15 Ebikade, E., Athaley, A., Fisher, B., Yang, K., Wu, C., Ierapetritou, M.G., et al. The future is garbage: repurposing of food waste to an integrated biorefinery. *ACS Sustain. Chem. Eng.* 2020. https://doi.org/10.1021/acssuschemeng.9b07479.

16 Tsegaye, B., Jaiswal, S., and Jaiswal, A.K. Food waste biorefinery: pathway towards circular bioeconomy. *Foods.* 2021. https://doi.org/10.3390/foods10061174.

17 Alagumalai, A., Devarajan, B., and Song, H. Unlocking the potential of catalysts in thermochemical energy conversion processes. *Catal. Sci. Technol.* 2023. https://doi.org/10.1039/d3cy00848g.

18 Carmona-Cabello, M., Garcia, I.L., Leiva-Candia, D., and Dorado, M.P. Valorization of food waste based on its composition through the concept of biorefinery. *Curr. Opin. Green Sustain. Chem.* 2018. https://doi.org/10.1016/j.cogsc.2018.06.011.

19 Saini, J.K., Kumar, G., Singh, S., Hemansi, and Kuhad, R.C. Municipal solid waste biorefinery for sustainable production of bioenergy. In: *Waste Biorefinery Value Addition Through Resource Utilization*. Elsevier, 2021, pp. 207–233. https://doi.org/10.1016/B978-0-12-821879-2.00008-9.

20 Rajendran, N., Gurunathan, B., Han, J., Krishna, S., Ananth, A., Venugopal, K., et al. Recent advances in valorization of organic municipal waste into energy using biorefinery approach, environment and economic analysis. *Bioresour. Technol.* 2021. https://doi.org/10.1016/j.biortech.2021.125498.

21 Escamilla-Alvarado, C., Poggi-Varaldo, H.M., and Ponce-Noyola, M.T. Bioenergy and bioproducts from municipal organic waste as alternative to landfilling: a comparative life cycle assessment with prospective application to Mexico. *Environ. Sci. Pollut. Res.* 2017. https://doi.org/10.1007/s11356-016-6939-z.

22 Molina-Peñate, E., Artola, A., and Sánchez, A. Organic municipal waste as feedstock for biorefineries: bioconversion technologies integration and challenges. *Rev. Environ. Sci. Biotechnol.* 2022. https://doi.org/10.1007/s11157-021-09605-w.

23 Usmani, Z., Sharma, M., Awasthi, A.K., Lukk, T., Tuohy, M.G., Gong, L., et al. Lignocellulosic biorefineries: the current state of challenges and strategies for efficient commercialization. *Renew. Sustain. Energy Rev.* 2021. https://doi.org/10.1016/j.rser.2021.111258.

24 Yogalakshmi, K.N., Mohamed Usman, T.M., Kavitha, S., Sachdeva, S., Thakur, S., Adish Kumar, S., et al. Lignocellulosic biorefinery technologies: a perception into recent advances in biomass fractionation, biorefineries, economic hurdles and market outlook. *Fermentation.* 2023. https://doi.org/10.3390/fermentation9030238.

25 Korberg, A.D., Mathiesen, B.V., Clausen, L.R., and Skov, I.R. The role of biomass gasification in low-carbon energy and transport systems. *Smart Energy.* 2021. https://doi.org/10.1016/j.segy.2021.100006.

26 Zhang, Y. ChemInform abstract: hydrothermal liquefaction to convert biomass into crude oil. *ChemInform*. 2011. https://doi.org/10.1002/chin.201137270.

27 Konur, O. Hydrothermal pretreatment of the biomass: review. In: *Bioethanol Fuel Production Processes. I Biomass Pretreat*. Taylor & Francis Group, 2023, p. 18. https://doi.org/10.1201/9781003226536-19.

28 Morales, A., Labidi, J., and Gullón, P. Hydrothermal treatments of walnut shells: a potential pretreatment for subsequent product obtaining. *Sci. Total Environ*. 2021. https://doi.org/10.1016/j.scitotenv.2020.142800.

29 Singh, N., Singhania, R.R., Nigam, P.S., Dong, C-D., Patel, A.K., and Puri, M. Global status of lignocellulosic biorefinery: challenges and perspectives. *Bioresour. Technol*. 2022. https://doi.org/10.1016/j.biortech.2021.126415.

30 Tabassum-Abbasi, Patnaik, P., Rahi, R., and Abbasi, S.A. A circular biorefinery-integrating wastewater treatment with the generation of an energy precursor and an organic fertilizer. *Sustain*. 2022. https://doi.org/10.3390/su14095714.

31 Pott, R., Johnstone-Robertson, M., Verster, B., Rumjeet, S., Nkadimeng, L., Raper, T., et al. Wastewater biorefineries: integrating water treatment and value recovery. In: *Green Energy and Technology*. Springer Nature, 2018, pp. 289–302. https://doi.org/10.1007/978-3-319-63612-2_18.

32 Zabaniotou, A. Redesigning a bioenergy sector in EU in the transition to circular waste-based Bioeconomy-A multidisciplinary review. *J. Clean. Prod*. 2018. https://doi.org/10.1016/j.jclepro.2017.12.172.

33 Agrawal, P.S., Belkhode, P.N., and Rokhum, S.L. Greener catalytic processes in biofuel production. In: *Sustainability in Biofuel Production Technology*. John Wiley & Sons, 2022, pp. 295–314. https://doi.org/10.1002/9781119888864.ch12.

34 Morales, A., Labidi, J., and Gullón, P. Integral valorisation of walnut shells based on a three-step sequential delignification. *J. Environ. Manage*. 2022. https://doi.org/10.1016/j.jenvman.2022.114730.

35 Jiao, Y., Xu, Q., Hu, Y., Wu, J., and Su, X. Pyrolysis co-activation synthesized magnetic biochar for efficient removal of Cr(VI) from aqueous solution. *Mater. Chem. Phys*. 2023. https://doi.org/10.1016/j.matchemphys.2022.127080.

36 Naderi, M. and Vesali-Naseh, M. Hydrochar-derived fuels from waste walnut shell through hydrothermal carbonization: characterization and effect of processing parameters. *Biomass Convers. Biorefin*. 2021. https://doi.org/10.1007/s13399-019-00513-2.

37 Chen, D., Zhang, K., Yang, J., Zhang, X., and Zhang, Y. Enhancement of properties of walnut shell/attapulgite/polyvinyl chloride composites with ultrasonic and hydrothermal treatment. *Text. Res. J*. 2023. https://doi.org/10.1177/0040517523 1156662.

38 Monteiro, C.R.M., Ávila, P.F., Pereira, M.A.F., Pereira, G.N., Bordignon, S.E., Zanella, E., et al. Hydrothermal treatment on depolymerization of hemicellulose of mango seed shell for the production of xylooligosaccharides. *Carbohydr. Polym*. 2021. https://doi.org/10.1016/j.carbpol.2020.117274.

39 Raj, T., Chandrasekhar, K., Naresh Kumar, A., Rajesh Banu, J., Yoon, J.J., Kant Bhatia, S., et al. Recent advances in commercial biorefineries for lignocellulosic ethanol production: current status, challenges and future perspectives. *Bioresour. Technol*. 2022. https://doi.org/10.1016/j.biortech.2021.126292.

40 Zhang, Y., Ding, Z., Shahadat Hossain, M., Maurya, R., Yang, Y., Singh, V., et al. Recent advances in lignocellulosic and algal biomass pretreatment and its biorefinery approaches for biochemicals and bioenergy conversion. *Bioresour. Technol.* 2023. https://doi.org/10.1016/j.biortech.2022.128281.

41 Kylili, A., Koutinas, M., Georgali, P.Z., and Fokaides, P.A. Lignin valorisation: life cycle Assessment (LCA) considerations for enabling Circular Bioeconomy. *Int. J. Sustain. Energy*. 2023. https://doi.org/10.1080/14786451.2023.2250869.

42 Lunze, A., Heyman, B., Chammakhi, Y., Eichhorn, M., Büchs, J., Anders, N., et al. Investigation of silphium perfoliatum as feedstock for a liquid hot water–based biorefinery process towards 2,3-butanediol. *Bioenergy Res.* 2021. https://doi.org/10.1007/s12155-020-10194-9.

43 Amidon, T.E. and Liu, S. Water-based woody biorefinery. *Biotechnol. Adv.* 2009. https://doi.org/10.1016/j.biotechadv.2009.04.012.

44 Hanum, F., Yuan, L.C., Kamahara, H., Aziz, H.A., Atsuta, Y., Yamada, T., et al. Treatment of sewage sludge using anaerobic digestion in Malaysia: current state and challenges. *Front. Energy Res.* 2019. https://doi.org/10.3389/fenrg.2019.00019.

45 Krátký, L., Ledakowicz, S., Slezak, R., Bělohlav, V., Peciar, P., Petrik, M., et al. Emerging sustainability in carbon capture and use strategies for V4 countries via biochemical pathways: a review. *Sustain.* 2024. https://doi.org/10.3390/su16031201.

46 Amulya, K., Morris, S., and Lens, P.N.L. Aquatic biomass as sustainable feedstock for biorefineries. *Biofuels, Bioprod. Biorefin.* 2023. https://doi.org/10.1002/bbb.2471.

47 Kalafatakis, S., Braekevelt, S., Carlsen, V., Lange, L., Skiadas I.V., and Gavala, H.N. On a novel strategy for water recovery and recirculation in biorefineries through application of forward osmosis membranes. *Chem. Eng. J.* 2017. https://doi.org/10.1016/j.cej.2016.11.092.

48 Zijlstra, D.S., de Korte, J., de Vries, E.P.C., Hameleers, L., Wilbers, E., Jurak, E., et al. Highly efficient semi-continuous extraction and in-line purification of high β-O-4 butanosolv lignin. *Front. Chem.* 2021. https://doi.org/10.3389/fchem.2021.655983.

49 Solarte-Toro, J.C. and Cardona Alzate, C.A. Sustainability of biorefineries: challenges and perspectives. *Energies.* 2023. https://doi.org/10.3390/en16093786.

50 Boccia, F., Di Donato, P., Covino, D., and Poli, A. Food waste and bio-economy: a scenario for the Italian tomato market. *J. Clean. Prod.* 2019. https://doi.org/10.1016/j.jclepro.2019.04.180.

51 Solarte-Toro, J.C. and Cardona Alzate, C.A. Biorefineries as the base for accomplishing the sustainable development goals (SDGs) and the transition to bioeconomy: technical aspects, challenges and perspectives. *Bioresour. Technol.* 2021. https://doi.org/10.1016/j.biortech.2021.125626.

52 Djukić-Vuković, A., Mladenović, D., Ivanović, J., Pejin, J., and Mojović, L. Towards sustainability of lactic acid and poly-lactic acid polymers production. *Renew. Sustain. Energy Rev.* 2019. https://doi.org/10.1016/j.rser.2019.03.050.

53 Yan, J. and Salman, C.A. Uncertainty in waste-to-energy process designs and techno-economic analysis. In: *Waste Biorefineries*. Elsevier, 2023, pp. 229–269. https://doi.org/10.1016/b978-0-323-91761-2.00010-6.

54 Guo, M., van Dam, K.H., Touhami, N.O., Nguyen, R., Delval, F., Jamieson, C., et al. Multi-level system modelling of the resource-food-bioenergy nexus in the global south. *Energy*. 2020. https://doi.org/10.1016/j.energy.2020.117196.

55 De la Torre-Bayo, J.J., Martín-Pascual, J., Torres-Rojo, J.C., and Zamorano, M. Characterization of screenings from urban wastewater treatment plants: alternative approaches to landfill disposal. *J. Clean. Prod.* 2022. https://doi.org/10.1016/j.jclepro.2022.134884.

56 Manhongo, T.T., Chimphango, A.F.A., Thornley, P., and Röder, M. Current status and opportunities for fruit processing waste biorefineries. *Renew. Sustain. Energy Rev.* 2022. https://doi.org/10.1016/j.rser.2021.111823.

57 Clauser, N.M., González, G., Mendieta, C.M., Kruyeniski, J., Area, M.C., and Vallejos, M.E. Biomass waste as sustainable raw material for energy and fuels. *Sustain.* 2021. https://doi.org/10.3390/su13020794.

58 Nizami, A.S., Rehan, M., Waqas, M., Naqvi, M., Ouda, O.K.M., Shahzad, K., et al. Waste biorefineries: enabling circular economies in developing countries. *Bioresour. Technol.* 2017. https://doi.org/10.1016/j.biortech.2017.05.097.

59 Kishor, N. and Agrawal, P.S. Nitration of glycerol-A byproduct of biodiesel production. *Mater. Today: Proc.* 2020. https://doi.org/10.1016/j.matpr.2020.03.065.

60 Guan, H., Wang, L., Liu, J., Evrendilek, F., Chen, Z., Huang, S., et al. Energetic and environmental optimizations and byproduct valorization of pyrolysis of textile dyeing sludge with FeCl3. *J. Clean. Prod.* 2024. https://doi.org/10.1016/j.jclepro.2023.139940.

61 Austen, L.I., Dugmore, T.I.J., Matharu, A.S., and Hurst, G.A. Byproduct valorization: from spent coffee grounds to fatty acid ethyl esters. *J. Chem. Educ.* 2023. https://doi.org/10.1021/acs.jchemed.2c00728.

62 Stewart, D.W., Cortés-Peña, Y.R., Li, Y., Stillwell, A.S., Khanna, M., and Guest, J.S. Implications of biorefinery policy incentives and location-specific economic parameters for the financial viability of biofuels. *Environ. Sci. Technol.* 2023. https://doi.org/10.1021/acs.est.2c07936.

63 Hassan, M.A., Farid, M.A.A., Zakaria, M.R., Ariffin, H., Andou, Y., and Shirai, Y. Palm oil expansion in Malaysia and its countermeasures through policy window and biorefinery approach. *Environ. Sci. Policy*. 2024. https://doi.org/10.1016/j.envsci.2024.103671.

64 Agrawal, P.S., Belkhode, P.N., and Rokhum, S.L. State of the art design and fabrication of a reactor in biofuel production. In: *Sustainability in Biofuel Production Technology*. John Wiley & Sons, 2022. https://doi.org/10.1002/9781119888864.ch9.

6

Algal Biorefinery

6.1 Introduction

The widespread reliance on non-renewable energy sources, leading to the depletion of coal and petroleum reserves, underscores the urgent need for alternative, efficient, affordable, and renewable energy options. In this context, algae emerge as promising candidates due to their high biomass productivity and potential for generating significant energy yields. Algal biorefining represents an innovative and sustainable strategy for harnessing the potential of algae to produce a diverse range of valuable products, encompassing biofuels, bioproducts, and bioactive compounds. The core principle of algal biorefining involves the integrated processing of biomass to yield multiple products, including fuels, chemicals, materials, and energy carriers. This approach seeks to maximize biomass value and minimize waste by processing feedstock components in a cascaded manner. The significance of the biorefinery concept, particularly when applied to algae, lies in its capacity to extract maximum value from algal biomass through a multi-product approach [1]. Rather than focusing solely on biofuel production, algal biorefineries can process different fractions of algal biomass to generate a variety of products for diverse applications. By implementing a biorefinery model, algal biomass can be separated and processed to obtain lipids for biodiesel, proteins for animal feed, carbohydrates for bioethanol, pigments for food and cosmetics, and bioactive compounds for pharmaceuticals and nutraceuticals. This integrated strategy enhances both the economic viability and environmental sustainability of algal biorefining operations.

6.1.1 Algae as a Versatile Feedstock for Biorefining: An Overview

Algae are a widespread and diverse group of organisms found in various aquatic habitats, including freshwater, marine, and even extreme environments like hot springs and polar regions. They are increasingly recognized as a highly potential feedstock for biorefineries due to their distinctive biological characteristics and environmental benefits. Unlike traditional terrestrial crops, algae do not compete for arable land, consume vast amounts of freshwater, or require intensive

Agro-Waste Management and Valorization, First Edition. Pratibha S. Agrawal, et al.

agricultural inputs, thereby mitigating numerous sustainability concerns associated with conventional biofuel and biochemical production. Algal species range from microscopic microalgae, such as *Chlorella* and *Spirulina*, to larger macroalgae (seaweeds) like *Kelp* and *Ulva*, each possessing unique traits and properties [2]. The various types of algae and their particular advantages are outlined as follows:

- **Microalgae:** Often described as microscopic algae, these are single-celled photosynthetic organisms. They exhibit rapid growth rates and can efficiently convert solar energy into biomass through photosynthesis. Microalgae are well-known for their ability to accumulate lipids, proteins, and carbohydrates, making them valuable feedstocks for biofuel production. Specific genera like *Chlorella, Scenedesmus, Nannochloropsis*, and *Spirulina* are particularly advantageous due to their high productivity, adaptability, and rich biochemical profiles.
- **Macroalgae:** Commonly referred to as seaweeds (e.g., *Kelp, Laminaria, Ulva*, and *Gracilaria*), macroalgae are multicellular organisms that thrive in marine or freshwater environments. They are rich in polysaccharides such as alginate, agar, and carrageenan, which have diverse industrial applications in food, pharmaceuticals, and biotechnology. While typically having lower lipid content compared to microalgae, certain seaweeds can still be processed to produce biofuels and other biochemicals [3]. Seaweed cultivation also offers environmental benefits, including nutrient removal from coastal waters and carbon sequestration. Table 6.1 summarizes the unique advantages of algae as a feedstock for biorefining [4].

Therefore, the potential of algae lies in their inherent capability to produce a broad spectrum of compounds. These include lipids for biodiesel production,

Table 6.1 Advantages of algae as a feedstock for biorefinery. Adapted from [4].

Feature	Microalgae	Macroalgae (seaweeds)
Growth rate	Very high (doubling time in hours)	High (but generally slower than microalgae)
Land use	Non-arable land, vertical systems	Marine environments, coastal areas
Water use	Freshwater, saline, wastewater	Seawater
Nutrient requirements	Can utilize inorganic and organic nutrients	Utilize nutrients from seawater, which can be used for remediation
Lipid content	High (up to 70% DW in some species)	Lower than microalgae, varies with species and conditions
Protein content	High (up to 70% DW in some species)	Moderate (10–30% DW), valuable protein and amino acids
Carbohydrate content	Moderate to high (starches, cellulose)	Very high (polysaccharides like alginate, agar, carrageenan)
Biomass productivity	Very high (tons/ha/year)	High (tons/ha/year, depends on species and cultivation)
Environmental impact	Carbon sequestration, nutrient remediation	Carbon sequestration, nutrient remediation, habitat provision

Table 6.1 (Continued)

Feature	Microalgae	Macroalgae (seaweeds)
Products	Biofuels, biochemicals, biopolymers, feed	Biopolymers, food, feed, biofuels, biochemicals
Versatility	Scalable in various systems	Large-scale ocean farming, integrated aquaculture
Scalability	(Ponds, PBRs)	—

carbohydrates for bioethanol, proteins for animal feed, pigments for food and cosmetics, and bioactive compounds for pharmaceutical and nutraceutical uses. Furthermore, algae can be cultivated on non-arable land and can utilize nutrients present in wastewater, positioning them as a sustainable and environmentally friendly resource for biorefinery applications.

6.1.2 The Algal Biorefinery Concept: Integrated Processes

An algal biorefinery is defined as a facility that combines the production of algal biomass with its subsequent conversion processes to generate a range of valuable products. The fundamental principle is to maximize the utilization of the various components within algal biomass – lipids, proteins, carbohydrates, and pigments – to yield multiple outputs, thereby enhancing economic viability and overall sustainability. This integrated approach contrasts with facilities focused on producing only a single product, which often leads to underutilization of biomass fractions and increased waste streams [5]. An algal biorefinery can be designed to produce a variety of products, as illustrated in Figure 6.1:

- **Biofuels:** These can include biodiesel derived from algal lipids, bioethanol and butanol produced from algal carbohydrates through fermentation, biogas from the anaerobic digestion of residual biomass, and bio-oil generated via thermochemical conversion methods.

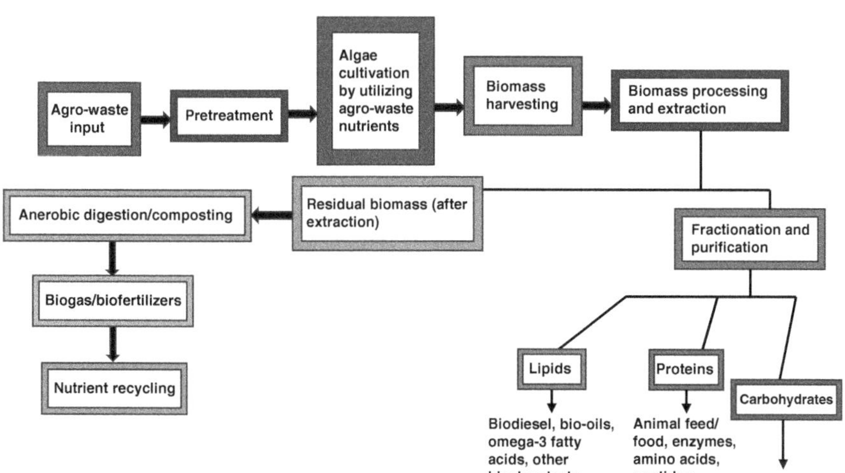

Figure 6.1 Schematic representation of an integrated algal biorefinery. Adapted from [6].

- **Biochemicals:** This category encompasses high-value compounds such as pigments (e.g., beta-carotene, astaxanthin, phycocyanin), omega-3 fatty acids (including eicosapentaenoic acid [EPA] and docosahexaenoic acid [DHA]), antioxidants, and enzymes, which find applications in food, nutraceuticals, pharmaceuticals, cosmetics, and industrial sectors.
- **Biopolymers:** Algal polysaccharides like alginate, agar, carrageenan, and exopolysaccharides can be used in biomaterials, packaging, hydrogels, and other applications.
- **Animal Feed and Food:** Algal biomass, being rich in protein and micronutrients, can serve as a sustainable and nutritious component in animal feed for aquaculture, poultry, and livestock, and is even explored as a direct food source for human consumption.
- **Biofertilizers:** Residual biomass remaining after product extraction can be composted or applied directly as biofertilizer, enriching soil and promoting sustainable agriculture, thus contributing to a circular nutrient economy.

6.1.3 Potential of Agro-waste as a Nutrient Source and Environmental Benefits in Algal Biorefining

Agricultural waste, encompassing crop residues, animal manure, and food-processing byproducts, represents a significant and underutilized resource. These waste streams are often rich in essential nutrients like nitrogen, phosphorus, potassium, and various micronutrients that are crucial for algal growth. Utilizing agro-waste as a nutrient source for algal cultivation offers a dual advantage: it reduces dependence on synthetic fertilizers, which are energy-intensive to produce and contribute to environmental pollution, while simultaneously lowering the operational costs of algae production, thereby making algal biorefineries more economically feasible and environmentally sustainable [7]. Integrating agro-waste treatment with algal biomass production creates a powerful synergy. Algal cultivation is effective in treating and remediating agro-waste streams by biologically removing excess nutrients, organic pollutants, and, in some cases, heavy metals. This bioprocess transforms waste into valuable algal biomass, aligning waste management with resource generation [8]. This synergistic approach not only addresses waste disposal challenges but also supports circular economy principles by converting waste materials into useful products. The environmental benefits of incorporating agro-waste into algal biorefineries are numerous:

- **Reduced Waste Disposal:** Diverting agro-waste from landfills decreases methane emissions from anaerobic decomposition, reduces land requirements for disposal, and alleviates pollution associated with improper waste management.
- **Carbon Sequestration:** Algae cultivation is inherently a carbon-negative process as algae consume atmospheric carbon dioxide (CO_2) during photosynthesis. When combined with agro-waste valorization, the net carbon footprint of product generation can be significantly reduced, contributing positively to climate change mitigation efforts.

- **Nutrient Recovery and Recycling:** Agro-waste is a rich source of essential nutrients. Algal cultivation enables the biological capture of these nutrients, converting them into algal biomass. Following product extraction, the residual algal biomass and digestate from anaerobic digestion can be further processed to recover and recycle nutrients back into agricultural systems, promoting nutrient circularity and reducing reliance on mined mineral fertilizers.
- **Reduced Eutrophication:** Using agro-waste (particularly nutrient-rich wastewater from agriculture and food processing) for algal cultivation helps remove excess nitrogen and phosphorus. This prevents these nutrients from entering water bodies, which can cause eutrophication – a major environmental issue leading to ecological imbalance and impaired water quality [9].

6.1.4 Chapter Objectives and Scope

The primary aim of this chapter is to provide a comprehensive overview of algal biorefining, emphasizing its potential, importance, and various applications. Specifically, the chapter intends to:

1. Introduce the concept of algal biorefining and highlight its significance in the context of sustainable biomass utilization.
2. Provide an overview of algae and their suitability as renewable feedstocks for biorefining applications.
3. Discuss the importance of the biorefining concept for maximizing the value extracted from algal biomass and minimizing waste generation.

The scope of this chapter covers a broad range of topics related to algal biorefining, including algal cultivation, biomass processing, product extraction, and their diverse applications. It will delve into various aspects of algal biorefining, such as technological advancements, challenges encountered, opportunities available, and future prospects. Additionally, the chapter will discuss the environmental and socio-economic implications of algal biorefining and its potential contributions to achieving sustainable development goals. Overall, the chapter serves as a foundational introduction to algal biorefining, setting the stage for more detailed discussions on specific areas like algal cultivation, biomass-processing techniques, and product development within the framework of sustainable biomass utilization. Through this comprehensive exploration of algal biorefining, the chapter aims to enhance understanding and contribute to the advancement of renewable energy and bioproduct production from algal sources. Consequently, the chapter provides an overview of algal biorefining, underscoring the importance of algae as a renewable resource, explaining the biorefinery concept, and outlining the objectives and scope of the discussion.

6.2 Algal Biomass Cultivation

Producing algal biomass is a fundamental step in algal biorefining, as it provides the raw material from which valuable compounds like lipids, proteins, carbohydrates, pigments, and bioactive substances are extracted. This section examines various aspects of algal biomass cultivation, including different cultivation methods, criteria for species selection, strategies for nutrient management, and techniques for harvesting [10].

6.2.1 Cultivation Systems: Comparing Open Ponds and Photobioreactors

a. **Open Ponds:** Open ponds, also known as raceways or ponds, are the most commonly used and cost-effective systems for cultivating algal biomass on a large scale. They typically consist of shallow, rectangular, or circular basins equipped with paddlewheels or other devices to circulate the culture and ensure uniform light exposure. Open ponds offer several advantages, including lower capital and operating costs, scalability, and operational simplicity [11]. A study by Yadala et al. [12] investigated the optimal raceway pond geometry for algal growth and biomass yield in a specific geographical location (Hyderabad, India). Their findings indicated that a raceway pond with a 30-cm channel depth, 300-m length, and 1-m width, harvested every 6 days, resulted in the lowest production cost for algal biomass. They suggested that longer ponds within considered ranges and smaller channel depths could further reduce production costs. However, open ponds also have drawbacks, such as susceptibility to contamination, sensitivity to fluctuating environmental conditions, and generally lower productivity compared to photobioreactors (PBRs).
b. **PBRs:** PBRs are closed, controlled cultivation systems designed to optimize algal growth conditions and maximize biomass productivity. They come in various configurations, including tubular, flat-panel, and airlift designs, each offering distinct advantages and limitations. PBRs allow for precise control over environmental parameters such as temperature, light intensity, pH, and nutrient concentration, leading to higher biomass yields and better quality control. However, PBRs are more expensive to construct and operate than open ponds and require sophisticated engineering and monitoring systems [13].

6.2.2 Strain Selection and Optimization of Algal Species and Nutrient Management

Selecting the appropriate algal species is a critical consideration in algal biomass production, as different strains exhibit varying growth rates, lipid content, nutrient requirements, and environmental tolerances [14]. The choice of strain depends on factors such as the target products, cultivation conditions, local climate, water availability, and relevant regulations. Commonly cultivated algal genera include *Chlorella, Spirulina, Nannochloropsis, Dunaliella*, and *Haematococcus*. Optimizing algal cultivation involves fine-tuning parameters like light intensity, temperature, pH, nutrient supply, and CO_2 concentration to maximize biomass productivity and quality. This process may involve screening and selecting suitable strains, optimizing the cultivation process using experimental design and statistical analysis, and employing advanced monitoring and control systems.

Algal biomass production necessitates essential nutrients such as nitrogen, phosphorus, potassium, micronutrients (e.g., iron, manganese, zinc), and CO_2 for photosynthesis and growth [15]. The availability and balance of these nutrients significantly influence algal productivity and biomass quality. Strategies for nutrient management involve optimizing the rates, ratios, and sources of nutrient dosing to meet the metabolic needs of the algal cells while minimizing nutrient losses and environmental impacts. This can include using inorganic fertilizers, organic amendments, wastewater effluents, and carbon capture technologies to supply nutrients and CO_2 to the algal cultures.

6.2.3 Harvesting and Dewatering Methods

Harvesting and dewatering are essential steps in algal biomass production. They involve separating algal cells from the culture medium and concentrating the biomass for subsequent processing. Several techniques exist for harvesting and dewatering algal biomass, each with its own advantages, limitations, and suitability for different cultivation systems and biomass characteristics. Mechanical methods such as centrifugation, filtration, and sedimentation are commonly used for large-scale biomass harvesting due to their effectiveness and scalability. Chemical flocculation and coagulation agents can improve the sedimentation and filtration of algal cells by promoting the aggregation of cells into larger flocs or precipitates. Other harvesting techniques include electrocoagulation, flotation, and membrane filtration, which offer benefits like energy efficiency, selectivity, and reduced chemical use. However, these methods may require specialized equipment and operating conditions, making them less suitable for certain applications or cultivation systems [16].

Dewatering harvested biomass involves reducing its water content to increase the solids concentration, which is necessary for further processing or storage. Mechanical dewatering techniques like pressing, drying beds, and screw presses are commonly employed to remove excess water from algal biomass.

Thermal drying methods, such as spray drying and freeze drying, can further decrease moisture content and produce dry algal powder or flakes suitable for long-term storage or downstream processing. Thus, efficient harvesting and dewatering techniques are vital for maximizing algal biomass recovery, minimizing energy and resource consumption, and improving the economic feasibility of algal biomass production for biorefinery applications [17].

Algal biomass production is a critical upstream component of algal biorefining, providing the raw material for extracting valuable compounds for various applications [19, 20]. As illustrated in Figure 6.2, cultivation methods, species selection, nutrient management, and harvesting techniques are key factors influencing biomass productivity, quality, and overall sustainability. By carefully selecting and integrating these elements, algal biorefinery operations can optimize efficiency, profitability, and environmental sustainability throughout the production process.

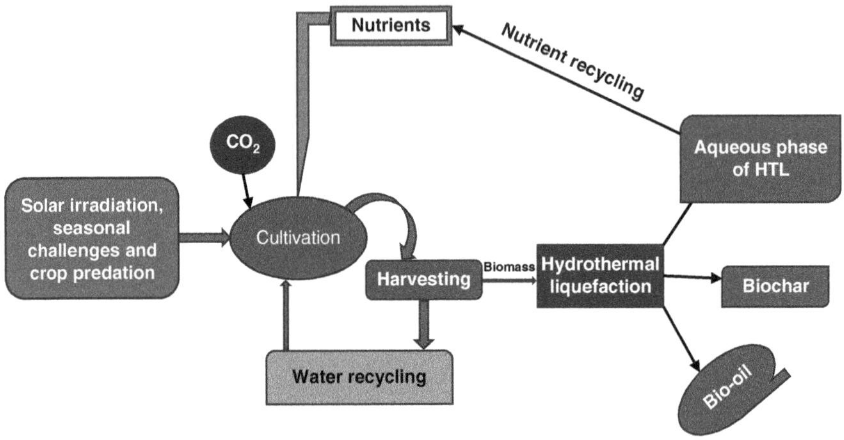

Figure 6.2 Factors responsible for algal biomass production. Adapted from [18].

6.3 Algal Biomass Processing

Algal biorefining involves crucial downstream steps to convert algal biomass into value-added products such as lipids, proteins, carbohydrates, pigments, and bioactive compounds [21]. Algal biomass processing is carried out through various steps, including pretreatment, extraction, fractionation, and purification, as well as co-product recovery and valorization techniques.

6.3.1 Pretreatment Techniques for Biomass Conversion

Pretreatment is a crucial initial step in processing algal biomass, as it helps to break down the rigid cell wall structure of algae and improves the accessibility of intracellular compounds for extraction and conversion. Various pretreatment methods are available for biomass conversion, including physical, chemical, biological, and thermal techniques.

- **Physical Pretreatment:** These methods utilize mechanical force to disrupt algal cells, thereby releasing their intracellular components [22]. Techniques such as milling, grinding, ultrasonication, and high-pressure homogenization can be employed to rupture algal cells and facilitate the extraction of lipids, proteins, and carbohydrates. While generally energy-intensive, physical pretreatment can effectively reduce particle size and increase the surface area for subsequent processing steps.
- **Chemical Pretreatment:** Chemical pretreatment methods involve using chemical agents to degrade the cell wall and enhance the release of intracellular compounds [23]. Acidic or alkaline treatments, enzymatic hydrolysis, and oxidative reactions are common chemical pretreatment approaches for algal biomass conversion. Acid hydrolysis can break down hemicellulose and cellulose into fermentable sugars, while alkaline treatment can dissolve proteins and disrupt cell wall structures. Enzymatic hydrolysis utilizes enzymes

like cellulases, hemicellulases, and proteases to degrade biomass components into soluble sugars, amino acids, and peptides [24].

- **Biological Pretreatment:** Biological pretreatment methods utilize microorganisms or enzymes to break down the cell wall and release intracellular compounds from algal biomass. Microbial fermentation, anaerobic digestion, and composting are examples of biological pretreatment methods that can improve the digestibility and biodegradability of algal biomass. Anukam et al. [25] reported that a residence time of approximately 10–14 days is required for the biological pretreatment process using microorganisms such as bacteria and fungi (including soft-rot, white, and brown fungi), and algae, which produce enzymes like cellulases, proteases, and lipases that degrade biomass components and aid conversion.
- **Thermal Pretreatment:** Thermal pretreatment methods involve applying heat to disrupt the cell wall and improve the efficiency of extracting intracellular compounds from algal biomass [26]. Techniques such as steam explosion, microwave irradiation, and hydrothermal treatment can break down biomass structures and increase the availability of biomass components for extraction and conversion. Thermal pretreatment methods can be energy-intensive but are effective in improving digestibility and enzymatic hydrolysis of algal biomass.

6.3.2 Extraction of Lipids, Proteins, and Carbohydrates

Following pretreatment, algal biomass can undergo extraction processes to recover valuable compounds such as lipids, proteins, and carbohydrates. The choice of extraction method for each component depends on factors like biomass composition, the target products, and processing requirements.

- **Extraction of Lipids:** Lipids are typically extracted from algal biomass using solvent-based methods such as Soxhlet extraction, Bligh and Dyer extraction, and supercritical fluid extraction (SFE) [27]. Solvents like hexane, ethanol, and supercritical CO_2 are commonly used to dissolve lipids from the biomass matrix. Zhou et al. [28] suggested that one of the most effective methods for extracting lipids from microalgae is Soxhlet extraction. Their study showed that an ethanol : chloroform (1 : 1) mixture resulted in a high extraction rate (11.8%). Folch extraction, introduced in 1957 for extracting lipids from animal fat, is another method. It involves a methanol : chloroform (1 : 2) mixture for lipid extraction, followed by adding water to separate phases and then using rotary evaporation to recover the extracted lipids. The Bligh and Dyer method is a variation of Folch extraction based on a two-phase solvent system. Compared to methods using non-polar solvents, this method yielded approximately double the mass of extracted lipids. Table 6.2 details various methods applied for lipid extraction, highlighting that extraction efficiency can be influenced by factors such as solvent polarity, temperature, pressure, and extraction time.

- **Extraction of Proteins and Carbohydrates:** Proteins can be extracted from algal biomass using various techniques, including solvent extraction, aqueous extraction, and enzymatic hydrolysis. Solvent extraction involves using organic

Table 6.2 Various methods applied for the extraction of lipids. Adapted from [28].

Extraction methods	Microalgae	Condition	Solvent	Inference
Subcritical fluid extraction	*Nannochloropsis salina*	Temperature 205–220 °C, t = 25 minutes	Water	Under subcritical water method, 70% lipid extraction; with microwave assistance, 100% extraction
Pulsed electric field-assisted extraction	*Chlorella*	Pulse repetition rate: 11 kHz, voltage 35 kV	Ethanol/ hexane	Increased extraction yield of lipids (166%)
Pressurized fluid extraction	*Nannochloropsis oculata*	Pressure 10–12 MPa, temperature 60 °C, t = 48 minutes	n-hexane, n-hexane/ 2-PrOH (2 : 1 vol%), C_2H_5OH (96%)	*n*-hexane showed lower extraction rate (6.1 ± 0.3 mass%); ethanol showed highest rate (36 ± 4 mass%) among the three solvents
Ultrasound-assisted extraction	*Scenedesmus* sp.	Ultrasonic power 20 W, time interval 2 seconds	Hexane	Total lipid yield increased from 0.76 to 1.31 g/L

solvents like ethanol, methanol, or water-miscible solvents to solubilize proteins from the biomass matrix. Aqueous extraction methods, such as adjusting pH, ultrasonication, and mechanical agitation, can also be used to extract proteins. Enzymatic hydrolysis employs proteases or carbohydrases to break down biomass proteins into soluble peptides and amino acids [29]. Carbohydrates can be extracted from algal biomass using methods such as acid hydrolysis, enzymatic hydrolysis, and mechanical disruption. Acid hydrolysis utilizes mineral acids like sulfuric acid or hydrochloric acid to hydrolyze polysaccharides into fermentable sugars. Enzymatic hydrolysis uses carbohydrases such as cellulases, hemicellulases, and amylases to break down biomass carbohydrates into soluble sugars. Mechanical disruption methods, including milling, grinding, and ultrasonication, can also break down biomass structures and release carbohydrates. Hammann et al. [30] evaluated physical/chemical pretreatments on *Chlorella vulgaris* for carbohydrate (glucose and galactose) retrieval. They found that temperature-assisted weak-acid extraction yielded the highest carbohydrate recovery (90 ± 1.1 wt.% at 95% confidence) and suggested that including carbohydrate recovery is feasible for improving the economic viability of microalgae-derived fuel and chemical production.

6.3.3 Fractionation and Purification Techniques

After extraction, algal biomass components can be subjected to fractionation and purification processes to isolate individual compounds or fractions for further processing and utilization. Fractionation techniques such as chromatography, membrane filtration, and centrifugation can be used to separate and purify target compounds based on their molecular size, polarity, or solubility.

Chromatographic techniques like column chromatography, high-performance liquid chromatography (HPLC), and gas chromatography (GC) can isolate and purify individual compounds from complex mixtures based on their differential partitioning between a stationary phase and a mobile phase. Membrane filtration techniques such as ultrafiltration, nanofiltration, and reverse osmosis can separate and concentrate target compounds based on their molecular size and charge. Membrane filtration can remove impurities, concentrate desired compounds, and fractionate biomass extracts. Centrifugation techniques, including differential centrifugation, density gradient centrifugation, and continuous flow centrifugation, can separate and purify biomass components based on their density and sedimentation rates [28]. Centrifugation can remove insoluble solids, concentrate biomass extracts, and isolate target compounds from complex mixtures. Garcia et al. [31] applied a fractionation process (bead milling, centrifugation, filtration, and diafiltration) after cultivating and harvesting two microalgae strains. They estimated that crude protein content reached 48.1 ± 1.8% for *T. suecica* and 43.5 ± 2.7% for *N. oleoabundans*, respectively.

6.3.4 Product Recovery and Valorization

In addition to major components like lipids, proteins, and carbohydrates, processing algal biomass can also generate co-products such as pigments, antioxidants, vitamins, and minerals [2, 10]. These co-products can be recovered and valorized for various applications, including use in food, feed, cosmetics, pharmaceuticals, and bioplastics. Pigments like chlorophyll, carotenoids, and phycobiliproteins can be extracted from algal biomass and used as natural colorants in food, cosmetics, and pharmaceuticals. Pigment extraction methods such as solvent extraction, SFE, and aqueous extraction are used to recover pigments from biomass matrices [32].

Antioxidants such as tocopherols, tocotrienols, phenolic compounds, and flavonoids can be extracted from algal biomass for use as natural antioxidants in food, nutraceuticals, and cosmetics. Methods like solvent extraction, solid-phase extraction (SPE), and microwave-assisted extraction (MAE) are employed to recover antioxidants from biomass matrices. Vitamins such as B12, K, and E, as well as minerals like calcium, magnesium, and iron, can also be extracted from algal biomass and used as nutritional supplements in food, feed, and pharmaceuticals. Vitamin and mineral extraction methods include solvent extraction, acid hydrolysis, and enzymatic hydrolysis [33].

Processing algal biomass is a critical step in algal biorefining, enabling the conversion of biomass into valuable added-value products like lipids, proteins, carbohydrates, pigments, antioxidants, vitamins, and minerals. Pretreatment methods (physical, chemical, biological, thermal) help break down the cell wall and improve the extractability of intracellular compounds. Extraction methods (solvent, aqueous, enzymatic, mechanical) are used to recover lipids, proteins, carbohydrates, and other compounds from biomass matrices. Fractionation and purification techniques (chromatography, membrane filtration, centrifugation) separate and purify target compounds. Co-product recovery and valorization techniques further enhance the economic and environmental sustainability of algal biomass processing by utilizing byproducts for various applications. Overall, algal biomass processing is fundamental to valorizing algae as a sustainable and renewable resource for biorefinery applications.

6.4 Biofuel Production from Algal Biomass

Biofuels derived from algal biomass represent a promising alternative to conventional fossil fuels, offering environmental, economic, and energy security advantages. This section explores various pathways for producing biofuels from algal biomass, including biodiesel, bioethanol, biobutanol, and bio-oil via hydrothermal liquefaction (HTL) [34, 35]. The importance of techno-economic analysis (TEA) and sustainability assessment in evaluating the feasibility and viability of biofuel production is also discussed.

6.4.1 Production of Biodiesel, Bioethanol, and Biobutanol from Algal Lipids

Biodiesel, a renewable fuel derived from triglycerides present in algal lipids, has attracted significant attention as a sustainable alternative to petroleum diesel. The typical production process involves the transesterification of algal lipids with methanol or ethanol to produce fatty acid methyl esters (FAMEs) or fatty acid ethyl esters (FAEEs), which are the primary components of biodiesel. The production of biodiesel from algal lipids [36] generally involves several steps:

a. **Algal Biomass Cultivation:** Algal biomass is cultivated using various systems, such as open ponds or PBRs, optimized for maximum lipid accumulation.
b. **Lipid Extraction:** Algal lipids are extracted from the biomass using solvent-based methods like Soxhlet extraction or SFE to obtain a lipid-rich feedstock for biodiesel production.
c. **Transesterification:** The extracted algal lipids are converted into biodiesel through transesterification. In this process, triglycerides react with an alcohol (e.g., methanol or ethanol) in the presence of a catalyst (e.g., sodium hydroxide or potassium hydroxide) to produce biodiesel and glycerol as a co-product.
d. **Product Purification:** The resulting crude biodiesel is purified to remove impurities and excess alcohol through processes such as washing, drying, and distillation.
e. **Product Refinement:** The purified biodiesel may undergo further refinement processes, like esterification, to improve its quality and meet fuel specifications [37].

Hasnain et al. [38] investigated the production of biodiesel from algal biomass using ultrasound and microwave methods and analyzed the fuel properties to optimize the algal energy feedstock efficiency. They studied six algal types (*Oscillatoria, Chlorella, Oedogonium, Ulothrix, Spirogyra*, and *Cladophora*) and found that ultrasonic treatment (80 kHz) with an extraction time of 8 minutes was effective in extracting lipids (21, 40, 32, 51, 20, and 36%, respectively). They observed improvements in cetane number, cold filter plugging point, high heating value, and oxidative stability compared to controls, suggesting that converting algal biomass to biodiesel is an effective approach for developing an environmentally friendly and sustainable product. Similarly, Karmarkar et al. [39] synthesized biodiesel by transesterifying oil extracted from algae using the Soxtherm solvent extraction method. They reported extraction yields of approximately 9.0 wt.% with hexane (costlier) and 8.0 wt.% with acetone (cheaper). The transesterification process involved an oil : methanol molar ratio of 1 : 6 with 3.0 wt.% KOH as a base catalyst at 333 K for 1 hour. The characteristic properties of the biodiesel (pour and cloud point of −6 and −1 °C, viscosity of 3.12 mm^2/s, carbon residue content of 0.03%, flash and fire point of 153 and 158 °C) were within standard ranges, making it suitable for blended use. Biodiesel derived from algal lipids offers several advantages, including high energy density and low emissions of greenhouse gases and pollutants, along with compatibility with existing diesel engines and infrastructure.

In addition to biodiesel, bioethanol and biobutanol are also renewable fuels that can be produced from fermenting sugars derived from algal biomass. While bioethanol is mainly produced through yeast or bacterial fermentation of sugars, biobutanol can be produced by fermentation using certain strains of *Clostridium* bacteria [40]. The production process for both bioethanol and biobutanol from algal biomass involves biomass pretreatment, sugar fermentation, product recovery, and co-product utilization. Initially, algal biomass is pretreated to release fermentable sugars from carbohydrates like cellulose and hemicellulose, using physical, chemical, or enzymatic methods. The pretreated biomass is then hydrolyzed into fermentable sugars, such as glucose and xylose, which are subsequently fermented by yeast or bacteria to produce ethanol or butanol, respectively. The fermented broth is then distilled to separate and concentrate the alcohol, followed by purification and dehydration to obtain fuel-grade bioethanol or biobutanol. Finally, co-products generated during fermentation, such as lignin or residual biomass, can be valorized for additional value-added products or applications [41]. Abomohra et al. [42] reported that algae serve as a superior raw material for biodiesel production, and additionally, biobutanol and bioethanol are attractive biofuels produced during the processing of microalgae after lipid extraction. Figure 6.3 illustrates the overall process.

Bioethanol and biobutanol produced from algal biomass offer several advantages, including compatibility with existing fuel infrastructure, reduced greenhouse gas emissions, and potential for higher energy yields compared to traditional biofuel feedstocks [43, 44].

6.4.2 HTL for Bio-oil Generation

HTL is a thermochemical conversion process that transforms wet biomass, such as algal biomass, into bio-oil under high temperature and pressure conditions, often with a catalyst. HTL provides a promising route for converting algal biomass into liquid fuels, specifically bio-oil, which can be upgraded and refined into

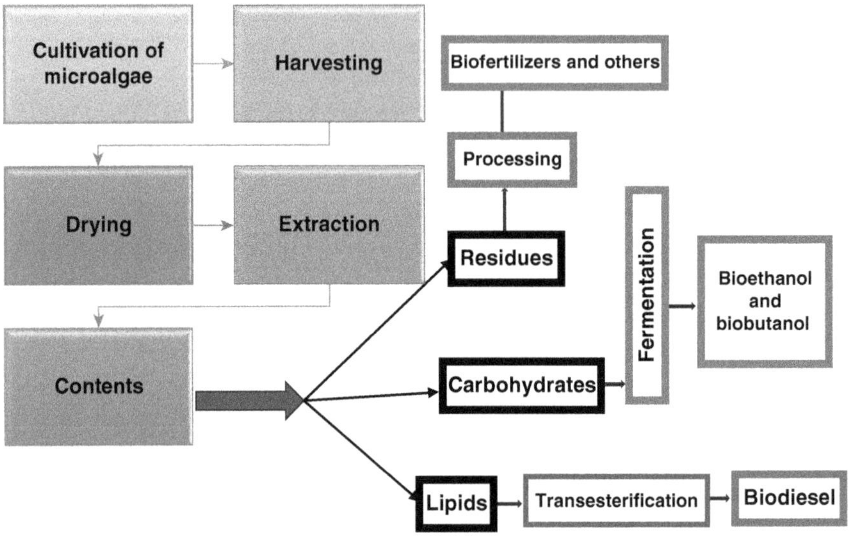

Figure 6.3 Schematic presentation of microalgae downstream processing into biofuels. Adapted from [42].

Table 6.3 Recapitulation of hydrothermal liquefaction conditions for algal biomass.

Microalgae	High heating value (MJ/kg)	Optimal conditions	Oil yield (dry-ash-free basis %)
Chlorella pyrenoidosa	21.4	Total solids 20%; 270 °C, 30 minutes, 0.7 MPa initial pressure (N_2); toluene extraction	39.2
Dunaliella tertiolecta	11.5	Total solids 5.6%; 360 °C, 60 minutes, K_2CO_3, $CHCl_3$ extraction	32.1
Chlorella vulgaris	22.3	Total solids 9.0%; 350 °C, 60 minutes, 20 MPa (N_2); CH_2Cl_2 extraction	35.2
Porphyridium cruentum	22.4	Total solids 9.0%; 350 °C, 60 minutes, Na_2CO_3, 20 MPa (N_2); CH_2Cl_2 extraction	29.3
Desmodesmus sp.	22	Total solids 9.0%; 375 °C, 5 minutes, 0.1–0.5 MPa initial pressure (He); CH_2Cl_2 extraction	49.0

transportation fuels or chemicals [45]. Table 6.3 outlines the operational conditions used for HTL processing of algal biomass.

The HTL process begins with pretreatment of algal biomass to reduce particle size and increase surface area, enhancing the efficiency of liquefaction. The pretreated biomass is then mixed with water and subjected to high temperatures (250–400 °C) and pressures (10–25 MPa) in a reactor vessel, where biomass components are converted into bio-oil, gas, and char [46, 47]. The reaction products are separated using methods like filtration, centrifugation, or solvent extraction to recover bio-oil, the aqueous phase, and solid residues. The resulting bio-oil can be further upgraded through processes such as hydrotreating, hydrocracking, or esterification to improve its quality, stability, and compatibility with existing fuel infrastructure.

Co-products from HTL, such as the aqueous phase or solid residues, can be valorized for additional products or applications, including fertilizer, animal feed, or biogas production. Ravichandran et al. [48] illustrated the reaction pathway of the HTL thermochemical process applied to biomass (high water content) at moderate pressure and temperature (100–200 bar, 200–400 °C) converting it into bio-crude oil, as shown in Figure 6.4.

Costa Magalhães et al. [49] synthesized bio-oil from microalgae using HTL with a sulfided NiW/Al_2O_3 catalyst at various temperatures (623, 648, and 673 K). They concluded that C16 and C18 lipid chains were converted into carboxylic acids and fatty acids (C16 and C18) constituting approximately 41.2 wt.% of the bio-oil. HTL offers several advantages for biofuel production from algal biomass, including high conversion efficiency, the ability to process wet biomass directly, and potential for co-product valorization.

6.4.3 Techno-economic and Sustainability Evaluation

TEA and sustainability assessment are crucial tools for evaluating the feasibility, viability, and environmental impacts of biofuel production from algal biomass. TEA involves assessing production costs, revenues, and profitability, considering factors such as capital investment, operating expenses, feedstock costs, and product

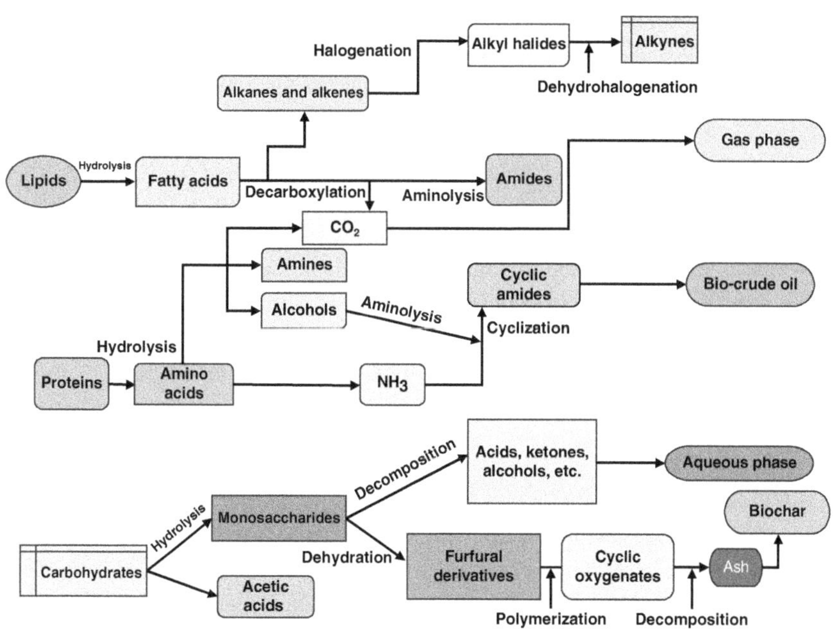

Figure 6.4 Reaction pathway of hydrothermal liquefaction to produce bio-oil.

prices. Sustainability assessment evaluates the environmental, social, and economic impacts of biofuel production processes, including greenhouse gas emissions, land use, water consumption, energy efficiency, and socio-economic benefits. The importance of conducting TEA and sustainability assessment in developing algal biofuel production processes lies in their ability to pinpoint key cost drivers, identify technical challenges, assess environmental impacts, and inform decision-making and policy development. By integrating TEA and sustainability assessment into the research and development process, stakeholders can make informed decisions regarding technology selection, process optimization, and investment strategies to promote the commercialization of algal biofuels sustainably and economically.

Algal biomass-derived biofuels offer a promising solution to growing challenges related to energy security, climate change, and sustainability. Production pathways like biodiesel, bioethanol, biobutanol, and bio-oil via HTL are viable options for converting algal biomass into renewable transportation fuels. TEA and sustainability assessment play vital roles in evaluating the feasibility and viability of these processes, guiding decision-making, and advancing the commercialization of algal biofuels sustainably and economically. Overall, biofuels from algal biomass demonstrate significant potential to contribute to a more sustainable and resilient energy future.

6.5 Bioproducts and Bio-compounds from Algae

Beyond biofuels, algal biomass can be utilized to produce a diverse array of high-value bioproducts and bio-compounds applicable across various industries, including food, pharmaceuticals, cosmetics, and nutraceuticals. These products can be categorized into three main types: high-value compounds like omega-3 fatty acids, pigments, and antioxidants; functional food ingredients and nutraceuticals; and bioactive compounds for pharmaceutical and cosmetic applications.

a. **High-value Products:**
   - **Omega-3 Fatty Acids:** Omega-3 fatty acids, such as EPA and DHA, are essential nutrients known for numerous health benefits, including promoting cardiovascular health, cognitive function, and possessing anti-inflammatory properties. Algae, particularly marine microalgae species like *Schizochytrium* sp. and *Crypthecodinium cohnii*, are rich sources of omega-3 fatty acids [50, 51]. These fatty acids can be extracted from algal biomass and used as dietary supplements, functional ingredients in food products, or additives in animal feed formulations.
   - **Pigments:** Algae synthesize a wide range of pigments, including chlorophylls, carotenoids, and phycobiliproteins, contributing to their diverse colors and photosynthetic capabilities. These pigments have various applications in food, cosmetics, and pharmaceuticals due to their antioxidant, anti-inflammatory, and photoprotective properties [52]. For example, astaxanthin, a carotenoid pigment found in certain microalgae species like *Haematococcus pluvialis*, is a potent antioxidant with potential health benefits for cardiovascular health, eye health, and skin protection.

- **Antioxidants:** Algal biomass contains a variety of antioxidant compounds, such as tocopherols, tocotrienols, phenolic compounds, and flavonoids, which help protect cells from oxidative damage and may reduce the risk of chronic diseases like cancer, diabetes, and cardiovascular diseases [53]. These antioxidants can be extracted from algal biomass and incorporated into functional foods, dietary supplements, or cosmetic products to enhance their health-promoting properties.

b. **Functional Food Ingredients and Nutraceuticals:** Functional food ingredients and nutraceuticals derived from algal biomass provide enhanced nutritional value and offer health benefits beyond basic nutrition. These products are gaining popularity among consumers seeking natural, plant-based alternatives to conventional food and dietary supplements.
   - **Proteins and Amino Acids:** Algal biomass is a rich source of proteins and essential amino acids, making it a valuable ingredient for developing functional foods and dietary supplements aimed at improving muscle growth, managing weight, and supporting overall health [54]. Algal proteins can be hydrolyzed or isolated to improve their digestibility, bioavailability, and functionality in food applications.
   - **Carbohydrates and Dietary Fibers:** Algal biomass contains carbohydrates and dietary fibers that offer potential health benefits such as improved digestion, satiety, and glycemic control. Alginate, a polysaccharide found in brown algae, has been shown to reduce cholesterol levels and promote weight loss when included in functional foods and beverages [55].
   - **Vitamins and Minerals:** Algal biomass is a natural source of essential vitamins (e.g., vitamins A, C, E, K) and minerals (e.g., calcium, magnesium, iron, zinc) necessary for maintaining optimal health and well-being. These vitamins and minerals can be extracted from algal biomass and added to fortified foods, beverages, and dietary supplements to address specific nutritional deficiencies and support overall health. According to Katiyar et al. [56], while microalgae were extensively explored as a raw material for fuel production, growing health awareness has led industries to shift focus toward microalgal bioproducts, as summarized in Table 6.4.

c. **Bioactive Compounds for Pharmaceutical and Cosmetic Applications:** Algal biomass contains a diverse range of bioactive compounds with potential applications in pharmaceuticals and cosmetics. These include compounds with anti-inflammatory, antimicrobial, anti-aging, and skin-whitening properties. These compounds are increasingly being incorporated into novel drugs, cosmeceuticals, and personal care products targeting various skin and health conditions.
   - **Anti-inflammatory Compounds:** Compounds such as polyunsaturated fatty acids (e.g., omega-3 fatty acids), phycocyanin, and astaxanthin found in algal biomass have demonstrated anti-inflammatory properties, making them promising candidates for treating inflammatory diseases like arthritis, asthma, and inflammatory bowel disease [57, 58].
   - **Antimicrobial Compounds:** Algal biomass contains compounds exhibiting antimicrobial activity against a wide range of pathogens, including bacteria, viruses, and fungi [59]. These compounds, such as sulfated polysaccharides

Table 6.4 Applications of microalgae products in different industries. Adapted from [56].

Bioproducts	Microalgal species	Configuration used in industries	Applications
γ-Linolenic acid (GLA)	*Spirulina* sp.	Capsules, syrups, and tablets	Infant formulas and nutritional supplements
Phycocyanin	*Spirulina platensis*	Pasta, ointments, shakes, noodles, chips	Health, food, and cosmetics
Proteins and fibers	*Chlorella* sp.	Shakes, noodles, chips, ointments, soaps, face washes, face packs, and shampoo	Health, food, and cosmetics
Arachidonic acid (AA)	*Parietochloris incise*	Capsules	Food additives and nutraceuticals
Vitamin B12	*Nostoc muscorum, Hapalosiphon fontinalis*	Capsules and tablets	Food and feed industry

(e.g., carrageenan) and antimicrobial peptides, hold potential for developing novel antibiotics, antiviral drugs, and antifungal agents.

- **Anti-aging Compounds:** Algal compounds with antioxidant and anti-inflammatory properties, such as polyphenols, tocopherols, and carotenoids, have been shown to protect against oxidative stress and DNA damage associated with aging [60]. These compounds are increasingly being incorporated into anti-aging skincare products and supplements to promote youthful-looking skin and overall well-being.

Algal biomass is a rich source of high-value bioproducts and bio-compounds with diverse applications in food, pharmaceuticals, cosmetics, and nutraceuticals. Products such as omega-3 fatty acids, pigments, antioxidants, functional food ingredients, and bioactive compounds derived from algal biomass offer numerous health benefits and environmental advantages compared to conventional sources, as discussed in Table 6.5.

As consumer demand for natural, sustainable, and plant-based products grows, the commercialization and utilization of algal bioproducts and bio-compounds are anticipated to expand significantly. This expansion is driving innovation and investment in the algal biotechnology industry. Overall, algal biomass represents a valuable and versatile resource for developing novel products and solutions to address global health, nutrition, and sustainability challenges.

6.6 Integrated Algal Biorefinery Approach

The integration of various algal biomass components and processing streams is central to the concept of an algal biorefinery. This synergistic approach maximizes resource efficiency, enhances product yields, and improves economic viability. Integrated biorefinery systems are designed to optimize the utilization of biomass

Table 6.5 Potential utilization of bioactive compounds from cyanobacteria and microalgae. Adapted from [61].

Cyanobacteria/Microalgae	Bioactive compounds	Uses in cosmetics
Odontella aurita	Chrysolaminarin	Antioxidant
Chlorella	β-1,3-Glucan	Free-radical collector, immune system booster
Skeletonema	β-1,3-Glucan	Anti-inflammatory
Haematococcus pluvialis	Astaxanthin	Antioxidant, sunscreen protection
Chlorella	Polysaccharides	Moisturizing and thickener agent
Porphyridium and *Rhodella reticulata*	Sulfated polysaccharides	Antioxidant
Spirulina porphyridium	Phycocyanobilin, phycoerythrobilin	Antioxidant pigment for eyeliner and lipsticks
Chlorella sp.	Chlorophyll	To mask odors in dentifrices and deodorants
Porphyridium cruentum	Phycocyanin	Eye shadows
Anabaena vaginicola	Lycopene	Antioxidant, anti-aging sunscreen
Halomonas elongata	Ectoine	Immune protection, UV protection, stress protection

feedstocks and processing technologies to produce a diverse portfolio of products and co-products while minimizing waste generation and environmental impact. Unlike conventional single-product biorefineries, integrated systems capitalize on the synergies between different processing streams and value chains to improve overall efficiency and profitability. Integrated biorefinery systems include:

a. **Feedstock Flexibility:** Integrated biorefineries can process a variety of feedstocks, such as algal biomass, lignocellulosic biomass, agricultural residues, and various waste streams, to generate multiple products and co-products.
b. **Process Integration:** Different processing technologies, including fermentation, extraction, fractionation, and catalysis, are integrated to maximize the efficiency of converting biomass feedstocks into value-added products.
c. **Product Diversification:** Integrated biorefineries produce a range of products, such as biofuels, bioproducts, and biochemicals, to meet diverse market demands and increase revenue streams.
d. **Waste Utilization:** Byproducts and waste streams generated during biorefinery operations are valorized to produce additional value-added products or used for energy generation, minimizing waste and improving overall resource efficiency.

Furthermore, co-production strategies involve the simultaneous production of multiple products from algal biomass feedstocks through integrated biorefinery approaches. By leveraging the synergies between different processing streams and value chains, co-production strategies aim to maximize resource utilization and product yields while minimizing costs and environmental impact. Some common co-production strategies in algal biorefining integration are as follows:

- **Simultaneous Extraction:** Algal biomass can be processed to extract multiple products concurrently, such as lipids for biodiesel production, proteins for food and feed, and pigments for nutraceuticals and cosmetics.
- **Sequential Processing:** Algal biomass can undergo a sequence of processing steps to produce different products at various stages of biorefinery operations. For example, biomass residues after lipid extraction can be further processed to produce bioethanol or biogas through fermentation or anaerobic digestion.
- **Co-cultivation:** Algal cultivation systems can be designed to co-culture multiple algal species with complementary characteristics, such as combining lipid- and protein-rich species, to enhance overall biomass productivity and product diversity.
- **Byproduct Valorization:** Byproducts and waste streams generated during biorefinery operations, such as spent biomass, extraction residues, and wastewater, can be valorized to produce additional value-added products or used for energy generation through anaerobic digestion or gasification. This valorization of microalgae biomass presents significant opportunities for enhancing the commercial viability of microalgae biorefineries (Figure 6.5).

As shown in Figure 6.5, cost savings and improvements can be achieved through integrated algal biomass processing. Similarly, integrating and optimizing processes are essential for maximizing the efficiency and economic viability of integrated biorefinery systems. By optimizing process parameters, resource utilization, and product yields, biorefineries can enhance overall performance and competitiveness in the market. Major aspects of process integration and optimization include selecting appropriate algal biomass feedstocks based on composition, growth characteristics, and availability, which is crucial for optimizing product yields and quality. Designing integrated biorefinery processes involves selecting and integrating various technologies (cultivation, extraction,

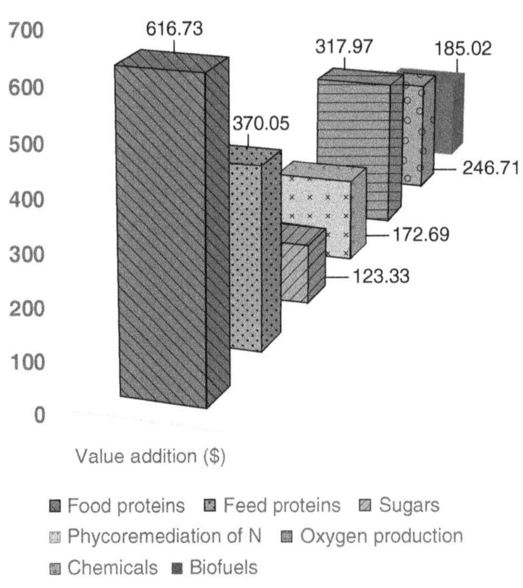

Figure 6.5 Microalgal biomass (per 10^3 kg) value addition in a biorefinery. Adapted from [62].

fermentation, purification) to maximize resource efficiency and product yields. Energy integration strategies, such as heat recovery, cogeneration, and process integration, can reduce energy consumption and environmental impact while improving overall process efficiency. Optimizing resource utilization, including water, nutrients, and chemicals, is essential for minimizing waste and environmental impact while maximizing product yields and profitability. Furthermore, real-world applications of integrated biorefinery systems in algal biotechnology demonstrate the feasibility and potential of co-production strategies and process integration approaches for enhancing resource efficiency, product yields, and economic viability. Some of the examples of integrated algal biorefineries are as follows:

- **Algenol Biorefinery:** Algenol, a biotechnology company, operates an integrated biorefinery producing biofuels (ethanol and biodiesel), feedstock (algae biomass), and co-products (omega-3 fatty acids and animal feed) from algal biomass. The facility integrates PBRs for cultivation, lipid extraction, and fermentation for biofuel production, along with downstream processing for co-product recovery [63].
- **Sapphire Energy:** Sapphire energy, a renewable energy company, utilizes an integrated biorefinery approach to produce renewable crude oil from algal biomass for biofuel production. Their facility integrates cultivation ponds, harvesting, dewatering, and HTL for oil extraction and conversion [64].
- **Neste:** Neste, a Finnish renewable fuels company, operates an integrated biorefinery that produces renewable diesel and renewable aviation fuel from algal biomass. The biorefinery combines cultivation ponds, lipid extraction, and hydrotreating for biofuel production, alongside co-production of high-value chemicals and bioproducts [65].

Integrated biorefinery systems offer a promising path to maximize resource efficiency, product yields, and economic viability in algal biorefining. By utilizing co-production strategies, process integration, and optimization, biorefineries can produce a diverse range of products and co-products from algal biomass feedstocks, contributing to the transition toward a sustainable and bio-based economy. Real-world case studies and applications demonstrate the feasibility and potential of integrated biorefinery systems in improving resource utilization, diversifying products, and enhancing environmental sustainability in algal biotechnology.

6.7 Genetic Engineering and Algal Strain Improvement

While naturally occurring algal species possess inherent advantages for biorefining, genetic engineering and other strain improvement techniques offer significant potential to further enhance their performance and tailor them for specific agro-waste streams and biorefinery objectives. Genetic engineering tools can be used to modify algal genomes to improve various traits relevant to biorefining. Overexpressing genes involved in nutrient transport (e.g., nitrate, phosphate transporters) can increase nutrient uptake rates and efficiency, leading to higher biomass productivity and improved nutrient removal from agro-waste. Metabolic pathway engineering can optimize nutrient assimilation pathways within the cell. Genetic manipulation of lipid biosynthesis pathways can enhance lipid accumulation, particularly TAGs [66]. This involves overexpressing key enzymes in lipid synthesis, downregulating genes involved in lipid degradation, or engineering regulatory pathways to promote lipid storage. Introducing or enhancing detoxification pathways can increase algal tolerance to inhibitory compounds present in agro-waste. This involves expressing enzymes capable of degrading or modifying inhibitors or strengthening cellular stress response mechanisms. Genetic engineering can be used to alter the protein or carbohydrate content of algal biomass, optimizing it for specific applications like feed, food, or bio-based materials. Engineering algal strains to flocculate spontaneously or produce larger cell sizes can improve harvesting efficiency and reduce associated costs [67]. Table 6.6 summarizes strategies related to genetic engineering and

Table 6.6 Genetic engineering and strain improvement strategies for algae. Adapted from [68, 69].

Target trait	Genetic engineering/ Strain improvement approach	Potential benefits	Challenges and research needs
Nutrient utilization	Overexpression of nutrient transporters, metabolic enzymes	Enhanced uptake and assimilation of complex nutrients from agro-waste hydrolysate	Identifying key genes, optimizing gene expression, ensuring metabolic balance
Biomass productivity	Enhancement of photosynthesis, carbon fixation pathways, and cell division	Faster growth rates, higher biomass yields, reduced cultivation time	Understanding regulatory mechanisms, balancing growth and product accumulation, maintaining genetic stability
Product accumulation	Metabolic engineering to redirect carbon flux toward target products (lipids, carbohydrates, proteins)	Increased yield of desired products, improved biorefinery efficiency	Detailed metabolic pathway mapping, precise gene editing, optimizing metabolic flux control
Stress tolerance	Introduction of genes conferring resistance to inhibitors and challenging environmental stresses	Improved survival and performance in agro-waste-based environments	Identifying stress-responsive genes, engineering robust tolerance mechanisms, maintaining productivity under stress
Strain robustness	Genome editing for strain stabilization, development of marker-free selection systems	Stable and high-performing strains over multiple generations, reduced contamination risk	Addressing genetic instability, developing efficient transformation methods, ensuring biosafety of modified strains

strain improvement for algae. Several successful examples of strain improvement in algae for biorefinery applications exist:

- **Increased Lipid Production:** Studies have shown that overexpressing genes involved in fatty acid synthesis, such as acyl-CoA carboxylase (ACCase) or diacylglycerol acyltransferase (DGAT), can significantly increase lipid accumulation in various algal species.
- **Enhanced Nutrient Uptake:** Genetic modification to enhance nitrogen assimilation pathways has been shown to improve growth and biomass productivity under nitrogen-limited conditions, which is relevant for nutrient-depleted agro-waste streams.
- **Improved Stress Tolerance:** Engineering algae to overexpress stress-responsive genes, such as those involved in antioxidant defense or heat shock proteins, can improve tolerance to environmental stresses.

6.8 Environmental Sustainability and Life Cycle Assessment in Algal Biorefining

As the world transitions toward renewable and sustainable energy sources and products, evaluating the environmental sustainability of processes like algal biorefining becomes crucial. Algal biorefinery processes can have both positive and negative environmental impacts. While algae cultivation can contribute to eutrophication if not managed properly, and energy-intensive processes for extraction and conversion can lead to greenhouse gas emissions and resource depletion, algal biorefineries also offer opportunities to mitigate environmental impacts. These include using wastewater and CO_2 as nutrient sources [70], reducing land and water use compared to conventional agriculture, and producing renewable products.

6.8.1 Life Cycle Assessment Methodology

Life Cycle Assessment (LCA) is a systematic methodology for evaluating the environmental impacts of a product or process throughout its entire life cycle, from the extraction of raw materials to disposal at the end of life. LCA involves four main stages: defining the goal and scope, inventory analysis, impact assessment, and interpretation of results. LCA considers various environmental indicators, including greenhouse gas emissions, energy use, water consumption, land use, and toxicity, to quantify the environmental footprint of algal biorefinery processes [71]. Analyzing the environmental footprint involves quantifying the environmental impacts of algal biorefining processes using LCA methodology. This analysis helps to identify "hotspots" or areas where environmental performance can be improved. Mitigation strategies aim to reduce or offset the environmental footprint through measures such as:

1. **Energy Efficiency Improvements:** Implementing energy-efficient technologies and optimizing processes to reduce energy consumption and greenhouse gas emissions.

2. **Resource Recycling and Reuse:** Recycling and reusing water, nutrients, and byproducts to minimize resource consumption and waste generation.
3. **Carbon Capture and Utilization (CCU):** Capturing CO_2 emissions from biorefinery processes and using them as feedstock for algae cultivation, thereby reducing carbon emissions and enhancing product sustainability.
4. **Sustainable Sourcing:** Ensuring sustainable sourcing of raw materials and inputs, such as algal strains, nutrients, and chemicals, to minimize environmental impacts throughout the supply chain [72].

6.8.2 Policy and Regulatory Considerations

Policy frameworks and regulations play a vital role in promoting environmental sustainability and mitigating the impacts of algal biorefining processes. Governments and regulatory bodies can implement policies and incentives to encourage the adoption of sustainable practices and technologies in the algal biorefinery sector [73]. Policy implications and regulatory considerations include:

- **Environmental Standards and Certification:** Establishing environmental standards and certification schemes ensures compliance with sustainability criteria and promotes best practices in algal biorefinery operations.
- **Financial Incentives:** Providing financial incentives, grants, and subsidies supports the research, development, and implementation of sustainable technologies and practices in the algal biorefinery sector.
- **Market-based Instruments:** Introducing market-based instruments such as carbon pricing, emissions trading schemes, and renewable energy incentives helps to internalize environmental costs and promote investments in low-carbon and renewable technologies.
- **Stakeholder Engagement:** Engaging stakeholders, including industry, academia, NGOs, and local communities, in policy development and decision-making processes is essential to build broad support for environmental sustainability initiatives in the algal biorefinery sector.

Maga et al. [74] applied the LCA method to an algal biorefinery to quantify its environmental impact. They reported that the algal biorefinery contributed to climate protection, ozone depletion, and fossil resource protection compared to conventional wastewater treatment methods. Thus, environmental sustainability is a crucial aspect in the development and deployment of algal biorefining processes. By conducting LCA and environmental footprint analysis, stakeholders can quantify the environmental impacts of algal biorefinery processes and identify opportunities for improvement. Mitigation strategies, such as energy efficiency improvements, resource recycling, CCU, and sustainable sourcing, can help reduce the environmental footprint. Policy frameworks and regulatory measures are essential for promoting environmental sustainability and incentivizing investments in sustainable technologies and practices. Overall, integrating environmental considerations into the planning, design, and operation of algal biorefinery processes is essential for achieving long-term environmental sustainability and mitigating climate change impacts.

6.9 Challenges and Future Perspectives

As the algal biorefinery sector continues to evolve, it faces various challenges and opportunities that will shape its future trajectory. Some technological challenges associated with algal biorefining include:

a. **Cultivation Efficiency:** Improving the efficiency and productivity of algal cultivation systems is crucial for reducing production costs and increasing biomass yields. Research is needed to develop novel cultivation methods, optimize growth conditions, and enhance strain selection to maximize biomass productivity.
b. **Downstream Processing:** Streamlining downstream processing steps, such as biomass harvesting, dewatering, and extraction, is essential for reducing energy consumption and production costs. Innovative technologies and scalable processes for biomass fractionation, lipid extraction, and product recovery are needed to improve process efficiency and product quality.
c. **Product Diversification:** Expanding the range of products derived from algal biomass beyond biofuels to include high-value bioproducts and bio-compounds is critical for enhancing the economic viability of algal biorefining. Research is needed to develop scalable processes for producing functional food ingredients, nutraceuticals, pharmaceuticals, and cosmetics from algal biomass.
d. **Resource Sustainability:** Ensuring the sustainable use of resources, such as water, nutrients, and land, is essential for the long-term viability of algal biorefinery operations. Research is needed to develop resource-efficient cultivation systems, closed-loop nutrient recycling technologies, and environmentally friendly cultivation practices to minimize environmental impacts and enhance resource sustainability [75].

Additionally, scaling up algal biorefinery processes from lab-scale to commercial-scale operations poses challenges related to cost-effectiveness and process scalability. Research is needed to identify cost-effective production technologies, optimize process parameters, and scale up production systems while maintaining product quality and consistency. Achieving market acceptance and penetration for algal biorefinery products requires overcoming barriers such as price competitiveness, regulatory hurdles, and consumer perception. Research is needed to demonstrate the economic and environmental benefits of algal biorefinery products, address regulatory concerns, and educate consumers about the sustainability and value of algal-derived products. Developing the necessary infrastructure, including cultivation facilities, processing plants, and distribution networks, is essential for scaling up operations [76]. Research is needed to assess infrastructure requirements, identify optimal facility locations, and establish partnerships with stakeholders to support infrastructure development and investment.

Emerging technologies such as PBRs, bioreactors, and closed-loop cultivation systems offer opportunities to improve the efficiency, productivity, and scalability of algal cultivation. Integrating automation, sensors, and control systems can enhance process control and optimization in algal cultivation. Advances in bioprocess engineering, including biocatalysis, fermentation, and downstream processing, enable the development of novel processes for converting algal biomass into value-added products. Integrating biorefinery processes and co-production strategies can maximize resource utilization and product yields. Genetic engineering and synthetic biology approaches can be used to engineer algal strains with enhanced traits, such as higher lipid content, faster growth rates, and tolerance to environmental stresses. Genetic tools like CRISPR-Cas9 enable precise genome editing and metabolic engineering of algal strains for improved biorefinery performance [77]. For further research and development, the following points are recommended:

- **Interdisciplinary Collaboration:** Foster collaboration among scientists, engineers, economists, and policymakers to address complex challenges and develop holistic solutions for sustainable algal biorefining.
- **Technology Transfer and Adoption:** Facilitate the transfer and adoption of research innovations by industry stakeholders through partnerships, pilot-scale demonstrations, and knowledge exchange initiatives.
- **Education and Training:** Invest in education and training programs to develop a skilled workforce capable of driving innovation and growth in the algal biorefinery sector. Provide opportunities for hands-on training, internships, and professional development in algal biotechnology and bioprocess engineering.
- **Policy Support:** Advocate for supportive policies and regulatory frameworks that incentivize investment, innovation, and commercialization in the algal biorefinery sector. Encourage policymakers to consider the environmental, economic, and social benefits of algal biorefinery products and support initiatives promoting sustainability and resource efficiency [78].

Henceforth, addressing the challenges and leveraging the opportunities facing the algal biorefinery sector require concerted efforts from academia, industry, and government to drive innovation, overcome barriers, and realize the full potential of algal biorefining. By tackling technological challenges, scaling up and commercialization issues, and embracing future trends and emerging technologies, the algal biorefinery sector can play a significant role in advancing sustainability, reducing dependence on fossil fuels, and promoting economic development. Continued research and development, coupled with supportive policies and investments, are essential for accelerating the transition toward a bio-based economy powered by algal biorefinery technologies.

6.10 Conclusion

Algal biorefining is considered a promising approach for reducing the operational costs of fuel production and managing waste and decontamination. Integrating algae cultivation with other sustainable practices such as wastewater treatment, carbon capture, and nutrient recycling creates closed-loop systems that minimize waste and resource consumption. There is a clear call for action from both the industry and the research community to advance the field of algal biorefining. Industry stakeholders must invest in research, development, and commercialization efforts to overcome technical challenges, scale up production, and bring algal biorefinery products to market. Collaboration among industry, academia, and government agencies is essential for driving innovation, developing sustainable technologies, and creating supportive policy frameworks. The research community must continue exploring novel cultivation techniques, bioprocess innovations, and product diversification strategies to unlock the full potential of algal biorefining. Additionally, education and outreach efforts are needed to raise awareness about the benefits of algal biorefining and inspire future generations of scientists, engineers, and entrepreneurs to contribute to its advancement. In conclusion, algal biorefining holds immense promise as a sustainable and versatile platform for producing renewable energy, bioproducts, and biochemicals. With concerted efforts and collaboration, we can harness the power of algae to drive sustainable development, mitigate climate change, and create a more prosperous and resilient future for generations to come.

References

1. Laurens, L.M.L., Nagle, N., Davis, R., Sweeney, N., Van Wychen, S., Lowell, A., et al. Acid-catalyzed algal biomass pretreatment for integrated lipid and carbohydrate-based biofuels production. *Green Chem.* vol. 17, pp. 1145–1158, 2015.
2. Sambusiti, C., Bellucci, M., Zabaniotou, A., Beneduce, L., and Monlau, F. Algae as promising feedstocks for fermentative biohydrogen production according to a biorefinery approach: a comprehensive review. *Renew. Sustain. Energy Rev.* vol. 44, pp. 20–36, 2015.
3. da Rosa, M.D.H., Alves, C.J., dos Santos, F.N., de Souza, A.O., Zavareze, E.D.R., Pinto, E., et al. Macroalgae and microalgae biomass as feedstock for products applied to bioenergy and food industry: a brief review. *Energies.* vol. 16, no. 4, p. 1820, 2023.
4. Suganya, T., Varman, M., Masjuki, H.H., and Renganathan, S. Macroalgae and microalgae as a potential source for commercial applications along with biofuels production: a biorefinery approach. *Renew Sustain Energy Rev.* vol. 55, pp. 909–941, 2016.
5. Das, D. *Algal Biorefinery: An Integrated Approach.* 2016.
6. Offei, F. Integrated biorefinery approach to lignocellulosic and algal biomass fermentation processes. In: *Fermentation - Processes, Benefits and Risks.* IntechOpen, 2021, 220–243.
7. Vaish, B., Srivastava, V., Singh, P.K., Singh, P., and Singh, R.P. Energy and nutrient recovery from agro-wastes: rethinking their potential possibilities. *Environ. Eng. Res.* vol. 25, no. 5, pp. 623–637, 2020.
8. Fenton, O. and Ó hUallacháin, D. Agricultural nutrient surpluses as potential input sources to grow third generation biomass (microalgae): a review. *Algal Res.* vol. 1, no. 1, pp. 49–56, 2012.
9. Talebi, S., Edalatpour, A., and Tavakoli, O. Algal biorefinery: a potential solution to the food-energy-water-environment nexus. *Sustain. Energy Fuels.* vol. 6, no. 11, pp. 2623–2664, 2022.

10 Santiago, A.F., Calijuri, M.L., Assemany, P.P., Calijuri, M.D.C., and Reis, A.J.D.D. Algal biomass production and wastewater treatment in high rate algal ponds receiving disinfected effluent. *Environ Technol* (United Kingdom). vol. 34, no. 13–14, pp. 1877–1885, 2013.

11 Slegers, P.M., Lösing, M.B., Wijffels, R.H., van Straten, G., and van Boxtel, A.J.B. Scenario evaluation of open pond microalgae production. *Algal Res.* vol. 2, no. 4, pp. 358–368, 2013.

12 Yadala, S. and Cremaschi, S. A dynamic optimization model for designing open-channel raceway ponds for batch production of algal biomass. *Processes.* vol. 4, no. 2, p. 10, 2016.

13 Johnson, M., Villani, T.S., Azmat, A., Simon, J.E., and Both, A.J. Evaluation of algal biomass production on vertical aeroponic substrates. *Algal Res.* vol. 10, pp. 240–248, 2015.

14 Zhuang, Y., Su, Q., Wang, H., Wu, C., Tong, S., Zhang, J., et al. Strain screening and conditions optimization in microalgae-based monosodium glutamate wastewater (MSGW) treatment. *Water* (Switzerland). vol. 15, no. 9, p. 1663, 2023.

15 Iqbal, M.M., Muhammad, G., Aslam, M.S., Hussain, M.A., Shafiq, Z., and Razzaq, H. Algal biofertilizer. In: *Biofertilizers: Study and Impact.* John Wiley & Sons, Ltd, 2021, pp. 287–310.

16 Fasaei, F., Bitter, J.H., Slegers, P.M., and van Boxtel, A.J.B. Techno-economic evaluation of microalgae harvesting and dewatering systems. *Algal Res.* vol. 31, pp. 347–362, 2018.

17 Pahl, S.L., Lee, A.K., Kalaitzidis, T., Ashman, P.J., Sathe, S., and Lewis, D.M. Harvesting, thickening and dewatering microalgae biomass. In: *Algae for Biofuels and Energy.* Springer Nature, 2013, pp. 165–185.

18 Rajvanshi, M. and Sayre, R. Recent Advances in algal biomass production. In: *Biotechnological Applications of Biomass.* Intechopen, 2021, 188–202.

19 Bhushan, S., Jayakrishnan, U., Shree, B., Bhatt, P., Eshkabilov, S., and Simsek, H. Biological pretreatment for algal biomass feedstock for biofuel production. *J. Environ. Chem. Eng.* vol. 11, no. 3, p. 109870, 2023.

20 Sarwer, A., Hamed, S.M., Osman, A.I., Jamil, F., Al-Muhtaseb, A.H., Alhajeri, N.S., et al. Algal biomass valorization for biofuel production and carbon sequestration: a review. *Environ. Chem. Lett.* vol. 20, pp. 2797–2851, 2022.

21 Singh, S., Singh, L., Kumar. V., Ali, W., Ramamurthy, P.C., Singh Dhanjal, D., et al. Algae-based approaches for Holistic wastewater management: a low-cost paradigm. *Chemosphere.* vol. 345, p. 140470, 2023.

22 Shetty, K. and Krishnakumar, G. Strategies to enhance lipid productivity in microalgae: role of cell disruption and media composition. *Fuel.* vol. 345, p. 128240, 2023.

23 Ocreto, J.B., Chen, W.H., Rollon, A.P., Chyuan Ong, H., Pétrissans, A., Pétrissans, M., et al. Ionic liquid dissolution utilized for biomass conversion into biofuels, value-added chemicals and advanced materials: a comprehensive review. *Chem. Eng. J.* vol. 445, p. 136733, 2022.

24 Shukla, A., Kumar, D., Girdhar, M., Kumar, A., Goyal, A., Malik, T., et al. Strategies of pretreatment of feedstocks for optimized bioethanol production:

distinct and integrated approaches. *Biotechnol. Biofuels Bioprod.* vol. 16, no. 1, p. 44, 2023.
25 Anukam, A. and Berghel, J. Biomass pretreatment and characterization: a review. In: *Biotechnological Applications of Biomass*. Intechopen, 2021,198–210.
26 Shah, T.A., Zhihe, L., Zhiyu, L., and Zhang, A. New development in renewable energy research. In: *Environmental Science and Engineering*. Springer International Publishing, 2023, pp. 269–285.
27 Pendyala, B., Hanifzadeh, M.M., Abel, G.A., Viamajala, S., and Varanasi, S. Production of organic acids via autofermentation of microalgae: a promising approach for sustainable Algal Biorefineries. *Ind. Eng. Chem. Res.* vol. 59, no. 5, p. 132236, 2020.
28 Zhou, J., Wang, M., Saraiva, J.A., Martins, A.P., Pinto, C.A., Prieto, M.A., et al. Extraction of lipids from microalgae using classical and innovative approaches. *Food Chem.* vol. 384, p. 132236, 2022.
29 Martins, P.L., Duarte, L.C., Pereira, H., Reis, A., and Carvalheiro, F. Evaluation of different fractionation methods for the simultaneous protein and carbohydrate extraction from microalgae. *Biomass Convers. Biorefin.* vol. 15, pp. 999–1011, 2024.
30 Hammann, W., Ross, A., and Seames, W. Sequential extraction of carbohydrates and lipids from *Chlorella vulgaris* using combined physical and chemical pre-treatments. *ChemEngineering*. vol. 8, no.1, p. 11, 2024.
31 Suarez Garcia, E., Suarez Ruiz, C.A., Tilaye, T., Eppink, M.H.M., Wijffels, R.H., and van den Berg, C. Fractionation of proteins and carbohydrates from crude microalgae extracts using an ionic liquid based-aqueous two phase system. *Sep. Purif. Technol.* vol. 204, pp. 56–65, 2018.
32 Obeid, S., Beaufils, N., Peydecastaing, J., Camy, S., Takache, H., Ismail, A., et al. Microalgal fractionation for lipids, pigments and protein recovery. *Process Biochem.* vol. 121, pp. 240–247, 2022.
33 Izanlou, Z., Akhavan Mahdavi, M., Gheshlaghi, R., and Karimian, A. Sequential extraction of value-added bioproducts from three *Chlorella* strains using a drying-based combined disruption technique. *Bioresour. Bioprocess.* vol. 10, p. 44, 2023.
34 Bhujade, R. and Agrawal, P.S. Algal biocrude as drop-In feedstock for green fuel and green chemicals. In: *The Production of Biodiesel and Related Fuel Additives [Internet]*. Bentham Science Publishers, 2024, pp. 178–198. Available from: https://www.benthamdirect.com/content/books/9789815196740.chapter-6
35 Agrawal, P.S., Belkhode, P.N., Rokhum, S.L. Microalgae – Biofuel production trends. In: *Sustainability in Biofuel Production Technology*. John Wiley & Sons, 2022, 215–226.
36 Sati, H., Mitra, M., Mishra, S., and Baredar, P. Microalgal lipid extraction strategies for biodiesel production: a review. *Algal Res.* vol. 38, p. 101413, 2019.
37 Seon, G., Kim, H.S., Cho, J.M., Kim, M., Park, W.K., and Chang, Y.K. Effect of post-treatment process of microalgal hydrolysate on bioethanol production. *Sci. Rep.* vol. 10, p. 16698, 2020.
38 Hasnain, M., Munir, N., Abideen, Z., Macdonald, H., Hamid, M., Abbas, Z., et al. Prospects for biodiesel production from emerging algal resource: process optimization and characterization of biodiesel properties. *Agric.* vol. 13, no. 2, p. 407, 2023.

39 Karmakar, R., Rajor, A., Kundu, K., and Kumar, N. Production of biodiesel from unused algal biomass in Punjab, India. *Pet. Sci.* vol. 15, pp. 164–175, 2018.
40 Kumar, V., Nanda, M., Joshi, H.C., Singh, A., Sharma, S., and Verma, M. Production of biodiesel and bioethanol using algal biomass harvested from fresh water river. *Renew. Energy*. vol. 116, no. Part A, pp. 606–612, 2018.
41 Elkatory, M.R., Hassaan, M.A., and El Nemr, A. Algal biomass for bioethanol and biobutanol production. In: *Handbook of Algal Biofuels: Aspects of Cultivation, Conversion, and Biorefinery*. Elsevier, 2021, pp. 251–279.
42 Abomohra, A.E.F. and Elshobary, M. Biodiesel, bioethanol, and biobutanol production from microalgae. In: *Microalgae Biotechnology for Development of Biofuel and Wastewater Treatment*. Springer Nature, 2019, pp. 293–321.
43 Sudhakar, M.P. and Viswanaathan, S. Algae as a sustainable and renewable bioresource for bio-fuel production. In: *New and Future Developments in Microbial Biotechnology and Bioengineering: Microbial Biotechnology in Agro-environmental Sustainability*. Elsevier, 2019, pp. 77–84.
44 Yaşar, F. Evaluation and advantages of Algae as an energy source. *J. Turk. Chem. Soc. Sect. A: Chem*. vol. 5, no. 3, pp. 1309–1318, 2018.
45 Ağbulut, Ü., Sirohi, R., Lichtfouse, E., Chen, W.H., Len, C., Show, P.L., et al. Microalgae bio-oil production by pyrolysis and hydrothermal liquefaction: mechanism and characteristics. *Bioresour. Technol*. vol. 376, p. 128860, 2023.
46 Agrawal, P. and Rokhum, S.L. Supported metal/metal oxide catalysts in biodiesel production: an overview. In: *Biodiesel Production: Feedstocks, Catalysts, and Technologies*. John Wiley & Sons, 2022, pp. 119–141.
47 Agrawal, P.S., Belkhode, P.N., and Rokhum, S.L. Greener catalytic processes in biofuel production. In: *Sustainability in Biofuel Production Technology*. Apple Academic Press, 2022, pp. 133–164.
48 Ravichandran, S.R., Venkatachalam, C.D., Sengottian, M., Sekar, S., Kandasamy, S., Ramasamy Subramanian, K.P., et al. A review on hydrothermal liquefaction of algal biomass on process parameters, purification and applications. *Fuel*. vol. 313, p. 122679, 2022.
49 da Costa Magalhães, B., Checa, R., Lorentz, C., Afanasiev, P., Laurenti, D., and Geantet, C. Catalytic hydroconversion of HTL micro-algal bio-oil into biofuel over NiWS/Al2O3. *Algal Res*. vol. 71, p. 103012, 2023;
50 Stokes, J., Tu, R., Peters, M., Yadav, G., Fabiano, L.A., and Seider, W.D. Omega-3 fatty acids from algae produced biodiesel. *Algal Res*. vol. 51, p. 102047, 2020.
51 Priyadarsini, M., Raichel Nivetha, X., Mathimani, T., Anto, S., Hareesh Krishnan, H., Glivin G., et al. Omega-3-fatty acids from algae for health benefits. *Mater. Today: Proc*. vol. 66, pp. 1514–1518, 2022.
52 Chen, Z., Wu, W., Wen, Y., Zhang, L., Wu, Y., Farid, M.S., et al. Recent advances of natural pigments from algae. *Food Prod. Process. Nutr*. vol. 5, p. 39, 2023.
53 Ngoc Boi, V., Thi My Trang, N., Xuan Cuong, D., and Thai Ha, H. Antioxidant phlorotannin from brown algae *Sargassum dupplicatum*: enzyme-assissted extraction and purification. *World J Food Sci Technol*. vol. 4, no. 2, pp. 62–68, 2020.
54 Arakaki, N., Flores Ramos, L., Oscanoa Huaynate, A.I., Ruíz Soto, A., and Ramírez, M.E. Biochemical and nutritional characterization of edible seaweeds from the Peruvian Coast. *Plants*. vol. 12, no. 9, p. 1795, 2023.

55 Dini, I. The potential of algae in the nutricosmetic sector. *Molecules*. vol. 28, no. 10, p. 4032, 2023.

56 Katiyar, R., Banerjee, S., and Arora, A. Recent advances in the integrated biorefinery concept for the valorization of algal biomass through sustainable routes. *Biofpr*. vol. 15, no. 3, pp. 879–898, 2021.

57 Ghallab, D.S., Ibrahim, R.S., Mohyeldin, M.M., and Shawky, E. Marine algae: a treasure trove of bioactive anti-inflammatory compounds. *Mar. Pollut. Bull.* vol. 199, p. 116023, 2024.

58 Khursheed, M., Ghelani, H., Jan, R.K., and Adrian, T.E. Anti-inflammatory effects of bioactive compounds from seaweeds, bryozoans, jellyfish, shellfish and peanut worms. *Mar. Drugs*. vol. 21, no. 10, p. 524, 2023.

59 Afzal, S., Yadav, A.K., Poonia, A.K., Choure, K., Yadav, A.N., and Pandey, A. Antimicrobial therapeutics isolated from algal source: retrospect and prospect. *Biologia*. vol. 78, pp. 291–305, 2023.

60 Kalasariya, H.S., Yadav, V.K., Yadav, K.K., Tirth, V., Algahtani, A., Islam, S., et al. Seaweed-based molecules and their potential biological activities: an eco-sustainable cosmetics. *Molecules*. vol. 26, no. 17, p. 5313, 2021.

61 Mourelle, M.L., Gómez, C.P., and Legido, J.L. The potential use of marine microalgae and cyanobacteria in cosmetics and thalassotherapy. *Cosmetics*. vol. 4, no. 4, p. 46, 2017.

62 Wijffels, R.H., Barbosa, M.J., and Eppink, M.H.M. Microalgae for the production of bulk chemicals and biofuels. *Biofpr*. vol. 4, no. 3, pp. 287–295, 2010.

63 Fal, S., Smouni, A., and El Arroussi, H. Integrated microalgae-based biorefinery for wastewater treatment, industrial CO2 sequestration and microalgal biomass valorization: a circular bioeconomy approach. *Environ. Adv*. vol. 12, p. 100365, 2023.

64 Taher, H., Giwa, A., Abusabiekeh, H., and Al-Zuhair, S. Biodiesel production from nannochloropsis gaditana using supercritical CO_2 for lipid extraction and immobilized lipase transesterification: economic and environmental impact assessments. *Fuel Process. Technol*. vol. 198, p. 106249, 2020.

65 Posada, J.A., Brentner, L.B., Ramirez, A., and Patel, M.K. Conceptual design of sustainable integrated microalgae biorefineries: parametric analysis of energy use, greenhouse gas emissions and techno-economics. *Algal Res*. vol. 17, pp. 113–131, 2016.

66 Sproles, A.E., Fields, F.J., Smalley, T.N., Le, C.H., Badary, A., and Mayfield, S.P. Recent advancements in the genetic engineering of microalgae. *Algal Res*. vol. 53, p. 102158, 2021.

67 Sreenikethanam, A., Raj S., J, R.B., Gugulothu, P., Bajhaiya, A.K. Genetic engineering of microalgae for secondary metabolite production: recent developments, challenges, and future prospects. *Front. Bioeng. Biotechnol*. vol. 10, p. 836056, 2022.

68 Geraldi, A., Syarafa, H.S., Febriana, E., Setianingsih, P.N.M., Tamamia, E., Al Hanif, A.K.R., et al. Application of genetic engineering for increasing biofuel yield from microalgae: a review. *AIP Conf. Proc*. vol. 2679, no. 1, p. 040002, 2023.

69 Fayyaz, M., Chew, K.W., Show, P.L., Ling, T.C., Ng, I.-S., and Chang, J.S. Genetic engineering of microalgae for enhanced biorefinery capabilities. *Biotechnol. Adv*. vol. 43, p. 107554, 2020.

70 San Juan, J.L.G., Ching, P.M.L., Mayol, A.P., Culaba, A.B., and Ubando, A. Envinronmental life cycle analysis of algal biorefineries for biofuel production under the circular economy concept. In: *2020 IEEE 12th International Conference on Humanoid, Nanotechnology, Information Technology, Communication and Control, Environment, and Management, HNICEM 2020*. 2020.
71 Ubando, A.T., Anderson, S., Ng, E., Chen, W.H., Culaba, A.B., and Kwon, E.E. Life cycle assessment of microalgal biorefinery: a state-of-the-art review. *Bioresour. Technol.* vol. 360, p. 127615, 2022.
72 Singh, J. and Dhar, D.W. Overview of carbon capture technology: microalgal biorefinery concept and state-of-the-art. *Front. Mar. Sci.* vol. 6, p. 29, 2019.
73 Solis, C.M.A., Juan, J.L.G.S., Mayol, A.P., Sy, C.L., Ubando, A.T., and Culaba, A.B. A multi-objective life cycle optimization model of an integrated algal biorefinery toward a sustainable circular bioeconomy considering resource recirculation. *Energies*. vol. 14, no. 5, pp. 1416, 2021.
74 Maga, D. Life cycle assessment of biomethane produced from microalgae grown in municipal waste water. *Biomass Convers. Biorefin.* vol.7, pp. 1–10, 2017.
75 Kumar, L., Mohan, L., Anand. R., Joshi, V., Chugh, M., and Bharadvaja, N. A review on unit operations, challenges, opportunities, and strategies to improve algal based biodiesel and biorefinery. *Front. Chem. Eng.* vol. 4, p. 998289, 2022.
76 Koyande, A.K., Show, P.L., Guo, R., Tang, B., Ogino, C., and Chang, J.S. Bio-processing of algal bio-refinery: a review on current advances and future perspectives. *Bioengineered*. vol. 10, no. 1, pp. 574–592, 2019.
77 Uma, V.S., Usmani, Z., Sharma, M., Diwan, D., Sharma, M., Guo, M., et al. Valorisation of algal biomass to value-added metabolites: emerging trends and opportunities. *Phytochem. Rev.* vol. 22, pp. 1015–1040, 2023.
78 Muthukrishnanm, L. Bio-engineering of microalgae: challenges and future prospects toward industrial and environmental applications. *J. Basic Microbiol.* vol. 62, no. 3–4, pp. 310–329, 2022.

7

Waste-to-bio-additive

7.1 Introduction

The relentless growth in global waste generation, fueled by rising population, rapid urbanization, and expanding industrial activity, represents a profound environmental and socio-economic issue. Traditional approaches to waste management, primarily landfilling and incineration, are increasingly recognized as unsustainable. They impose significant environmental burdens, including greenhouse gas (GHG) emissions, contamination of soil and water from leachate and ash, and the depletion of limited natural resources. This situation necessitates developing and implementing innovative and sustainable waste management methods that prioritize resource recovery, minimize environmental impact, and foster a circular economy.

The increasing global challenges of waste management and the growing demand for sustainable materials have sparked significant interest in transforming waste streams into valuable bio-additives. Waste-to-bio-additive involves the process of converting various types of waste materials into valuable bio-based additives using biological, thermochemical, or physicochemical processes [1]. These additives can serve multiple purposes, such as generating energy, improving soil, and industrial applications. The concept centers on turning waste, which would otherwise be sent to landfills or incinerated, into useful and sustainable products that aid environmental preservation and resource efficiency.

The term "waste" encompasses a wide variety of materials, including organic waste from homes, agricultural residues, industrial byproducts, and municipal solid waste (MSW). These waste materials contain abundant organic matter and can potentially be transformed into bio-based additives using various conversion technologies. The additives produced from waste can substitute for or complement traditional additives made from fossil fuels, helping to decrease GHG emissions and reliance on limited resources. Therefore, waste-to-bio-additive is significant because it can help address several urgent environmental and socio-economic issues. First, it provides a sustainable way to manage and decrease waste volumes, redirecting organic matter away from landfills and incineration, and lessening environmental harm and GHG emissions linked to waste disposal. Second, it supports the move

Agro-Waste Management and Valorization, First Edition. Pratibha S. Agrawal, et al.

toward a circular economy by giving value to waste materials and converting them into useful resources [2]. By capturing the energy and nutrients found in waste streams, waste-to-bio-additive facilitates resource recovery and promotes the efficient use of resources, thereby reducing the consumption of new materials and mitigating environmental degradation.

Additionally, waste-to-bio-additive has important implications for energy security and generating renewable energy. By converting organic waste into bio-based additives like biogas, bioethanol, and biodiesel, waste-to-bio-additive creates a renewable and sustainable energy source. This reduces reliance on fossil fuels and helps diversify the energy supply. From a socio-economic standpoint, waste-to-bio-additive can generate new economic chances and jobs, especially in rural and underserved areas where waste management infrastructure might be lacking. Setting up local facilities to convert waste into bio-additives allows communities to earn income from waste valorization and develop products with market value. Therefore, the main goals of waste-to-bio-additive [3] are as follows:

- **Waste Reduction and Management:** The primary goal of waste-to-bio-additive is to lessen the volume of waste going to landfills or incinerators by turning organic materials into useful bio-based additives. By rerouting waste from disposal, WtBA eases the burden on current waste management systems and lowers environmental pollution [4].
- **Resource Recovery and Valorization:** WtBA seeks to reclaim valuable resources like energy, nutrients, and organic matter from waste and convert them into commercially viable bio-based additives. By adding value to waste, WtBA supports the efficient use of resources and aligns with circular economy principles.
- **Energy Production and Security:** Another aim of WtBA is to capture the energy contained in organic waste and produce renewable energy sources such as biogas, bioethanol, biodiesel, or biochar. Turning waste into energy via WtBA improves energy security, decreases GHG emissions, and aids the shift to a low-carbon economy.
- **Environmental Protection and Sustainability:** WtBA aims to lessen the environmental harm linked to waste disposal and traditional methods of making additives. Using renewable resources and lessening dependence on fossil fuels, WtBA helps combat climate change, save natural resources, and safeguard ecosystems and biodiversity.
- **Economic Development and Innovation:** Lastly, WtBA intends to spur economic growth and innovation by creating new markets for bio-based additives, encouraging entrepreneurship, and creating jobs in waste management and bioenergy industries.

Waste-to-bio-additive represents a multifaceted approach to waste management, resource recovery, and sustainable development (Figure 7.1). The tons of yield as summarized in Table 7.1 from various feedstocks can surely help in economic development of the country.

a. **Top of Form:** By converting waste materials into valuable bio-based additives, waste-to-bio-additive offers environmental, economic, and social benefits,

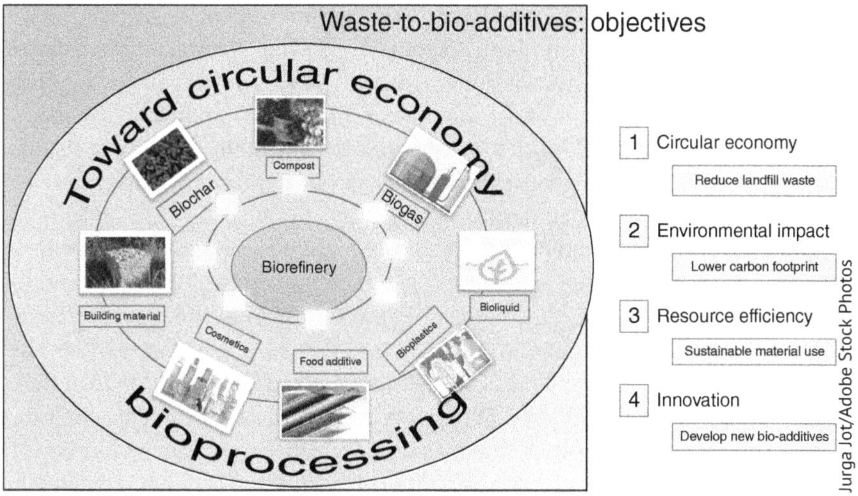

Figure 7.1 Objectives of waste-to-bio-additive conversion.

Table 7.1 Bio-additive yields from various waste streams.

Feedstock	Process	Bio-additive product	Yield (kg/ton)	Ref.
Food waste	Anaerobic digestion	Biogas	90–120 m^3	[5]
Rice husk	Pyrolysis	Biochar	250–350	[6]
Plastic waste (PE/PP)	Pyrolysis	Liquid fuel	400–600	[7]
Sewage sludge	Composting	Compost	500–700	[8]

supporting a more sustainable and robust future. Therefore, this chapter looks at the potential of different waste types – organic waste, agricultural residues, industrial byproducts, and plastic waste – for conversion into bio-additives using various biological and thermochemical techniques.

7.2 Types of Waste Utilized for Bio-additive Production

a. **Organic Waste:** Organic waste is defined as material derived from plants or animals that is naturally biodegradable. Examples include food waste, yard waste, sewage sludge, and other organic matter. With high carbon and nutrient content, organic waste is well-suited for conversion into bio-based additives using various biological and thermochemical methods. Anaerobic digestion (AD) is a common biological technique in this area, utilizing microbial communities to break down organic matter in an oxygen-free setting. This produces biogas (mainly methane [CH_4]) and a digestate high in nutrients. The digestate can be refined to create bio-fertilizers or soil conditioners, providing a sustainable substitute for synthetic fertilizers and supporting circular nutrient flow. Similarly, composting, an aerobic method, facilitates the breakdown of organic waste into a humus-like product suitable for improving soil quality and structure [9]. However, challenges like feedstock heterogeneity, varying moisture content, and potential presence of pathogens exist, requiring specific pretreatment methods and strict quality checks.

b. **Agricultural Residues:** These are secondary products created during farming operations like harvesting, processing, and handling crops. Examples include crop residues (e.g., straw, stalks, and husks), animal manure, and residues from agro-industries (e.g., bagasse, corn stover, and rice husks). These residues are widely available in farming areas, positioning them as excellent raw materials for making bio-based additives. They contain a large amount of lignocellulosic material that can be converted into various bio-additives. Techniques like pyrolysis and gasification, which use heat, can turn agricultural residues into bio-oil, syngas, and biochar. Bio-oil can be refined further into biofuels and chemical feedstocks, while syngas can be used for generating power or synthesizing valuable chemicals. Biochar, a carbon-rich solid residue, has great qualities for improving soil, capturing carbon, and holding water. Biological approaches, such as enzymatic hydrolysis and fermentation, can also be used to break down lignocellulose into sugars that can be fermented. These fermentable sugars can then be used to make bioplastics, biofuels (like ethanol), and other bio-based products [10]. The economic feasibility of using agricultural residues depends on factors like how available the feedstock is, the cost of transport, and how efficient the processing is. Pretreatment methods such as milling, delignification, and enzymatic hydrolysis are essential for making lignocellulose more accessible for later conversion steps.

c. **Industrial Byproducts:** These are materials produced during industrial processes and manufacturing. This includes organic waste streams (like wastewater, sludge, and organic residues) and non-organic waste streams (such as spent solvents, co-product gases, and ash). Their composition and properties differ significantly based on the industry and production method. Using industrial byproducts creates value, lowering disposal costs and contributing to a circular economy by substituting fossil resources with renewable options. However, the varied composition and quality require processing strategies and purification steps specific to each type [11].

d. **MSW:** MSW is composed of waste from households, businesses, and institutions in urban areas. Typically, MSW contains a mix of organic, recyclable, and non-recyclable items like food waste, paper, plastics, glass, and metals. Although MSW is challenging due to its varied composition and contamination, it does contain valuable organic parts. These can be turned into bio-based additives using modern waste-to-energy and waste-to-product techniques.

e. **Plastic Waste:** The buildup of plastic waste is a significant environmental problem. Recycling is the preferred approach, but a large amount of plastic waste cannot be recycled because of contamination or complex polymer combinations. Pyrolysis and catalytic methods provide promising ways to convert

Table 7.2 Typical feedstock properties for bio-additive production. Adapted from [13, 14].

Waste type	Moisture (%)	Volatile solids (%)	Ash (%)	C/N ratio	Notable contaminants
Food waste	70–85	80–95	2–5	15–25	Plastics, metals
Agricultural residue	10–20	70–85	5–15	30–80	Soil, pesticides
Sewage sludge	80–95	60–75	15–30	7–10	Heavy metals
Plastic waste (PE/PP)	<1	99–100	<1	–	Additives, dyes

specific kinds of plastic waste into bio-additives like fuel oils, waxes, and monomers. Pyrolysis heats plastic waste without oxygen, causing the polymer chains to break down into smaller molecules. These products can be processed further and refined into valuable chemicals and fuels. However, the resulting composition largely depends on the kind of plastic waste used and the process settings [12]. Additionally, worries about the environmental effects and safety of pyrolysis and its final product must be thoroughly examined using detailed life cycle assessment (LCA) and toxicity testing. Table 7.2 lists typical properties of waste feedstocks suitable for bio-additive production.

7.3 Waste-to-bio-additive Conversion Technologies

Producing bio-based additives from waste involves various technologies, each designed for specific waste types and target products. These methods can be broadly classified as biological, thermochemical, and physicochemical processes.

7.3.1 Biological Processes

Biological methods are important for transforming different waste materials into valuable bio-additives. They utilize the metabolic actions of microorganisms, enzymes, and biological catalysts to break down organic matter and convert it into bio-based products. Table 7.3 summarizes various biological processes along with their typical operating temperature and pH.

7.3.1.1 Anaerobic Digestion

AD is a biological process where organic waste is converted into biogas and digestate in an environment without oxygen. This happens naturally in places lacking oxygen, like wetlands, marshes, and the digestive tracts of animals. Dedicated AD facilities create carefully controlled, oxygen-free conditions crucial for microbes to convert various organic wastes, including food waste, agricultural residues, and sewage sludge, into useful biogas and nutrient-rich digestate. These facilities maintain controlled conditions to enhance the breakdown of organic matter by anaerobic microbes. The process involves several steps [16].

Table 7.3 Biological conversion process parameters. Adapted from [15].

Process	Temp. (°C)	pH	HRT (days)	Microbial community	Main product
Anaerobic digestion	35–55	6.8–7.5	15–30	Methanogens, acidogens	Biogas, digestate
Fermentation	28–37	4.5–6	1–5	Yeasts, lactic bacilli	Ethanol, acids
Composting	50–65	6.0–8	20–40	Thermophilic bacteria	Compost

As reported by Kang et al. [17], AD starts with essential feedstock preparation. This can include physical reduction in size, homogenization, or chemical/thermal treatments. These steps improve how easily the substrate is digested and remove substances that could inhibit microbial activity, thereby improving the performance of anaerobic microorganisms in later stages. Inside the digester, organic matter breaks down in four specific, sequential biochemical phases. First, hydrolysis breaks down complex organic compounds (carbohydrates, proteins, lipids) into simpler soluble substances such as sugars, amino acids, and fatty acids. Fermentative bacteria then convert these simpler molecules during acidogenesis into volatile fatty acids (VFAs), alcohols, hydrogen (H_2), and carbon dioxide (CO_2). In the third stage, acetogenesis, acetogenic bacteria convert VFAs and alcohols (excluding acetate) into acetate, H_2, and CO_2. This occurs via syntrophic relationships with hydrogenotrophic methanogens. Finally, methanogenesis, carried out by methanogenic archaea, completes the process. They transform acetate or H_2 and CO_2 into CH_4 and CO_2, which are the main constituents of biogas. The resulting biogas, usually containing 50–75% CH_4, serves as a renewable energy source. It can be burned directly to produce heat and power or upgraded to biomethane for use as vehicle fuel or injected into the gas grid. The remaining material, called digestate, is abundant in nutrients and organic matter. It is often further treated, for example, by dewatering to separate solids and liquids or post-composting for stability. After treatment, it can be used safely and effectively as a valuable organic fertilizer or soil improver, completing nutrient cycles and supporting circular economy goals.

7.3.1.2 Fermentation

Fermentation is a metabolic process where microorganisms like bacteria, yeast, or fungi convert organic materials into biochemicals or biofuels. While fermentation has a long history in food and beverage production, it is also applied in producing biofuels, pharmaceuticals, and chemicals. This process involves preparing and pretreating organic materials like sugars, starches, or lignocellulosic biomass to free up fermentable sugars or carbohydrates. Methods include enzymatic hydrolysis, acid or alkali treatment, or mechanical pretreatment. The treated material is then inoculated with specific microorganisms and kept under controlled conditions (temperature, pH, oxygenation) to allow fermentation to occur. The microbes process the sugars or carbohydrates in the material, creating desired biochemicals or biofuels as a result of their metabolism. This includes the following fermentation processes:

- **Ethanol Fermentation:** Yeast species, such as *Saccharomyces cerevisiae*, convert sugars into ethanol and CO_2 through alcoholic fermentation.
- **Lactic Acid Fermentation:** Lactic acid bacteria change sugars into lactic acid via anaerobic fermentation, commonly used in food preservation and manufacturing.
- **H_2 Fermentation:** Some bacteria generate H_2 gas via dark fermentation of organic materials, which could potentially be used as a renewable energy carrier.

Following fermentation, the bio-based product is separated from the liquid mixture using techniques like filtration, centrifugation, or distillation. The collected product might be further purified or processed before it is used [18].

7.3.1.3 Composting

Composting is a natural biological process that transforms organic waste into compost, a soil amendment rich in nutrients. It happens aerobically, with microbes like bacteria, fungi, and actinomycetes breaking down organic matter into stable, humus-like substances. Composting is a managed biological process using aerobic microbial decomposition to turn organic waste into a stable, humus-like soil improver. Pajura et al. [19] suggest that composting starts with choosing the feedstock. Organic waste materials like food waste, yard waste, and agricultural residues are selected based on important factors such as their carbon-to-nitrogen (C/N) ratio and moisture content; a proper balance of carbon-rich (e.g., straw, leaves) and nitrogen-rich (e.g., food scraps, manure) materials is essential for best microbial activity. The chosen materials are then arranged in layers or mixed to create a pile or windrow. This is actively aerated to ensure enough oxygen reaches all parts, maintaining necessary aerobic conditions throughout the process. Following this, biological breakdown happens as various microbes (bacteria, fungi, actinomycetes) process the organic matter using a sequence of biochemical reactions involving hydrolysis, oxidation, and mineralization. This metabolic activity generates heat, greatly increasing the temperature of the compost pile and speeding up the decomposition rate. Temperatures often reach thermophilic levels, which helps kill pathogens and weed seeds. The maturation and curing phase involves the ongoing conversion of leftover organic matter into stable humic substances over weeks or months. Factors like temperature, moisture, and feedstock makeup influence this stage. Regular turning or forced air circulation is usually done during this phase to promote even breakdown and avoid areas without oxygen.

Once the main composting phase is finished, the material is left to cure for several weeks. This improves its stability and nutrient content. Finally, the finished compost is often screened to remove larger pieces or impurities. This results in a high-quality product suitable for many uses in agriculture, horticulture, and landscaping. Table 7.4 outlines various advantages and disadvantages of these biological conversion processes.

Table 7.4 Advantages and disadvantages of biological conversion technologies. [20]/Elsevier.

Technology	Feedstock felicitous	Bio-additive	Advantages	Disadvantages
Anaerobic digestion	Organic waste	Digestate, biogas	Renewable energy for heat and power generation, nutrient recovery, reduction in waste volume, mitigation of GHGs	Slow process, requirement of biogas purification
Fermentation	Organic waste	Various chemicals	Biofuel and biochemical production, utilization of diverse organic substrates, reduction in GHGs, and potential for biorefinery integration	Require sterile conditions and product purification
Composting	Organic waste	Compost	Nutrient-rich soil amendment production, diversion of organic waste, enhancement in agricultural ecosystems, cost-effective, and environmentally sustainable	Space requirement, potential odor issues

Table 7.5 Thermochemical conversion technologies. Adapted from [24].

Process	Temp. (°C)	Pressure (bar)	Atmosphere	Main product	Energy yield (%)
Pyrolysis	400–700	1	Inert (N_2/Ar)	Bio-oil, biochar	60–75
Gasification	700–1000	1–30	Steam/O_2	Syngas	65–80
Hydrothermal carbonization	180–250	10–40	Subcritical H_2O	Hydrochar	50–70

7.3.2 Thermochemical Processes

Thermochemical processes use heat and/or pressure on organic waste to cause chemical reactions and create bio-based additives. They provide efficient and flexible ways to turn many different waste materials into valuable products like biochar [21, 22], syngas, and bio-oil. Table 7.5 summarizes three main thermochemical processes: pyrolysis, gasification, and hydrothermal processing [23]. These are discussed in the following sections.

7.3.2.1 Pyrolysis

Pyrolysis is a thermochemical process where organic materials are broken down into biochar, bio-oil, and syngas with no oxygen or very little oxygen present. This method heats the raw material to high temperatures (usually 300–800 °C) in an oxygen-poor setting. This causes the material to break down thermally and release volatile substances. The process uses organic waste such as biomass, agricultural residues, and plastic waste. These materials are prepared and pre-processed to make the pyrolysis process more efficient. Pretreatment might include shredding, grinding, or drying the feedstock to improve how it is handled and how well it converts. The prepped material is then put into a pyrolysis reactor, where it is heated to high temperatures without oxygen. As the feedstock breaks down under heat, volatile compounds like gases and vapors are released from the solid material. The volatile compounds produced are condensed and collected as bio-oil. The solid material left behind, called biochar, is recovered from the reactor. Based on the pyrolysis settings and the type of feedstock, syngas (a mix of H_2, carbon monoxide [CO], and CH_4) can also be generated as a secondary product. Biochar from pyrolysis can serve as a soil additive, a way to capture carbon, or a raw material for different industrial uses.

Bio-oil can be processed further and refined into transport fuels, chemicals, or specialized products. Syngas can be burned to create heat and power, or it can be converted catalytically into fuels and chemicals. Jha et al. [25] suggested that high-temperature pyrolysis of waste biomass without oxygen produces non-condensable gases (CO_2, CO, CH_4, and H_2), bioproducts, and bio-additives, as illustrated in Figure 7.2.

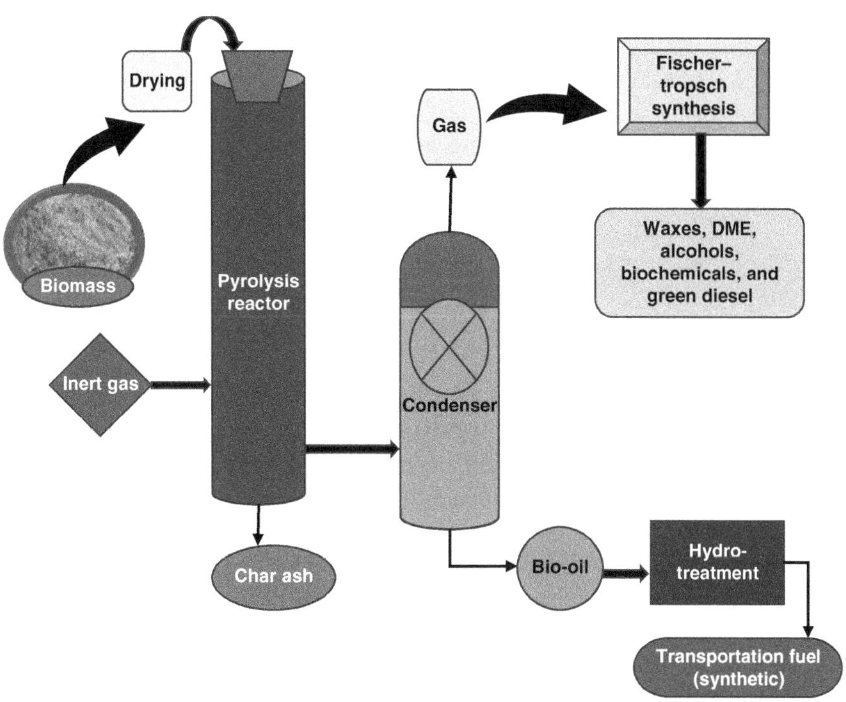

Figure 7.2 Schematic representation of pyrolysis process of converting biomass into bioproducts and bio-additives. Adapted from [25].

Therefore, this process includes the following advantages:

- Produces biochar, bio-oil, and syngas from various feedstocks, such as biomass, agricultural residues, and plastic waste.
- Decreases GHG emissions by sequestering carbon in biochar and replacing fossil fuels with renewable biofuels.
- Adds value to waste materials, creating valuable products with potential market uses.
- Offers flexibility to adjust process settings and product output according to the feedstock used and the desired final products.

7.3.2.2 Gasification

Gasification is a thermochemical technique that turns organic materials into syngas (synthesis gas), a combustible gas mixture, in a hot environment with little oxygen. This process is a flexible and effective way to convert many types of feedstocks into a clean, versatile fuel gas for producing heat, power, and chemicals. Santos et al. [26] reported that, similar to pyrolysis, gasification involves preparing and pre-processing organic waste materials like biomass, MSW, and coal to improve gasification results. Pretreatment can include drying, reducing size, and removing contaminants like metals and inert substances. The prepared material is fed into a gasifier and heated to high temperatures (usually 700–1,500 °C) with a controlled amount of oxygen or steam. Inside the gasifier, thermochemical reactions break the feedstock down into a gas mixture mostly containing H_2, CO, and CH_4. The raw syngas from gasification may have impurities like tars, particles, and sulfur compounds. These must be removed to meet quality standards. Syngas-conditioning methods like cooling, scrubbing, and filtration are used to clean and purify the syngas before it is used. The cleaned syngas can fuel heat and power generation in gas turbines, boilers, or internal combustion engines. Alternatively, syngas can undergo further processing through catalytic conversion or Fischer–Tropsch synthesis to create liquid fuels, chemicals, or H_2 gas. The process offers the following advantages:

- Produces clean, versatile syngas from a range of feedstocks, including biomass, waste, and coal.
- Allows flexibility in adjusting parameters to get the best syngas composition and energy content.
- Can be integrated with existing gas networks and conversion technologies for generating power and synthesizing chemicals.
- Potential for using carbon capture and utilization (CCU) to reduce GHG emissions and improve process sustainability.

7.3.2.3 Hydrothermal Processing

Hydrothermal processing is a thermochemical method using high temperature and pressure in water to convert organic materials into bio-oil, biochar, and aqueous phase products. This method provides a distinct way to add value to waste, especially for wet and biomass-heavy feedstocks. Feedstock preparation involves mixing organic wastes like sewage sludge, algae, and food waste with water to form a slurry or suspension [27]. Zhou et al. [28] reported that the feedstock might be pretreated with acids, bases, or catalysts to improve its solubility and encourage chemical reactions. The prepared slurry is then placed in a hydrothermal reactor and heated to 150–400 °C under high pressure (typically 5–25 MPa). In this hydrothermal setting, organic compounds undergo thermal breakdown, hydrolysis, and solvolysis reactions, yielding bio-oil, gases, and solid residues. The reaction products are cooled, and pressure is reduced to recover bio-oil, water-based products, and solid residues. Bio-oil is separated from the water phase and processed further to remove water and impurities. Solid residues, called hydrochar or hydrothermal carbon, can be collected and used to improve soil or as a carbonaceous material. Bio-oil from hydrothermal processing can be used as renewable fuel for heat and power or refined into transportation fuels via hydrotreating or catalytic cracking. The water-based products may contain valuable chemicals like organic acids, sugars, and phenolic compounds, which can be recovered and used for different purposes. Advantages linked to this process are as follows:

- Converts wet and biomass-heavy feedstocks into bio-based products under relatively mild conditions.
- Uses water effectively as a reaction medium, reducing wastewater production and resource use.
- Yields bio-oil high in energy and low in oxygen, suitable for direct burning or upgrading to more valuable fuels.
- Possibility to co-produce biochar- and water-based products, increasing resource recovery and efficiency.

Thermochemical processes like pyrolysis, gasification, and hydrothermal processing are efficient and adaptable ways to convert various waste materials into valuable bio-based additives. They allow for making biochar, bio-oil, and syngas from a variety of feedstocks, supporting waste valorization, renewable energy generation, and resource reclamation.

7.3.3 Physiochemical Processes

Physicochemical processes use physical and chemical techniques to separate, extract, or change organic waste materials into bio-based additives. They employ different chemical reactions and catalysts to turn waste into valuable products like biofuels, biochemicals, and platform chemicals. The three primary physicochemical processes are hydrolysis, transesterification, and acid/base catalysis.

7.3.3.1 Hydrolysis

Hydrolysis is a chemical process that uses water molecules to break chemical bonds in organic compounds. For converting waste-to-bio-additives, hydrolysis is often used to break down complex organic molecules such as polysaccharides, proteins, and lipids into simpler substances such as sugars, amino acids, and fatty acids. Organic wastes like lignocellulosic biomass, animal fats, and protein-heavy residues are prepared and pretreated to make them easier to hydrolyze. Pretreatment can involve reducing size, chemical treatment (e.g., using acid or alkali), or enzymatic hydrolysis. The treated material is mixed with water and then hydrolyzed under acidic, alkaline, or enzymatic conditions. For acid hydrolysis, acids like sulfuric acid or hydrochloric acid catalyze the breaking of glycosidic bonds in cellulose and hemicellulose, producing sugars like glucose and xylose. Alkaline hydrolysis uses bases such as sodium hydroxide or potassium hydroxide to hydrolyze ester bonds in fats and proteins, freeing up fatty acids and amino acids. Enzymatic hydrolysis utilizes enzymes like cellulase, hemicellulose, or protease to catalyze specific biochemical reactions and help break down the substrate. The hydrolysis reaction yields a mix of products, including sugars, amino acids, and fatty acids, plus soluble secondary products like salts and organic acids. The products are separated from the reaction mix using techniques including filtration, centrifugation, or chromatography. The collected sugars, amino acids, and fatty acids can be processed further into biofuels, biochemicals, or platform chemicals via subsequent conversion stages. This process offers the following advantages [29]:

- Breaks down complex organic compounds efficiently into simpler, more usable forms.
- Can hydrolyze a variety of feedstocks, such as lignocellulosic biomass, fats, oils, and proteins.
- Works with acid, alkaline, and enzymatic catalysis, allowing adjustment of process conditions and product yields.
- Integrates with later processing methods for making bio-based fuels, chemicals, and materials.

7.3.3.2 Transesterification

Transesterification is a chemical process exchanging ester groups between organic molecules, usually between an alcohol (like methanol or ethanol) and a triglyceride (like vegetable oil or animal fat). For converting waste-to-bio-additives, transesterification produces biodiesel from feedstocks high in lipids, such as waste cooking oil, animal fats, and algae biomass [30–32]. Lipid-heavy feedstocks like waste cooking oil, grease, or algae biomass are collected and pretreated to remove impurities such as water, free fatty acids, and particles. Pretreatment can involve filtration, centrifugation, or acid treatment to improve the quality and purity of the feedstock. The treated lipid feedstock is mixed with an alcohol (e.g., methanol or ethanol) and a catalyst (e.g., sodium hydroxide or potassium hydroxide) inside a reactor [33]. At controlled temperatures and pressures, the triglycerides in the feedstock react with the alcohol and catalyst, forming fatty acid alkyl esters (biodiesel) and glycerol. The reaction proceeds through an intermediate phase where fatty acid alkyl ester intermediates are formed, which then react further to produce biodiesel and glycerol. Once the reaction is finished, the mixture is left to settle, and the biodiesel layer is separated from the glycerol layer. The biodiesel is washed to remove leftover catalyst and impurities, producing a high-quality fuel [34]. The glycerol co-product [35], containing unreacted alcohol and catalyst, can be processed further to purify glycerol for applications in pharmaceuticals, cosmetics, or chemical synthesis [36]. Major advantages linked to this process include:

- Makes biodiesel from renewable sources, lowering reliance on fossil fuels and reducing GHG emissions.
- Works with many lipid-rich feedstocks, such as waste cooking oil, animal fats, and algae biomass.
- Uses simple conditions and relatively low temperatures, decreasing energy use and processing costs.
- Produces glycerol as a valuable co-product, improving process efficiency and economic feasibility.

7.3.3.3 Acid/Base Catalysis

This process uses acids or bases as catalysts to speed up chemical reactions and help convert organic compounds into desired products. For converting waste-to-bio-additives, acid/base catalysis is used in processes like esterification, dehydration, and hydrolysis to make bio-based fuels, chemicals, and materials. Acid or base catalysts can be chosen for catalytic conversion [37], depending on the needed reaction and feedstock makeup. Acid catalysts like sulfuric acid or hydrochloric acid are used in esterification, alcohol dehydration, and cellulose hydrolysis [38–40]. Base catalysts such as sodium hydroxide or potassium hydroxide are used in transesterification, saponification of fats and oils, and aldol condensation. The catalyst is added to the reaction mixture in a suitable solvent or medium, and the reaction takes place under controlled temperature, pressure, and time. Acid catalysis usually happens at lower temperatures and might need more time to reach conversion. Base catalysis works at higher temperatures with faster reaction rates. The reaction catalyzed by acid or base goes through intermediate species and transition states, resulting in desired products like esters, ethers, aldehydes, or ketones. The catalyst helps the reaction by lowering the energy barrier and stabilizing intermediates, thus speeding up the overall conversion process [41]. Once the reaction is finished, the product mixture is separated from the catalyst and solvent using techniques like extraction, distillation, or chromatography. Collected products might be further purified or processed to meet quality standards and market needs. Advantages of acid/base catalysis include:

- Converts organic waste into value-added products with high selectivity and efficiency using catalysis.
- Allows control over reaction conditions and catalyst properties for best process performance and product output.
- Can catalyze many chemical reactions, such as esterification, transesterification, dehydration, and hydrolysis.
- Potential for recycling and reusing the catalyst to minimize waste and improve process sustainability.

Physicochemical processes like hydrolysis, transesterification, and acid/base catalysis are effective and adaptable ways to convert organic waste into bio-based additives. They make it possible to produce biofuels, biochemicals, and platform chemicals from various feedstocks, supporting waste valorization, resource recovery, and sustainable development. These products are useful due to their valuable chemical composition, as summarized in Table 7.6.

Table 7.6 Chemical composition of selected bio-additives.

Additive	C (%)	H (%)	N (%)	O (%)	Ash (%)	Typical use	Ref.
Biochar	50–80	1–4	0.5–2	10–30	5–15	Soil amendment	[42]
Digestate	30–40	4–6	2–6	30–50	10–20	Fertilizer	[43]
Compost	35–45	4–6	1–3	25–40	10–15	Soil conditioner	[44]

7.4 Applications of Bio-additives

The conversion of waste into bio-additives opens up many applications across various sectors, supporting sustainable development, resource efficiency, and environmental protection. There are several key applications and uses of bio-additives derived from waste materials:

1. **Energy Generation:** Bio-additives from waste can be used as renewable energy sources for generating heat and power. Biofuels like biodiesel, bioethanol, and biogas can be burned in boilers, engines, or turbines to create heat, electricity, or mechanical energy. Energy from biofuels can replace fossil fuels, lower GHG emissions, and boost energy security and independence. Furthermore, making bioenergy from waste helps valorize waste and recover resources, turning waste into a valuable energy source [45].
2. **Transportation Fuels:** Biofuels made from waste offer sustainable options to traditional fossil fuels in transportation. Biodiesel, made from waste cooking oil, animal fats, or algae biomass, can be mixed with diesel or used alone in diesel engines. Bioethanol, derived from agricultural residues, food waste, or lignocellulosic biomass, can be mixed with gasoline or used in vehicles designed for flex fuel. It reduces carbon emissions, improves air quality, and increases energy resilience, aiding climate change efforts and sustainable transportation [46].
3. **Soil Amendment:** Bio-additives made from organic waste, like compost, biochar, and organic fertilizers, are valuable soil improvers that enhance soil health, fertility, and productivity. Compost, made by biological breakdown of organic waste, adds organic matter, nutrients, and helpful microbes to soil, improving soil structure, water-holding capacity, and nutrient cycling. Biochar, created from biomass pyrolysis, enhances soil structure, nutrient retention, and carbon storage, supporting sustainable farming and helping to combat climate change. Organic fertilizers, sourced from waste materials like animal manure, food waste, or sewage sludge, supply key nutrients for plants while lessening the need for synthetic fertilizers and chemicals [47].
4. **Waste Management Solutions:** Bio-additives are vital in waste management. They divert organic waste from landfills and support resource recovery and circular economy ideas. AD of organic waste yields biogas and digestate. These can be used for producing energy and improving soil, respectively. Composting organic waste results in compost, a valuable soil conditioner and source of nutrients (Figure 7.3) used in agriculture and landscaping [48]. Thermochemical

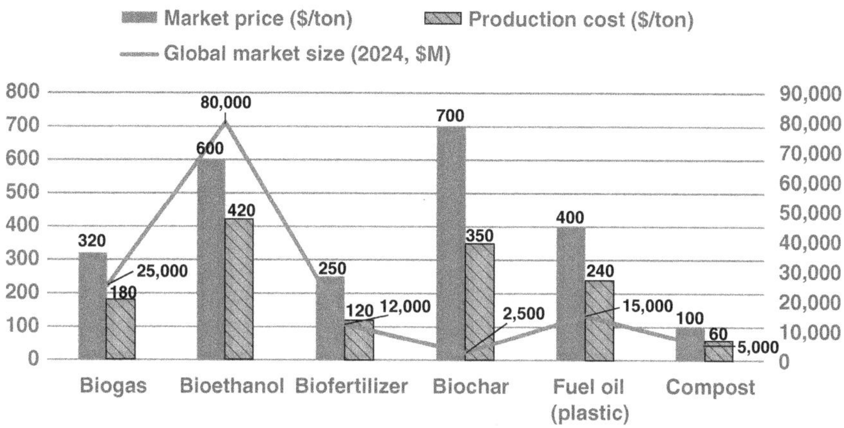

Figure 7.3 Economic and market assessment of waste-to-bio-additive technologies. Adapted from [49].

conversion of waste materials such as biomass, plastics, and organic residues creates biochar, bio-oil, and syngas. These can be used for generating energy, synthesizing chemicals, and valorizing waste.

By transforming waste into bio-additives, these methods help reduce waste, prevent pollution, and support sustainable waste management.

5. **Industrial Processes:** Bio-additives made from waste materials are used in various industrial processes, such as chemical synthesis, making materials, and treating wastewater. Chemicals sourced from waste feedstocks offer sustainable substitutes for petrochemical-based products in industries like pharmaceuticals, cosmetics, and food processing. Bio-based materials like bioplastics, bio-composites, and other biomaterials provide eco-friendly options instead of traditional plastics and synthetics. This lowers reliance on fossil resources and lessens environmental effects. Moreover, bio-additives are applicable in wastewater treatment for removing nutrients, controlling odor, and managing sludge, helping to improve water quality and environmental sustainability.

In summary, bio-additives from waste materials have diverse and numerous applications, covering energy generation, transportation fuels, soil amendment, waste management, and industrial processes. By using the potential of waste materials as valuable resources, bio-additives support sustainable development, protect the environment, and aid the shift toward a circular economy.

7.5 LCA of Waste-of-building Applications Technologies

LCA is a useful method for evaluating how environmentally sustainable waste-to-building application technologies are. LCA assesses a product's or process's environmental effects across its whole life cycle, from getting raw materials to disposal. The growing area of Waste-to-Building Applications technologies shows potential for recovering resources and making sustainable building materials [50]. However, their real environmental feasibility depends on a thorough LCA that carefully examines the many effects at each step of their lifecycle. This review examines crucial factors that should be included in WtBA LCA studies to get accurate and dependable evaluations of environmental performance.

1. **Energy Consumption:** A detailed LCA needs to measure the total energy required for WtBA processes. This includes not only the direct energy needed for the main conversion technology (e.g., pyrolysis, gasification, enzymatic hydrolysis) [51] but also the energy used for supporting operations. These include:
   - Thermal Energy: Heating needed for feedstock pretreatment (drying, grinding), running the reactor (maintaining temperatures), and purifying the product (e.g., distillation, evaporation). Detailed thermal modeling is essential for correctly estimating energy needs and finding potential ways to integrate heat for better efficiency.
   - Mechanical Energy: Energy used for mixing, stirring, and moving feedstock within the process equipment. Motor efficiency and pump features should be included to estimate energy use accurately.
   - Separation Processes: Energy-heavy separation methods, like membrane filtration, adsorption, and solvent extraction, are often important in WtBA processes for purifying the desired material. A full assessment of the energy used by these separation steps is vital.
   - Feedstock Pre-processing: Energy used for collecting, transporting, sorting, and reducing the size of the waste material.
2. **Emissions:** Detailed accounting of emissions is vital for assessing the environmental impact of waste-to-building application technologies. The LCA needs to include emissions to air, water, and soil, accounting for direct and indirect sources throughout the lifecycle [52]. Important emission categories that need careful evaluation are as follows:
   - GHGs: Measuring the amounts of CO_2, CH_4, N_2O, and other GHGs released from the conversion process, burning of co-products, and energy production. Carbon accounting must include both biogenic and fossil carbon sources. Air Pollutants: Assessing criteria air pollutants (e.g., NOx, SOx, particulate matter, VOCs) released from the conversion process and related energy sources. This needs detailed modeling of the process and emission data to predict effects on air quality. Additionally, specific toxic air pollutants generated based on the waste material and technology used should not be ignored. Water Pollutants: Describing the wastewater produced during conversion, including organic pollutants, heavy metals, and other contaminants. The environmental effect of discharging wastewater, even after treatment, needs to be assessed. Soil Contaminants: Analyzing possible soil pollution from disposing of solid waste or accidental leaks of process chemicals.Figure 7.4 illustrates the environmental impact of various processes.
3. **Waste Generation:** The LCA needs to carefully record all waste streams (solid, liquid, and gaseous [54]) produced by the WtBA process. This includes:

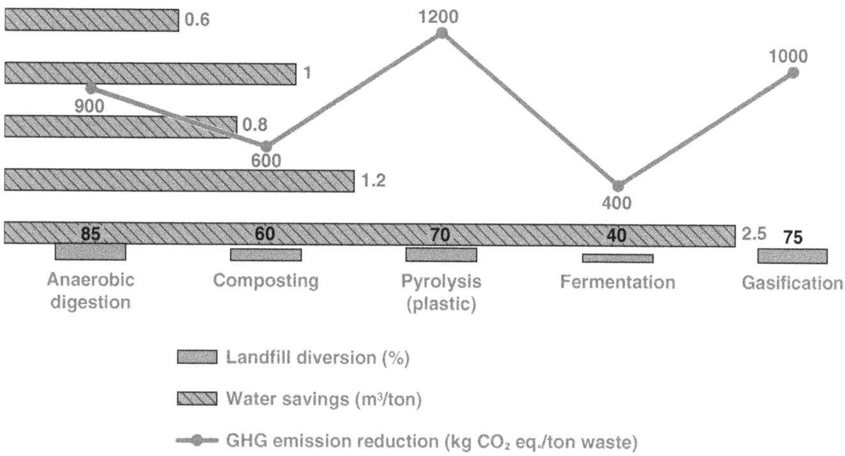

Figure 7.4 The environmental impact assessment of various processes. Adapted from [53].

- Solid Waste: Assessing ash, sludge, used catalysts, and other solid waste produced during the conversion. The disposal method (landfill, incineration, etc.) and related environmental impacts must be evaluated.
- Liquid Waste: Describing wastewater streams, including their volume, makeup, and potential for treatment and reuse. How well wastewater treatment methods remove pollutants should be considered.
- Gaseous Waste: A detailed understanding of what makes up and how to manage off-gases produced during conversion. Capturing and using these gases (e.g., for energy or chemical synthesis) is important, while also considering the lifecycle environmental impacts of the related technologies.
- Residue Management: Evaluating the environmental and economic viability of using co-products and residues for other applications, reducing waste disposal.

4. **Resource Depletion:** The LCA needs to account for natural resources used [55] throughout the WtBA process, focusing on:
   - Water Consumption: Measuring water used for cooling, process operations, and cleaning. The effects of water scarcity in the area where the WtBA facility is located should be considered.
   - Mineral Resources: Assessing the use of minerals and metals in building equipment, catalysts, and other process materials. The depletion of crucial raw materials should be emphasized.
   - Land Use: Evaluating the land needed for the WtBA facility, waste storage, infrastructure, and potential effects on biodiversity and ecosystem services.
   - Fossil Fuel Depletion: While WtBA often seeks to replace fossil fuels, it's important to account for fossil fuels used within the process, either directly for energy or indirectly for making supporting materials.

By carefully considering these factors within a full LCA framework, the environmental sustainability of waste-to-building application technologies can be accurately judged, helping to guide informed decisions and promote the creation of truly eco-friendly building materials. Also, sensitivity analysis should be used to pinpoint key factors affecting LCA results. This allows for focused optimization strategies to boost environmental performance. Including social and economic factors, along with environmental ones, will give a more complete sustainability assessment of waste-to-building application technologies.

7.6 Challenges and Limitations

Converting waste into bio-additives provides many environmental advantages. However, several challenges and limitations prevent widespread adoption and scaling up of waste-to-bio-additive technologies. Here, we cover the main challenges concerning feedstock availability and quality, the maturity and efficiency of the technology, market acceptance and infrastructure, as well as policy and regulatory hurdles.

1. **Feedstock Availability and Quality:** Issues with the availability and quality of feedstock materials pose considerable challenges for waste-to-bio-additive conversion processes. Getting a steady and sufficient supply of waste feedstocks, including agricultural residues, organic waste, and biomass, can be restricted by seasonal changes, location limitations, and competition from other industries [56]. Furthermore, differences in feedstock composition, moisture levels, and contaminants can impact process efficiency, the quality of the product, and its later uses. However, challenges like insufficient feedstock supply and quality control can lead to operational problems, inefficient production, and reduced product performance. Difficulty accessing appropriate feedstocks can limit how scalable and economically feasible waste-to-bio-additive projects are, especially in areas with limited resources or poor infrastructure.
2. **Technological Maturity and Efficiency:** The maturity and efficiency of technologies for converting waste-to-bio-additives differ greatly among various technologies and applications. While some conversion methods, like AD and composting, are commercially mature and widely used, others, such as thermochemical and biochemical processing, may still encounter technical difficulties, scaling problems, and performance limits. Lower technological maturity and efficiency can lead to high setup and running costs, low conversion rates, and product quality that is not optimal. Technical obstacles like designing reactors, optimizing processes, and developing catalysts can slow down the commercialization [57] and use of new waste-to-bio-additive technologies, delaying their adoption by the market and industry. The technology performance is depicted in Table 7.7.

   Major engineering challenges along with their solutions are presented in Table 7.8.
3. **Market Acceptance and Infrastructure:** Market acceptance and the readiness of infrastructure for bio-additives made from waste pose significant challenges for efforts to valorize waste. Despite the environmental and economic

Table 7.7 Process performance indicators.

Process	Energy efficiency (%)	GHG reduction (%)	Typical conversion rate (%)	Ref.
Anaerobic digestion	70–85	60–80	60–75	[58]
Pyrolysis	60–75	50–70	45–65	[59]
Composting	40–60	30–50	50–70	[60]

Table 7.8 Engineering challenges and solutions. Adapted from [61].

Challenge	Impact	Engineering solution
Feedstock heterogeneity	Process instability	Pre-sorting, blending, pretreatment
High moisture content	Low energy efficiency	Dewatering, drying
Contaminant removal	Product quality issues	Advanced filtration, separation
Scale-up	Economic feasibility	Modular reactor design

benefits of bio-based products, market demand and consumer acceptance may differ based on factors like product performance, cost competitiveness, and regulations [62]. Low market demand and insufficient infrastructure for waste-to-bio-additive products can slow down market entry and make them less attractive for investment. Weak market demand and distribution networks can discourage investors and industry players from putting money into waste valorization projects, resulting in available technologies and resources not being fully used.

4. **Policy and Regulatory Barriers:** Policy and regulatory hurdles can impede the progress, use, and market entry of waste-to-bio-additive technologies. Regulations that are not consistent or are out of date, permitting rules, and environmental standards can create uncertainty and make it hard for waste valorization projects to comply. Furthermore, insufficient supporting policies, incentives, and funding schemes can deter investment and innovation in the waste-to-bio-additive sector. Uncertainty in regulations and the cost of compliance can be major barriers for waste valorization projects to start, especially for small- and medium-sized enterprises (SMEs) and new companies. Lack of sufficient policy support and market incentives can slow down industry growth and investment in waste-to-bio-additive technologies, thereby limiting how much they contribute to sustainable development goals and circular economy aims [63]. To summarize, while waste-to-bio-additive technologies hold promise for managing waste, recovering resources, and producing sustainably, they face significant challenges and limitations that need to be overcome to fully achieve their potential. Addressing these hurdles will require cooperation among industry, government, and academia. Efforts are needed to improve feedstock availability and quality, increase technological maturity and

efficiency, encourage market acceptance and build infrastructure, and put supportive policies and regulations in place. By tackling these challenges, waste-to-bio-additive technologies can be crucial in moving toward a more sustainable and circular economy.

7.7 Case Studies and Success Stories

7.7.1 Successful Waste-to-bio-additive Projects:

- ***Case Study 1:*** *The Waste to Energy Plant in Vienna, Austria*
  The Waste to Energy Plant in Vienna is an example of a successful waste-to-bio-additive project. The plant uses advanced thermal treatment methods to turn MSW into heat, electricity, and district heating. Additionally, Kladnik et al. [64] reported that the plant makes biochar from organic waste, which is used to improve soil in farming and landscaping. The project has greatly decreased the city's dependence on landfills, reduced GHG emissions, and offered a sustainable way to handle municipal waste.
- ***Case Study 2****: The Zero Waste Program in San Francisco, USA*
  San Francisco's Zero Waste Program is another successful waste-to-bio-additive initiative. With thorough efforts in waste reduction, recycling, and composting, the city has reached a diversion rate over 80%, redirecting most of its waste away from landfills. Organic waste from homes and businesses is composted at modern facilities, yielding high-quality compost used for urban farming, community gardens, and restoration. The program has changed waste management methods, lessened environmental effects, and generated economic opportunities in the local green economy [65].
- ***Case Study 3:*** *The Biogas Plant in Güssing, Austria*
  The Biogas Plant in Güssing, Austria, demonstrates practical uses of waste-to-bio-additive technologies. The plant employs AD to transform organic waste, agricultural residues, and energy crops into biogas, a renewable energy source. The biogas is upgraded to biomethane and either put into the natural gas network or used as a fuel for vehicles. Furthermore, the digestate resulting from the AD process is used as a fertilizer rich in nutrients in farming [66]. The project has lowered GHG emissions, improved energy security, and aided rural development locally.
- ***Case Study 4:*** *The Waste-to-Energy Initiative in Singapore*
  Singapore's Waste-to-Energy Initiative offers useful insights and effective methods for waste-to-bio-additive projects. The city-state has put in place a thorough waste management approach that includes waste-to-energy incineration plants, recycling facilities, and initiatives to divert waste from landfills. Through public education, involving stakeholders, and using regulatory incentives, Singapore has successfully managed its waste stream, minimized landfilling, and maximized the recovery of resources. The initiative shows the importance of using holistic methods, involving stakeholders, and having policy support to achieve sustainable waste management results [67].

7.7.2 Future Directions and Emerging Trends

1. **Technological Innovations:** Advances in waste conversion technologies include ongoing research and development that can result in technological innovations in waste-to-bio-additive conversion. New technologies like hydrothermal liquefaction, microwave-assisted pyrolysis, and enzymatic hydrolysis show potential for converting various waste feedstocks into bio-additives more efficiently and affordably. Incorporating new reactor designs, catalysts, and process optimization methods will improve process efficiency, product quality, and how easily the technology can be scaled up [68]. Moreover, using digital and automation technologies in waste-to-bio-additive projects is expected to simplify operations, enhance process control, and make better use of resources. Advanced sensors, data analysis, and machine learning algorithms can offer real-time monitoring and predict maintenance needs, which lowers downtime and running costs. Robotics and automated systems can improve how waste materials are sorted, handled, and processed, boosting throughput and efficiency.
2. **Integration with Circular Economy Models:** As summarized in Table 7.9, WtBA can fit well into a circular economy framework. Combining waste-to-bio-additive projects with circular economy models will be more and more crucial as we move toward a more sustainable society that uses resources efficiently. Systems that close the loop by maximizing resource recovery, minimizing waste, and promoting material reuse and recycling will become more popular. Potential synergies between waste valorization, making renewable energy, and sustainable farming will be investigated to build circular value chains and improve how resources are used in a loop [69]. Using a lifecycle approach for designing, making, and managing products at the end of their life will encourage the inclusion of bio-additives made from waste into circular economy models. Creating products with features like recyclability, biodegradability, and cradle-to-cradle principles will make it easier to include bio-based materials and additives sourced from waste. Collaborative efforts among industry, academia, and policymakers will be necessary to develop standards, certifications, and regulations that support circularity and sustainability.
3. **Research and Development Priorities:** Feedstock diversification means research will concentrate on using a wider variety of feedstock sources and

Table 7.9 Process integration and circular economy potential.

Integration strategy	Waste input	Output streams	Circular economy benefit	Ref.
Biogas + compost	Organic waste	Biogas, digestate	Energy + fertilizer	[70]
Pyrolysis + bio-oil upgrading	Agri/plastic waste	Bio-oil, biochar, syngas	Renewable fuel + soil health	[71]
Fermentation + bioplastics	Food/agro-waste	Ethanol, lactic acid	Bioplastics, green solvents	[72]

improving supply chains to make waste-to-bio-additive projects more robust and sustainable. Looking into less common feedstocks like algae biomass, lignocellulosic residues, and industrial byproducts will increase the available feedstocks and lessen competition with food crops. Regarding process optimization, research will keep working to refine waste conversion technologies and improve efficiency, yield, and how easily they can be scaled up. Developing integrated biorefinery methods that produce several bio-additives from different waste streams will maximize resource use and create more value [73]. Improving pretreatment, conversion, and separation processes will reduce energy use, waste produced, and environmental effects.

4. **Market Opportunities and Growth Potential:** Increasing demand for sustainable products, renewable energy, and circular solutions will open up market opportunities for bio-additives made from waste. Industries like agriculture, construction, packaging, and consumer goods will look for bio-based options instead of materials and chemicals from fossil sources. This will drive market growth and diversification. Emerging markets in developing countries will offer significant growth potential for waste valorization technologies, driven by factors like urbanization, industrialization, and environmental concerns [74]. Creating high-value bio-additives with specific properties and uses will open up new market opportunities and sources of income from value-added products. Bio-based additives for functional materials, specialized chemicals, and advanced composites will serve niche markets and high-end segments, allowing for higher prices and profits. Working together with end-users and partners along the value chain will be crucial for identifying market needs, confirming product performance, and speeding up market acceptance.

Therefore, the future of waste-to-bio-additive projects depends on technological innovation, integration into the circular economy, research and development, and market expansion. By adopting these trends and priorities, stakeholders can fully realize the potential of waste valorization, promote sustainable development, and establish a circular and regenerative economy.

7.8 Conclusion and Recommendations

This study of waste-to-bio-additive projects has identified several key findings. Waste-to-bio-additive technologies provide sustainable ways to manage organic waste, decrease GHG emissions, and support resource recovery. Successful projects globally show that waste valorization initiatives are feasible and effective, bringing environmental, economic, and social benefits. Technological advances, integration with circular economy models, and research focus areas are influencing the future of waste-to-bio-additive projects, promoting market growth and sustainability. Challenges like feedstock availability, technological maturity, market acceptance, and policy barriers must be addressed to fully realize the potential of waste valorization. Waste-to-bio-additive projects have important consequences for sustainable development:

- **Environmental Benefits:** Waste-to-bio-additive projects help conserve the environment and combat climate change by diverting organic waste from landfills, reducing GHG emissions, and promoting resource recovery.
- **Economic Opportunities:** Creating bio-based products and renewable fuels from waste materials generates economic opportunities, encourages innovation, and supports job creation in the green economy.
- **Social Impacts:** Waste valorization projects increase community resilience, improve public health, and empower local people through methods that encourage participation and inclusive development.

To support the growth and scaling of waste-to-bio-additive projects, policymakers should implement supportive policies, regulations, and incentives. These measures would promote waste valorization, renewable energy production, and circular economy initiatives, invest in building infrastructure, research and development, and capacity to make waste-to-bio-additive projects more resilient and sustainable, and encourage cooperation among government, industry, academia, and civil society to tackle regulatory hurdles, market issues, and knowledge gaps in waste valorization. To move the field of waste-to-bio-additive projects forward, the following research areas are recommended:

- Investigate new feedstock sources, conversion technologies, and product uses to broaden the scope and influence of waste valorization efforts.
- Improve process efficiency, product quality, and how easily technologies can be scaled up through technological innovations, optimizing processes, and integrating digital tools and automation.
- Perform research across different fields on the environmental, economic, and social effects of waste-to-bio-additive projects. This research should guide policy decisions, investment plans, and how stakeholders are involved.

Waste-to-bio-additive projects offer a promising way to achieve sustainable development goals, support circular economy principles, and address global environmental issues. By adopting innovation, fostering collaboration, and gaining policy support, stakeholders can fully use the potential of waste valorization and build a future that is more resilient, inclusive, and sustainable for future generations.

References

1 Mitra, P. and Boruah, A. Synthesizing bio-additive from bio-wastes for the effective cement settling process. *Mater. Today: Proc.* vol. 60, pp. 2250–2253, 2022.

2 Brenner, M. and Weichold, O. Poultry feather waste as bio-based cross-linking additive for ethylene propylene diene rubber. *Polymers (Basel).* 2021. https://doi.org/10.3390/polym13223908.

3 Romani, A., Suriano, R., and Levi, M. Biomass waste materials through extrusion-based additive manufacturing: a systematic literature review. *J. Clean. Prod.* vol. 386, p. 135779, 2023.

4 Tiwari, R., Agrawal, P.S., Belkhode, P.N., Ruatpuia, J.V.L., and Rokhum, S.L. Hazardous effects of waste transformer oil and its prevention: a review. *Next Sustain*. vol. 3, p. 100026, 2024.
5 Getahun, T., Gebrehiwot, M., Ambelu, A., Van Gerven, T., and Van Der Bruggen, B. The potential of biogas production from municipal solid waste in a tropical climate. *Environ. Monit. Assess*. 2014. https://doi.org/10.1007/s10661-014-3727-4.
6 Bushra, B. and Remya, N. Biochar from pyrolysis of rice husk biomass—characteristics, modification and environmental application. *Biomass Conv. Bioref.* vol. 14, pp. 5759–5770, 2024.
7 Valizadeh, S., Valizadeh, B., Seo, M.W., Choi, Y.J., Lee, J., Chen, W.H., et al. Recent advances in liquid fuel production from plastic waste via pyrolysis: emphasis on polyolefins and polystyrene. *Environ. Res*. 2024. https://doi.org/10.1016/j.envres.2024.118154.
8 Siles-Castellano, A.B., López-González, J.A., Suárez-Estrella, F., López, M.J., Jurado, M.M., and Estrella-González, M.J. Compost quality and sanitation on industrial scale composting of municipal solid waste and sewage sludge. *Appl. Sci*. vol. 11, p. 7525, 2021.
9 Ariunbaatar, J., Panico, A., Esposito, G., Pirozzi, F., and Lens, P.N.L. Pretreatment methods to enhance anaerobic digestion of organic solid waste. *Appl. Energy*. vol. 123, pp. 143–156, 2014.
10 Sánchez, M.E., Lindao, E., Margaleff, D., Martínez, O., and Morán, A. Pyrolysis of agricultural residues from rape and sunflowers: production and characterization of bio-fuels and biochar soil management. *J. Anal. Appl. Pyrolysis*. vol. 85, pp. 142–144, 2009.
11 Sarode, D.D. Use of industrial waste for value-added products. In: S. Verma, R. Khan, M. Mili, S.A.R. Hashmi, and A.K. Srivastava (Eds.) *Advanced Materials from Recycled Waste*. Elsevier, 2022, pp. 179–198.
12 Papari, S., Bamdad, H., and Berruti, F. Pyrolytic conversion of plastic waste to value-added products and fuels: a review. *Materials (Basel)*. 2021. https://doi.org/10.3390/ma14102586.
13 Rashwan, A.K., Bai, H., Osman, A.I., Eltohamy, K.M., Chen, Z., Younis, H.A., et al. Recycling food and agriculture by-products to mitigate climate change: a review. *Environ. Chem. Lett*. vol. 21, pp. 3351–3375, 2023.
14 Mishra, K., Siwal, S.S., Nayaka, S.C., Guan, Z., and Thakur, V.K. Waste-to-chemicals: green solutions for bioeconomy markets. *Sci. Total Environ*. 2023. https://doi.org/10.1016/j.scitotenv.2023.164006.
15 Kataya, G., Cornu, D., Bechelany, M., Hijazi, A., and Issa, M. Biomass waste conversion technologies and its application for sustainable environmental development—a review. *Agronomy*. vol. 13, p. 2833, 2023.
16 Piadeh, F., Offie, I., Behzadian, K., Rizzuto, J.P., Bywater, A., Córdoba-Pachón, J.R., et al. A critical review for the impact of anaerobic digestion on the sustainable development goals. *J. Environ. Manage*. 2024. https://doi.org/10.1016/j.jenvman.2023.119458.
17 Kang, A.J. and Yuan, Q. Enhanced anaerobic digestion of organic waste. In: F.C. Mihai (Ed.) *Solid Waste Management in Rural Areas*. InTechOpen, 2017.

18 Esercizio, N., Lanzilli, M., Vastano, M., Landi, S., Xu, Z., Gallo, C., et al. Fermentation of biodegradable organic waste by the family thermotogaceae. *Resources*. vol. 10, p. 34, 2021.
19 Pajura, R. Composting municipal solid waste and animal manure in response to the current fertilizer crisis - a recent review. *Sci. Total Environ.* 2024. https://doi.org/10.1016/j.scitotenv.2023.169221.
20 Hussain, C. and Hait, S. (Editors) *Advanced Organic Waste Management: Sustainable Practices and Approaches*. Elsevier, 2022.
21 Agrawal, P. and Rokhum, S.L. Supported metal/metal oxide catalysts in biodiesel production. In: *Biodiesel Production* [Internet]. John Wiley & Sons, Ltd, 2022, pp. 119–141. Available from: https://onlinelibrary.wiley.com/doi/abs/10.1002/9781119771364.ch7
22 Ao, S., Rashid, U., Shi, D., Rokhum, S.L., Thuy, L.T., Alahmadi, T.A., et al. Synthesis and utilization of biomass-derived sulfonated heterogeneous catalyst-BT-SO3H for microalgal biodiesel production. *Environ. Res.* 2024. https://doi.org/10.1016/j.envres.2023.118025.
23 Tiwari, R., Agrawal, P.S., Belkhode, P.N., Brijpuriya, D., Devase, A., and Samarth, P. Utilization of waste transformer oil for the synthesis of biodiesel and glyptal. In: K.D. Yadav, N.D. Jariwala, A. Kumar, and A. Sinha (Eds.) *Recent Advances in Sustainable Waste Management Practices*. Lecture Notes in Civil Engineering. Springer, 2024, pp. 85–97.
24 Chandraratne, M.R. and Daful, A.G. Recent advances in thermochemical conversion of biomass. In: *Recent Perspectives in Pyrolysis Research*. IntechOpen, 2022, 155–164.
25 Jha, S., Nanda, S., Acharya, B., and Dalai, A.K. A review of thermochemical conversion of waste biomass to biofuels. *Energies*. vol. 15, p. 6352, 2022.
26 dos Santos, R.G. and Alencar, A.C. Biomass-derived syngas production via gasification process and its catalytic conversion into fuels by Fischer Tropsch synthesis: a review. *Int. J. Hydrogen Energy*. vol. 45, pp. 18114–18132, 2020.
27 Basar, I.A., Liu, H., Carrere, H., Trably, E., and Eskicioglu, C. A review on key design and operational parameters to optimize and develop hydrothermal liquefaction of biomass for biorefinery applications. *Green Chem.* vol. 23, pp. 1404–1446, 2021.
28 Zhou, Y., Remón, J., Pang, X., Jiang, Z., Liu, H., and Ding, W. Hydrothermal conversion of biomass to fuels, chemicals and materials: a review holistically conneccting product properties and marketable applications. *Sci. Total Environ.* vol. 886, p. 163920, 2023.
29 Lee, S.Y., Sankaran, R., Chew, K.W., Tan, C.H., Krishnamoorthy, R., Chu D.-T., et al. Waste to bioenergy: a review on the recent conversion technologies. *BMC Energy*. vol. 1, 2019, 78–91.
30 Agrawal, A., Karadbhajne, V., Agrawal, P., Bawa, S., Tiwari, R., and Murthy, S.S. Formulation of biogreases derived from Mahua and Karanja Oil. *Res Sq*. vol. 5, 226–234, 2022
31 Agrawal, A.J., Karadbhajne, V.Y., Agrawal, P.S., Arekar, P.S., and Chakole, N.P. Synthesis of biolubricants from non edible oils. *Int. Res. J. Eng. Technol.* vol. 4, pp. 1753–1757, 2017.

32 Rajkumari, K. and Rokhum, L. A sustainable protocol for production of biodiesel by transesterification of soybean oil using banana trunk ash as a heterogeneous catalyst. *Biomass Conv. Bioref.* vol. 10, pp. 839–848, 2020.

33 Katre, G., Raskar, S., Zinjarde, S., Ravi Kumar, V., Kulkarni, B.D., and RaviKumar, A. Optimization of the in situ transesterification step for biodiesel production using biomass of Yarrowia lipolytica NCIM 3589 grown on waste cooking oil. *Energy*. vol. 142, pp. 944–952, 2018.

34 Singh, D., Sharma, D., Soni, S.L., Sharma, S., Kumar Sharma, P., and Jhalani, A. A review on feedstocks, production processes, and yield for different generations of biodiesel. *Fuel*. vol. 262, p. 116553, 2020.

35 Kishor, N. and Agrawal, P.S. Nitration of glycerol-A byproduct of biodiesel production. *Mater. Today: Proc.* vol. 32, pp. 524–527, 2020.

36 Perez, F.M., Gatti, M.N., Nichio, N.N., and Pompeo, F. Bio-additives from glycerol acetylation with acetic acid: chemical equilibrium model. *Results Eng*. vol. 15, p. 100502, 2022.

37 Sheldon, R.A. Green chemistry, catalysis and valorization of waste biomass. *J. Mol. Catal. A. Chem*. vol. 422, pp. 3–12, 2016.

38 Ibrahim, A., Rashid, U., Hazmi, B., Moser, B.R., Alharthi, F.A., Rokhum, S.L., et al. Biodiesel production from waste cooking oil using magnetic bifunctional calcium and iron oxide nanocatalysts derived from empty fruit bunch. *Fuel*. vol. 317, p. 123525, 2022.

39 Rajkumari, K., Changmai, B., Meher, A.K., Vanlalveni, C., Sudarsanam, P., Wheatley, A.E.H., et al. A reusable magnetic nanocatalyst for bio-fuel additives: the ultrasound-assisted synthesis of solketal. *Sustain. Energy Fuels*. vol. 5, no. 8, pp. 2362–2372, 2021.

40 Ao, S. and Rokhum, S.L. Recent advances in the valorization of biodiesel by-product glycerol to solketal. *J. Chem*. Vol. 8, 225–239, 2022.

41 Bodachivskyi, I., Kuzhiumparambil, U., and Williams, D.B.G. Biomass processing via acid catalysis. In: *Biomass Valorization: Sustainable Methods for the Production of Chemicals*. Wiley, 2021, 175–189.

42 Mohan, D., Abhishek, K., Sarswat, A., Patel, M., Singh, P., and Pittman, C.U. Biochar production and applications in soil fertility and carbon sequestration-a sustainable solution to crop-residue burning in India. *RSC Adv*. vol. 8, pp. 508–520, 2018.

43 Appels, L., Baeyens, J., Degrève, J., and Dewil, R. Principles and potential of the anaerobic digestion of waste-activated sludge. *Prog Energy Combust Sci*. vol. 34, pp. 755–781, 2008.

44 Haug, R.T. *The Practical Handbook of Compost Engineering*. Routledge, 2018.

45 Kumar, A., Bhattacharya, T., Mozammil Hasnain, S.M., Kumar Nayak, A., and Hasnain, M.S. Applications of biomass-derived materials for energy production, conversion, and storage. *Mater. Sci. Energy Technol*. vol. 3, pp. 905–920, 2020.

46 Raspolli Galletti, A.M., Caposciutti, G., Pasini, G., Antonelli, M., and Frigo, S. Bio-additives for CI engines from one-pot alcoholysis reaction of lignocellulosic biomass: an experimental activity. *E3S Web Conf*. vol. 197, p. 08005, 2020.

47 Rashmi, I., Kumawat, A., Munawery, A., Sreekumar Karthika, K., Kumar Sharma, G., Kala, S., et al. Soil amendments: an ecofriendly approach for soil

health improvement and sustainable oilseed production. In: M. Hasanuzzaman and K. Nahar (Eds.) *Oilseed Crops - Uses, Biology and Production*. IntechOpen, 2023, 121–134.
48 Chavan, S., Yadav, B., Atmakuri, A., Tyagi, R.D., Wong, J.W.C., and Drogui, P. Bioconversion of organic wastes into value-added products: a review. *Bioresour. Technol.* 2022. https://doi.org/10.1016/j.biortech.2021.126398.
49 World Bioenergy Association. *Global Bioenergy Statistics 2020 World Bioenergy Association*. World Bioenergy Association. 2020.
50 Abdeljaber, A., Mostafa, O., and Abdallah, M. Applications of life cycle costing in waste-to-energy projects. In: S.S. Muthu (Ed.) *Life Cycle Costing*. Environmental Footprints and Eco-Design of Products and Processes. Springer, 2023, pp. 77–115.
51 Ulkir, O. Energy-consumption-based life cycle assessment of additive-manufactured product with different types of materials. *Polymers (Basel)*. 2023. https://doi.org/10.3390/polym15061466.
52 Lewerenz, S., Sailer, G., Pelz, S., and Lambrecht, H. Life cycle assessment of biowaste treatment – considering uncertainties in emission factors. *Clean. Eng. Technol.* vol. 15, p. 100651, 2023.
53 Arafat, H.A., Jijakli, K., and Ahsan, A. Environmental performance and energy recovery potential of five processes for municipal solid waste treatment. *J. Clean. Prod.* vol. 105, pp. 233–240, 2015.
54 Bisinella, V., Schmidt, S., Varling, A.S., Laner, D., and Christensen, T.H. Waste LCA and the future. *Waste Manag.* 2024. https://doi.org/10.1016/j.wasman.2023.11.021.
55 Finnveden, G., Arushanyan, Y., and Brandão, M. Exergy as a measure of resource use in life cycle assessment and other sustainability assessment tools. *Resources.* vol. 5, p. 23, 2016.
56 Zhou, C. and Wang, Y. Recent progress in the conversion of biomass wastes into functional materials for value-added applications. *Sci. Technol. Adv. Mater.* 2020. https://doi.org/10.1080/14686996.2020.1848213.
57 Albahnasawi, A., Eyvaz, M., Alazaiza, M.Y.D., Özdoğan, N., Gurbulak, E., Alhout, S., et al. Biomass waste and bioenergy production: challenges and alternatives. In: A.L. Srivastav, A.K. Bhardwaj, and M. Kumar (Eds.) *Valorization of Biomass Wastes for Environmental Sustainability*. Springer, 2024, pp. 51–67.
58 Kleerebezem, R. Biochemical conversion: anaerobic digestion. In: W. De Jong and R. Van Ommen (Eds.) *Biomass as a Sustainable Energy Source for the Future: Fundamentals of Conversion Processes*. Wiley Blackwell, 2014, pp. 441–467.
59 Igliński, B., Kujawski, W., and Kiełkowska, U. Pyrolysis of waste biomass: technical and process achievements, and future development—a review. *Energies.* vol. 16, p. 1829, 2023.
60 Waqas, M., Hashim, S., Humphries, U.W., Ahmad, S., Noor, R., Shoaib, M., et al. Composting processes for agricultural waste management: a comprehensive review. *Processes.* vol. 11, p. 731, 2023.
61 Berenguer, C.V., Andrade, C., Pereira, J.A.M., Perestrelo, R., and Câmara, J.S. Current challenges in the sustainable valorisation of agri-food wastes: a review. *Processes.* vol. 11, p. 20, 2023.

62 Araujo, L. and Mason, K. Markets, infrastructures and infrastructuring markets. *AMS Rev.* vol. 11, pp. 240–251, 2021.

63 United Nations Environment Programme International Solid Waste Association. *Global Waste Management Outlook 2024 – Beyond an Age of Waste: Turning Rubbish Into a Resource.* 2024.

64 Kladnik, V., Dworak, S., and Schwarzböck, T. Composition of public waste – a case study from Austria. *Waste Manag.* 2024. https://doi.org/10.1016/j.wasman.2024.02.031.

65 Zaman, A. Zero-waste: a new sustainability paradigm for addressing the global waste problem. In: K. Edvardsson Björnberg, S.O. Hansson, M.Å. Belin, and C. Tingvall (Eds.) *The Vision Zero Handbook.* Springer, 2022, pp. 1195–1218.

66 Hermann, H., Reinhard, R., Klaus, B., Reinhard, K., and Christian, A. Biomass CHP plant güssing – a success story. In: *Expert Meet Pyrolysis Gasif Biomass Waste.* 2002.

67 Alao, M.A., Popoola, O.M., and Ayodele, T.R. Waste-to-energy nexus: an overview of technologies and implementation for sustainable development. *Clean Energy Syst.* vol. 3, p. 100034, 2022.

68 Czekała, W., Drozdowski, J., and Łabiak, P. Modern technologies for waste management: a review. *Appl. Sci. (Switzerland).* vol. 13, p. 8847, 2023.

69 Tanveer, M., Khan, S.A.R., Umar, M., Yu, Z., Sajid, M.J., and Haq, I.U. Waste management and green technology: future trends in circular economy leading towards environmental sustainability. *Environ. Sci. Pollut. Res. Int.* 2022. https://doi.org/10.1007/s11356-022-23238-8.

70 Jain, S. *Global Potential of Biogas.* World Biogas Association. 2019.

71 Inayat, A., Ahmed, A., Tariq, R., Waris, A., Jamil, F., Ahmed, S.F., et al. Techno-economical evaluation of bio-oil production via biomass fast pyrolysis process: a review. *Front. Energy Res.* vol. 9, p. 770355, 2022.

72 Melchor-Martínez, E.M., Macías-Garbett, R., Alvarado-Ramírez, L., Araújo, R.G., Sosa-Hernández, J.E., Ramírez-Gamboa, D., et al. Towards a circular economy of plastics: an evaluation of the systematic transition to a new generation of bioplastics. *Polymers* 2022. https://doi.org/10.3390/polym14061203.

73 Mashudi, Sulistiowati, R., Handoyo, S., Mulyandari, E., and Hamzah, N. Innovative strategies and technologies in waste management in the modern era integration of sustainable principles, resource efficiency, and environmental impact. *Int. J. Sci. Soc.* vol. 5, pp. 87–100, 2023.

74 Hull, C.E., Millette, S., and Williams, E. Challenges and opportunities in building circular-economy incubators: stakeholder perspectives in Trinidad and Tobago. *J. Clean. Prod.* vol. 296, p. 126412, 2021.

8

Agro-waste to Compost

8.1 Introduction

Agriculture, the bedrock of human civilization, is intrinsically linked to waste generation. Agricultural waste, or agro-waste, encompasses a vast array of organic materials discarded or left over from agricultural production, processing, and consumption [1]. As populations grow and agricultural intensification increases to meet global food demands, the volume of agro-waste generated globally is reaching unprecedented levels. Unmanaged agro-waste poses significant environmental threats, contributing to greenhouse gas emissions, soil and water pollution, and health hazards. Open burning, landfilling, and improper disposal methods are still prevalent in many regions, exacerbating these problems and resulting in the loss of valuable resources contained within this waste stream [2, 3]. Agricultural activities generate vast quantities of agro-waste, estimated at over 5 billion tons globally each year [4]. This includes crop residues, animal manures, and agro-industrial byproducts. Improper management – such as open burning or landfilling – leads to greenhouse gas emissions, leachate generation, and loss of soil nutrients. Conversely, composting transforms agro-waste into a stabilized, humified organic amendment, closing nutrient loops and supporting the circular economy [5, 6].

However, within this challenge lies a significant opportunity, agro-waste is a rich source of organic matter and nutrients. Composting, a biological process that converts organic waste into a stable, humus-like material, offers a sustainable and environmentally benign solution for managing agro-waste while simultaneously producing a valuable soil amendment. Compost not only improves soil physical, chemical, and biological properties, enhancing soil health and fertility but also reduces reliance on synthetic fertilizers [7]. In the context of growing concerns about soil degradation, climate change, and the need for sustainable agricultural practices, agro-waste composting emerges as a critical strategy for resource recovery, waste minimization, and building resilient agricultural systems [8].

This research review focuses specifically on the scientific literature concerning agro-waste to compost. It aims to provide an in-depth and critical analysis of the research conducted in this domain, encompassing the diverse facets of agro-waste

Agro-Waste Management and Valorization, First Edition. Pratibha S. Agrawal, et al.

Table 8.1 Environment consideration in landfilling and composting.

Parameter	Landfilling (kg CO_2-eq/ton)	Composting (kg CO_2-eq/ton)	% Reduction
GHG emissions (CO_2-eq)	120–200	20–40	70–85
Organic matter loss	>50%	<30%	–
Nitrogen loss	30–50%	10–20%	–

composting. This review will meticulously explore the types of agro-waste suitable for composting, the underlying scientific principles of the composting process, the methodologies employed, the factors influencing compost quality, the assessment of compost characteristics, and the diverse applications of agro-waste-derived composts [9] (Table 8.1). Furthermore, it will identify research gaps and suggest future research directions to advance the field and promote the wider adoption of agro-waste composting as a cornerstone of sustainable agricultural and waste management practices as environmental benefits compost [10].

8.2 Defining and Categorizing Agro-waste for Composting

"Agro-waste" is a broad term encompassing a diverse range of organic residues originating from various agricultural sectors [11]. For the purpose of composting, understanding the specific types and characteristics of agro-waste is crucial, as it directly influences the composting process and the quality of the final compost product. Agro-waste can be categorized based on its source, nature, and physical form [12] (Table 8.2). Research literature presents various classifications, often overlapping, but broadly, agro-waste for composting can be categorized as follows:

8.2.1 Crop Residues

Crop residues represent a principal and globally abundant component of agro-waste, comprising the non-harvested plant parts left in agricultural fields or generated during initial on-farm processing after crops are harvested [13]. This broad category encompasses diverse materials depending on the crop type. Cereal straw and Stover, such as rice straw, wheat straw, maize stover, and barley straw, constitute vast volumes annually and are primarily composed of lignocellulosic biomass (cellulose, hemicellulose, and lignin) [8, 14]. They typically exhibit a high carbon-to-nitrogen (C : N) ratio, which influences their suitability for biological conversion processes like anaerobic digestion and composting unless co-digested with nitrogen-rich feedstocks [15, 16]. Legume residues, including pea haulms, bean stalks, and soybean residues, are generally distinct from cereals due to their higher nitrogen content, resulting in a lower C : N ratio that makes them more suitable for direct biological conversion or co-digestion [17]. Vegetable and fruit wastes arising from cultivation, encompassing discarded leaves, stems, peels, and rejected produce, are

Table 8.2 Classification of agro-waste for composting based on source and nature.

Category	Sub-categories	Nature of waste	Examples	Characteristics for composting
Crop residues	Cereal straw and stover	High cellulose, lignin, low nitrogen	Rice straw, wheat straw, maize stover	Bulky, high C : N ratio, requires nitrogen amendment, slow decomposition
	Legume residues	Moderate cellulose, lignin, moderate nitrogen	Pea haulms, bean stalks, soybean residues	Less bulky than cereal straw, moderate C : N ratio, moderate decomposition
	Vegetable and fruit waste	High moisture, readily biodegradable, high nutrients	Vegetable leaves, fruit peels, rejected produce	Rapid decomposition, potential odor issues if not managed properly, high nutrient content
	Oilseed crop residues	Moderate cellulose, lignin, moderate oil content	Mustard stalks, sunflower stalks, rapeseed straw	Moderate decomposition, may require pretreatment for oil content
	Sugarcane trash and bagasse	High cellulose, lignin, high fiber	Sugarcane leaves, bagasse	Bulky, high C : N ratio, slow decomposition, requires pretreatment
Livestock manure	Cattle manure	Moderate moisture, moderate nutrients, moderate C : N	Cow dung, cattle urine	Good source of microorganisms, balanced C : N ratio when mixed with bedding
	Poultry manure	Low moisture, high nutrients, low C : N	Chicken droppings, poultry litter	High nitrogen and phosphorus content, rapid decomposition, potential ammonia odor
	Swine manure	High moisture, moderate nutrients, variable C : N	Pig slurry, swine manure solids	Requires careful management due to odor and pathogen concerns
Agro-industrial byproducts	Fruit and vegetable processing waste	High moisture, readily biodegradable, high sugar content	Fruit pomace, vegetable peels, processing rejects	Rapid decomposition, potential odor issues, high sugar content can lead to anaerobic conditions
	Food-processing industry waste	Variable depending on source	Brewery waste, distillery waste, oilseed cake, sugar press mud	Highly diverse in composition, requires characterization for composting suitability
	Slaughterhouse waste	High protein, High fat, pathogen risk	Blood, bones, Rumen content	Requires pretreatment (rendering, composting) due to pathogen and odor concerns

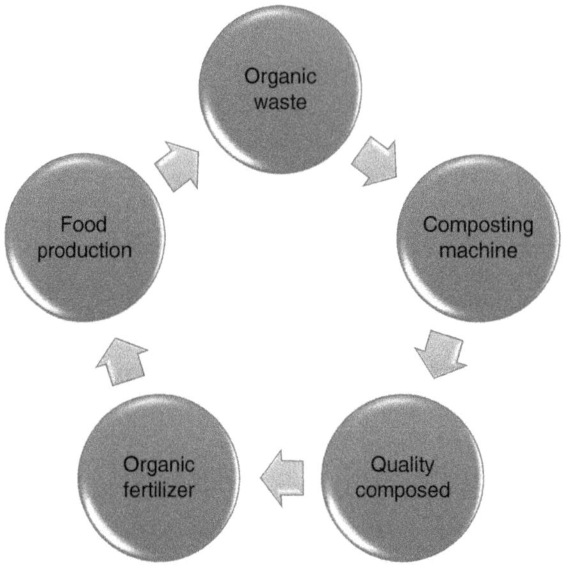

Figure 8.1 Organic waste conversion.

characterized by high moisture content and high biodegradability, making them highly amenable to anaerobic digestion and composting [18, 19]. Oilseed crop residues, such as mustard and sunflower stalks, are also significant lignocellulosic sources [20]. Finally, sugarcane trash (leafy tops) and bagasse (fibrous residue from milling) represent major lignocellulosic byproducts from sugarcane cultivation and processing, extensively explored for bioenergy and biochemical production [12, 15]. The lignocellulosic nature of many crop residues makes them prime feedstocks for thermochemical conversion processes like pyrolysis and gasification, yielding biochar, bio-oil, and syngas [7]. Understanding the specific composition and properties of different crop residues is crucial for tailoring effective pretreatment strategies (e.g., hydrolysis for fermentation) and selecting optimal conversion technologies to maximize valorization potential as shown in Figure 8.1 [21].

8.2.2 Livestock Manure: A Nutrient-rich Bioresource

Livestock manure, consisting of animal excreta, bedding materials, and spilled feed from animal husbandry operations, constitutes another substantial and nutrient-rich source of agro-waste globally, particularly in regions with intensive livestock farming [3]. The composition and characteristics of manure vary considerably depending on the animal type, diet, age, and the type of bedding used. Cattle manure, a significant volume stream from beef and dairy farming, is typically characterized by high moisture content and a moderate C : N ratio, making it highly suitable for anaerobic digestion to produce biogas and digestate [18]. Poultry manure, from chicken, turkey, and duck farming, is generally drier (especially layer manure) and significantly richer in nitrogen and phosphorus compared to cattle manure, resulting in a lower C : N ratio; its high nutrient density makes it valuable as a fertilizer, but high nitrogen can lead to ammonia (NH_3) inhibition in anaerobic digestion if not managed [22]. Swine manure is often managed as a liquid slurry with high organic content and nutrient load, primarily suitable for anaerobic digestion. Manure from sheep and goats tends to be drier and pelleted. Horse manure, often mixed with straw or wood shavings as bedding, presents a higher lignocellulosic content from the bedding material. All types of manure are rich in organic matter and essential plant nutrients, positioning them as valuable feedstocks for biological conversion processes like anaerobic digestion (producing biogas and nutrient-rich digestate) and composting [23]. Digestate and compost from manure are widely used as organic fertilizers and soil conditioners [18]. Challenges in utilizing manure include high moisture content (for many types), potential presence of pathogens and antibiotics, significant variability in composition, and logistical issues related to collection and transport from distributed farms [24].

8.2.3 Agro-industrial Byproducts: Residues from Processing

Agro-industrial byproducts are the diverse waste streams generated during the processing of agricultural commodities into food, feed, fibers, and other products [8]. Unlike field residues, these byproducts often originate from centralized processing facilities, potentially offering advantages in terms of collection logistics, though their composition can be highly variable and specific to the industry. Fruit and vegetable processing waste includes materials like pomace from juice production, peels, seeds, cores, and trimming waste [25]. These wastes are typically high in moisture, readily biodegradable sugars and pectin, and organic acids, making them excellent feedstocks for anaerobic digestion, fermentation (e.g., for organic acids), and composting. Food-processing industry waste is a broad category encompassing residues from breweries (spent grain, spent hops), distilleries, starch production, sugar refining (press mud), and oilseed processing (oilseed cake) [26]. These streams vary significantly in composition (e.g., spent grain is lignocellulosic, press mud is high in moisture and organic matter, oilseed cake is protein/lipid rich), dictating their suitability for different WtBA processes, including AD, fermentation, and thermochemical conversion [27]. Slaughterhouse waste, consisting of blood, bones, rendering waste, and rumen content, is high in protein, lipids, and moisture [15]. Due to its high organic content and potential pathogen load, it is primarily suited for anaerobic digestion (often thermophilic) or rendering, with strict regulations governing its handling and treatment [28]. While potentially valuable feedstocks, agro-industrial byproducts often contain high moisture, specific organic compounds (e.g., high lipid content), and sometimes contaminants or inhibitors (e.g., cleaning chemicals), necessitating tailored pretreatment and conversion strategies [25].

8.2.4 Forestry Residues and the Integrated Waste Basket

While not strictly classified under "agro-waste" in some contexts, forestry residues are closely related organic waste streams, particularly relevant in agricultural landscapes interspersed with forests or linked through biomass supply chains [13]. These residues are generated during logging operations (e.g., branches, tops, forest floor litter) and timber processing (e.g., wood chips, sawdust, bark) [18]. Forestry residues are predominantly lignocellulosic biomass with a high C : N ratio and relatively low moisture content (especially processed residues), making them excellent feedstocks for thermochemical conversion processes like pyrolysis and gasification to produce biochar, bio-oil, and syngas. They are also a source of lignocellulose for fermentation after pretreatment [22]. Integrating forestry residues into broader agro-waste valorization strategies expands the available feedstock base for bio-additive production. Managing this diverse "waste basket" from agricultural and related sectors presents overarching challenges, including the inherent variability in composition, moisture content, and physical form across different waste types and sources [12]. The potential for contamination with inerts (plastic, metals) and hazardous substances requires robust sorting and pretreatment [29]. The geographical dispersion of many agro-waste streams, particularly field residues, creates significant logistical challenges and costs for collection and transportation to centralized processing facilities. Despite these complexities, the sheer volume and diverse potential of these combined organic waste streams underscore the critical opportunity for developing integrated waste management and valorization systems that convert this complex basket of residues into a range of valuable bio-additives within a circular economy framework as shown in Figure 8.2 [10].

The diverse nature of agro-waste necessitates a tailored approach to composting [30]. Understanding the specific characteristics of each type of agro-waste, especially in terms of C : N ratio, moisture content, lignin content, nutrient composition, and potential presence of contaminants is crucial for designing effective composting strategies and achieving high-quality compost. Research often focuses on specific types or combinations of agro-waste to optimize composting processes and compost quality for particular applications. A detailed understanding of agro-waste types is essential for process optimization. Table 8.3 summarizes typical characteristics.

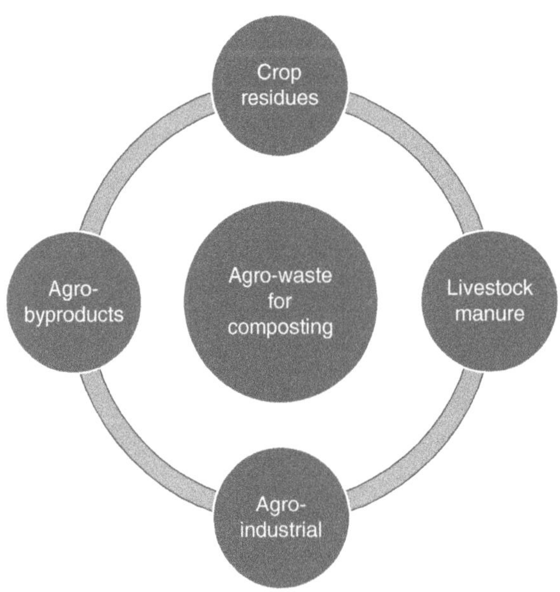

Figure 8.2 Classification of agro-waste for composting.

Table 8.3 Major agro-waste types and key composting parameters. Adapted from [31].

Agro-waste type	C : N ratio	Moisture (%)	Lignin (%)	Notable elements
Rice straw	60–80 : 1	10–15	12–15	Si, K
Wheat straw	80–100 : 1	10–15	15–18	K, P
Vegetable waste	15–25 : 1	75–90	2–5	N, K, Ca
Cattle manure	15–25 : 1	75–80	2–5	N, P, K
Poultry manure	10–15 : 1	60–70	1–3	N, P, K, Ca
Sugarcane bagasse	40–50 : 1	50–60	18–22	Si, K
Fruit processing waste	20–30 : 1	80–90	1–3	N, K, Ca

Optimal C : N ratio for composting: 25–30 : 1.

8.3 The Science of Composting: Biochemical Processes and Microbial Ecology

Composting is a complex and dynamic biological process driven by a diverse community of microorganisms that decompose organic matter under controlled aerobic conditions [26]. It is a natural process mimicking the decomposition of organic matter in forest floors and soil ecosystems, accelerated and managed for waste treatment and resource recovery. Understanding the biochemical processes and microbial ecology underpinning composting is essential for optimizing the process and ensuring efficient conversion of agro-waste into valuable compost as shown in Figure 8.3.

8.3.1 Biochemical Processes in Composting

The composting process involves a series of biochemical reactions primarily catalyzed by enzymes produced by microorganisms. The main biochemical processes are as follows:

- **Hydrolysis:** Complex organic polymers (cellulose, hemicellulose, lignin, proteins, fats) are broken down into simpler monomers (sugars, amino acids, fatty acids) by hydrolytic enzymes (cellulases, hemicellulases, proteases, lipases). This initial breakdown makes the organic matter readily available for microbial assimilation.
- **Oxidation:** Simple organic molecules are further oxidized through respiration, releasing energy and producing carbon dioxide (CO_2), water (H_2O), and heat. This is the primary energy-generating process for the microbial community.
- **Humification:** A crucial process in composting where complex and resistant organic molecules, collectively termed humic substances, are formed from the breakdown products of lignin and other organic matter. Humic substances contribute significantly to the beneficial properties of compost, including soil structure improvement, nutrient retention, and enhanced plant growth.

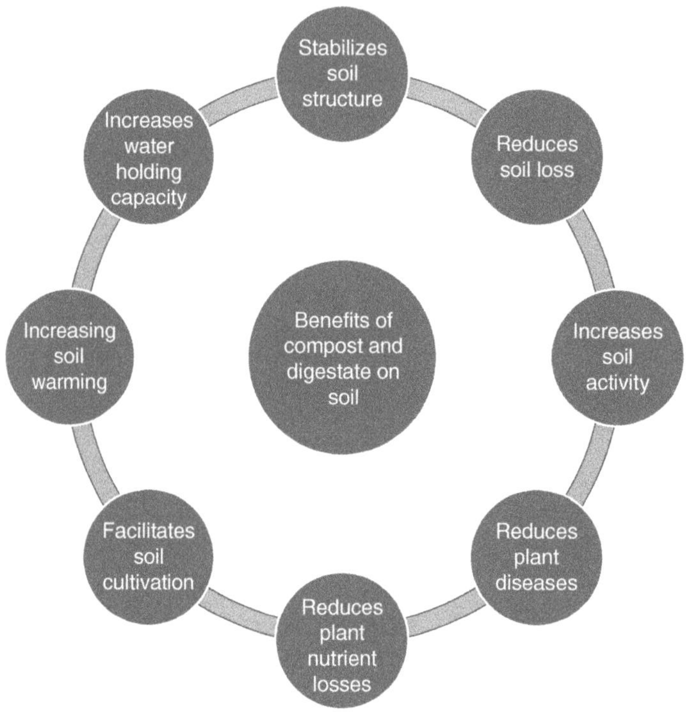

Figure 8.3 Compost and digestate for a circular economy.

- **Mineralization:** Organic forms of essential nutrients (nitrogen, phosphorus, potassium, etc.) are converted into inorganic, plant-available forms through microbial activity. This process is crucial for releasing nutrients from the agro-waste and making them available in the final compost product.

8.3.2 Microbial Ecology of Composting

A diverse array of microorganisms orchestrates the composting process [32]. These microbial communities undergo dynamic shifts in composition and activity throughout the different phases of composting. Key groups of microorganisms involved are as follows:

- **Bacteria:** Bacteria are the dominant microbial group in composting, responsible for the initial stages of decomposition, particularly the breakdown of readily biodegradable organic matter. Diverse bacterial genera, including Bacillus, Pseudomonas, Streptomyces, Thermus, and Acidovorax, are commonly found in compost.
- **Fungi:** Fungi play a critical role in decomposing more recalcitrant organic polymers like cellulose and lignin, especially during the thermophilic (high-temperature) phase. Genera such as Aspergillus, Trichoderma, Penicillium, and Thermoascus are important fungal decomposers in compost.

- **Actinomycetes:** Actinomycetes, filamentous bacteria, are particularly active in the later stages of composting, contributing to the breakdown of more resistant organic compounds and humification. They are also known for their ability to produce antibiotics, potentially suppressing plant pathogens in compost.
- **Protozoa and Nematodes:** These microscopic eukaryotes play roles in the composting process, primarily through grazing on bacteria and fungi, regulating microbial populations, and contributing to nutrient cycling.

8.3.3 Phases of Composting and Microbial Succession

The composting process typically progresses through distinct phases (Table 8.4), each characterized by specific temperature ranges, microbial communities, and biochemical activities:

- **Mesophilic Phase (Initial Phase):** Temperature rises slightly as mesophilic microorganisms (active at moderate temperatures) begin to decompose readily biodegradable organic matter. Bacteria, particularly mesophilic bacteria, are dominant during this phase.

Table 8.4 Microbial groups and biochemical processes dominating composting phases.

Composting phase	Temperature range (°C)	Dominant microbial groups	Biochemical processes	Characteristics
Mesophilic phase I	20–40	Mesophilic bacteria, some fungi	Hydrolysis of readily biodegradable organic matter (sugars, starch)	Initial decomposition, slight temperature increase
Thermophilic phase	50–70	Thermophilic bacteria, thermophilic fungi	Hydrolysis of cellulose, hemicellulose, lignin, proteins, fats, oxidation	Rapid decomposition, pathogen inactivation, weed seed destruction, high heat generation
Cooling phase (mesophilic II)	40–50 to ambient	Mesophilic bacteria, actinomycetes, fungi	Continued hydrolysis, oxidation, initiation of humification	Temperature decrease, stabilization of organic matter, humification begins
Maturation phase	Ambient	Actinomycetes, fungi	Humification, stabilization of compost, maturation	Formation of stable humic substances, compost maturity, reduced microbial activity

- **Thermophilic Phase (Heating Phase):** Temperature rapidly increases to 50–70 °C due to intense microbial activity and heat generation. Thermophilic bacteria and fungi become dominant, actively decomposing more complex organic polymers. This phase is crucial for pathogen inactivation and weed seed destruction.
- **Cooling Phase (Mesophilic Phase II):** As easily degradable organic matter is depleted, microbial activity and heat generation decrease, and the temperature gradually declines. Mesophilic microorganisms, including bacteria and actinomycetes, become active again, further decomposing remaining organic matter and initiating humification.
- **Maturation Phase (Curing Phase):** Temperature stabilizes at ambient levels. Actinomycetes and fungi become dominant, focusing on humification and stabilization of the compost. This phase is crucial for achieving a mature and stable compost product.

Understanding the microbial dynamics and biochemical processes in each phase is critical for managing the composting process effectively. Factors like aeration, moisture content, C : N ratio, and temperature need to be carefully controlled to support the desired microbial activity and optimize decomposition and humification [12]. The process control parameters are as follows:

Aeration rate: 0.2–0.5 L air/kg waste/minutes for optimal O_2 (>10%).
Moisture: 50–60% for maximal microbial activity; <40% slows decomposition, >65% risks anaerobiosis.
pH: 6.0–8.0 optimal; acidic conditions (<5.5) inhibit microbes.

Research in microbial ecology and biochemistry of composting is continuously advancing our understanding and enabling the development of more efficient and targeted composting strategies.

8.4 Composting Methodologies for Agro-waste: From Traditional to Advanced Techniques

In the face of mounting environmental concerns and the imperative for sustainable resource management, composting has emerged as a viable and environmentally sound solution for agro-waste management. Composting, fundamentally, is a naturally occurring biological process wherein microorganisms, primarily bacteria and fungi, decompose organic matter under controlled aerobic conditions. This process transforms complex organic waste into a stable, humus-like product known as compost. Compost is a nutrient-rich material that can be utilized as a soil amendment, enhancing soil fertility, improving soil structure, water retention, and nutrient availability.

Beyond its waste management capabilities, composting offers a multitude of benefits. It reduces the volume of waste requiring landfill disposal, diverts organic matter from incineration, and minimizes greenhouse gas emissions compared to anaerobic decomposition in landfills [3]. By producing compost, agro-waste is transformed from a liability into a valuable resource, contributing to a circular economy in agriculture. Compost application to agricultural lands can reduce the need for synthetic fertilizers, which are energy-intensive to produce and can contribute to environmental pollution through nutrient runoff and eutrophication. Furthermore, compost enhances soil biodiversity, suppresses soil-borne plant diseases, and improves overall soil health, leading to more sustainable and resilient agricultural systems.

This comprehensive research review aims to provide an in-depth analysis of composting methodologies specifically tailored for agro-waste management. Embark journey from traditional composting techniques, practiced for centuries, to modern, technologically advanced systems [4]. The analysis delves into the scientific principles underlying each method, compares their advantages and disadvantages, analyzes critical process parameters, and examines the quality and applications of compost derived from agro-waste. Ultimately, evaluation intends to provide a holistic understanding of composting as a sustainable and vital technology for agro-waste management, highlighting its role in promoting environmental protection, resource conservation, and sustainable agricultural practices.

8.4.1 Traditional Composting Methodologies for Agro-waste

Traditional composting methods, practiced for centuries in various cultures and agricultural systems, often rely on simple techniques and locally available resources. While these methods may be less efficient than advanced techniques in terms of speed and process control, they are often low-cost, easily adaptable to small-scale operations, and require minimal infrastructure [33]. This section explores the common traditional composting methods used for agro-waste management.

8.4.1.1 Pit Composting

Pit composting represents a fundamental and historically significant method for decomposing organic waste materials, distinguished by its inherent simplicity and minimal technological requirements compared to more controlled composting techniques [32]. As one of the oldest known composting practices, it primarily involves excavating a pit or trench directly within the ground, with dimensions varying according to the volume of waste materials and disadvantages.

8.4.1.2 The Pit Composting Process: A Simple, Ground-based Approach

Pit composting represents a rudimentary yet historically significant method for decomposing organic waste, distinguished by its inherent simplicity and minimal technological reliance [18]. The process fundamentally involves the excavation of a pit or trench directly within the ground, the dimensions of which are typically determined by the volume of organic material designated for decomposition [12]. Waste materials are then introduced into this pit, often deposited in sequential layers that alternate between "green" (nitrogen-rich, readily decomposable organic waste like vegetable scraps and fresh green manure) and "brown" (carbon-rich, more resistant materials such as dry leaves, straw, twigs, and shredded paper) components to achieve a balanced C : N ratio conducive to microbial activity [22]. While microbial populations inherent in the waste initiate decomposition, animal manure or other nitrogen-rich sources may be intentionally added to the pit as activators to potentially accelerate the initial breakdown phase [34]. Maintaining adequate moisture levels within the pit is crucial for supporting microbial metabolism, and this is typically achieved through periodic watering of the material. Following the layering and moistening, the pit is often left largely undisturbed for a considerable duration, allowing decomposition to occur naturally over several months to a year. Although passive aeration is limited within the pit structure, occasional physical turning of the material may be performed, depending on the specific practice and the desired rate of decomposition and homogeneity of the final product [35]. This simple, ground-based process contrasts significantly with more engineered composting systems that involve above-ground piles, windrows, or in-vessel reactors with active aeration mechanisms, highlighting its fundamental approach rooted in natural decomposition within an excavated space [36].

8.4.1.3 Advantages of Implementing Pit Composting

Despite its technological simplicity, pit composting offers several distinct advantages that contribute to its historical prevalence and continued use in certain contexts [8]. A primary benefit is its exceptionally low cost and simplicity of implementation, requiring minimal investment in specialized tools, equipment, or engineered infrastructure beyond basic excavation implements; this makes it highly accessible for individuals, small farms, and communities with limited financial resources [11]. Relatedly, it is easy to implement and requires minimal technical expertise, making it readily adoptable in backyards, gardens, or rural agricultural settings without the need for complex training. Furthermore, the pit structure provides an inherent advantage in moisture retention, as the surrounding soil walls help contain water, particularly beneficial in arid or semi-arid climates where moisture loss from open composting systems can be a significant challenge. This natural moisture management can reduce the frequency and volume of watering required [18]. While not as effective as sealed anaerobic digesters or actively aerated systems, burying organic material in a pit can offer some degree of odor control compared to open composting heaps, as the surrounding soil can help absorb or mitigate the dispersion of volatile organic compounds produced during decomposition, potentially reducing nuisance to nearby areas. These advantages, rooted in the low technological and financial requirements, position pit composting as a viable option for basic organic waste management where resources and technical capacity are limited.

8.4.1.4 Disadvantages and Limitations of Pit Composting Performance

While simple and low-cost, pit composting is subject to significant disadvantages and limitations that impede its efficiency, control, and suitability for large-scale or regulated waste management applications. A major drawback is the slow decomposition rate experienced in pit composting. Due to the inherently limited passive aeration within the confined pit structure, oxygen availability is often insufficient for optimal aerobic microbial activity throughout the entire waste volume, slowing down the rate of organic matter breakdown and prolonging the composting process to several months or even a year [32]. This lack of adequate aeration frequently leads to the development of anaerobic conditions within the pit. Anaerobic decomposition generates noxious odors (e.g., hydrogen sulfide [H_2S], volatile organic acids) and results in incomplete stabilization of organic matter, potentially leading to lower quality compost and the loss of valuable nutrients through leaching or volatilization [28]. A critical concern is pathogen survival; the temperatures achieved in pit composting, often remaining below the thermophilic range necessary for pathogen inactivation (e.g., 55 °C for several days), may not be sufficient to effectively eliminate disease-causing microorganisms and weed seeds present in the waste feedstock, raising concerns about the safety of the final compost product for use in food production [2]. Operationally, physically turning the material within a pit to improve aeration is cumbersome, labor-intensive, and less effective compared to turning surface piles or using mechanical equipment. Finally, pit composting requires dedicated space requirement for digging and managing pits, which can be a constraint in areas with limited land availability and necessitates rotating pit locations or managing multiple pits simultaneously for continuous waste-processing. These limitations in speed, control, safety, and operational efficiency constrain the applicability of pit composting compared to more advanced and controlled composting technologies, particularly for large volumes of waste or feedstocks with significant pathogen loads.

8.4.1.5 Heap Composting

Heap composting, also known as passive composting, is perhaps the most rudimentary method. Agro-waste materials are piled into a heap, often in an open area. Decomposition occurs naturally through the action of indigenous microorganisms and insects present in the waste and the surrounding environment. The important characteristics of traditional heap composting are represented in Table 8.5.

8.4.1.6 Windrow Composting

Windrow composting is a scaled-up version of heap composting, where agro-waste is formed into elongated piles or rows, known as windrows. This method facilitates better aeration and temperature control compared to heap composting due to

Table 8.5 Characteristics of traditional heap composting.

Feature	Description	Advantages	Disadvantages	Scale of application
Process	Passive, uncontrolled decomposition	Simple, low-cost, minimal equipment required	Slow process, potential for odor, nutrient loss,	Small-scale, farm level
Management	Minimal intervention (occasional turning if any)	Low management input	Difficult to control temperature and aeration	Inconsistent compost quality
Input materials	Diverse agro-waste, often unsorted	Adaptable to various waste streams	Limited ability to process large volumes	Not suitable for pathogen inactivation without active management

Table 8.6 Characteristics of traditional windrow composting.

Feature	Description	Advantages	Disadvantages	Scale of application
Process	Aerobic decomposition in long piles (windrows)	Improved aeration compared to heap, better temperature control with turning	Still relatively slow, potential for odor, requires turning equipment	Small to medium scale
Management	Requires periodic turning for aeration and mixing	Enhanced control over decomposition parameters	Labor-intensive or equipment dependent on turning method	More consistent compost quality than heap
Input materials	Can handle larger volumes of agro-waste	Scalable for larger farms or community composting projects	Space requirement for windrow formation	Can achieve some pathogen reduction with proper turning and temperature management

the increased surface area and shape. Windrows are typically turned periodically (mechanically or manually) to enhance aeration, moisture distribution, and accelerate decomposition. The important characteristics of traditional windrow composting are represented in Table 8.6.

8.5 Advanced Composting Methodologies

Advanced composting techniques have been developed to overcome the limitations of traditional methods, aiming for faster decomposition rates, improved compost quality, reduced environmental impact, and enhanced process control.

8.5.1 Vermicomposting

Vermicomposting is a bio-oxidative process employing earthworms to accelerate the decomposition of organic waste [37]. Earthworms ingest and digest organic matter, breaking it down into finer particles and enriching it with beneficial microbial communities and enzymes in their gut. The resultant vermicast (worm castings) is a high-quality compost rich in nutrients, enzymes, and plant growth hormones. The important characteristics of traditional vermicomposting are represented in Table 8.7.

8.5.2 In-vessel Composting

In-vessel composting systems are enclosed, controlled environments that optimize composting conditions. These systems utilize reactors or containers where

Table 8.7 Characteristics of vermicomposting.

Feature	Description	Advantages	Disadvantages	Scale of application
Process	Accelerated decomposition using earthworms	Rapid decomposition, high-quality vermicast production, nutrient enrichment	Requires controlled environment, specific worm species, sensitive to input material composition	Small to medium scale
Management	Maintaining optimal moisture, temperature, and bedding material	Reduced odor and pathogen presence in vermicast	Worm management and separation from vermicast can be labor-intensive	Premium compost production
Input materials	Primarily pre-composted or easily digestible agro-waste	Effective for processing diverse agro-waste (excluding highly lignocellulosic)	Not suitable for large volumes directly, pre-processing often needed	High nutrient content compost

temperature, aeration, and moisture are precisely regulated. In-vessel systems offer faster composting cycles, better odor control, and enhanced pathogen inactivation due to high temperatures. The important characteristics of in-vessel composting are represented in Table 8.8.

8.5.3 Thermal Composting (Aerated Static Pile with Forced Aeration)

Thermal composting, particularly aerated static pile composting, is an advanced method that combines the simplicity of pile composting with controlled aeration. Agro-waste is piled, and forced aeration is provided through perforated pipes embedded within the pile. This maintains aerobic conditions and elevates pile temperatures to thermophilic ranges (55–65 °C), effectively killing pathogens and weed seeds while accelerating decomposition [38]. The important characteristics of thermal composting are represented in Table 8.9.

8.6 Factors Influencing Composting Efficiency and Compost Quality

The efficiency of composting and the quality of the final compost product are influenced by several factors, regardless of the composting methodology employed. Composting efficiency and the resulting compost quality are governed by a complex interplay of physical, chemical, and biological factors, each of which must be

Table 8.8 Characteristics of in-vessel composting.

Feature	Description	Advantages	Disadvantages	Scale of application
Process	Controlled environment composting in enclosed reactors	Rapid decomposition, excellent odor control, high pathogen inactivation, consistent compost quality	High initial investment, higher operational costs (energy for aeration, temperature control), requires technical expertise	Medium to large scale
Management	Automated control of process parameters (temperature, aeration, moisture)	Precise control over composting process, reduced labor for turning and handling	More complex operation, potential for mechanical failures, requires regular maintenance	Suitable for municipalities, large farms, processing industries
Input materials	Varied agro-waste, can handle larger volumes and diverse types	Can handle a wider range of agro-waste compared to vermicomposting	Pre-processing (shredding, mixing) often required to ensure efficient operation	Reduced land footprint compared to windrow composting

Table 8.9 Characteristics of thermal composting (aerated static pile).

Feature	Description	Advantages	Disadvantages	Scale of application
Process	Static pile composting with controlled forced aeration	Relatively rapid decomposition, good pathogen inactivation due to thermophilic temperatures, reduced turning requirement	Requires aeration system installation and energy consumption, potential for odor if aeration is not properly managed	Medium to large scale
Management	Monitoring temperature and adjusting aeration rates	Lower labor costs compared to windrow turning, better control of decomposition environment	Potential for anaerobic zones if aeration is insufficient, requires monitoring and adjustment of aeration rates	Efficient pathogen reduction, good quality compost
Input materials	Suitable for various agro-waste types, larger volumes	Scalable for larger quantities of agro-waste, adaptable to different waste streams	Initial investment for aeration system, operational costs for aeration system, requires space for static piles	Can achieve higher throughput than windrow but less complex than in-vessel

Table 8.10 Factors influencing composting efficiency and compost quality.

Factor	Optimal range/value	Importance in composting
C : N ratio	25 : 1 to 30 : 1	Provides balanced nutrients for microbial growth and decomposition; too high C leads to slow decomposition, too high N can lead to ammonia volatilization and odor
Moisture content	50–60%	Essential for microbial activity and transport of nutrients; too low inhibits microbial activity, too high leads to anaerobic conditions and odor
Aeration/ oxygen	>5% (ideally >10%)	Crucial for aerobic decomposition and prevents anaerobic conditions and odor production; influences microbial respiration and decomposition rate
Temperature	Mesophilic (25–40 °C) and Thermophilic (55–65 °C) phases	Mesophilic phase initiates decomposition, thermophilic phase accelerates decomposition and pathogen inactivation; maintaining optimal temperature ranges is critical
pH	6.0–8.0 (slightly acidic to neutral)	Influences microbial activity and nutrient availability; extreme pH values can inhibit microbial growth
Particle size	Smaller particle size (shredded)	Increases surface area for microbial attack and decomposition; too fine can limit aeration
Turning/ mixing	Frequency depends on method (windrow, ASP)	Improves aeration, moisture distribution, temperature homogeneity, and accelerates decomposition; less critical in in-vessel and Aerated Static Pile (ASP) due to forced aeration

carefully managed to optimize the process [9]. The major parameters include the C : N ratio, moisture content, temperature, aeration, particle size, and pH as represented in Table 8.10.

An optimal C : N ratio (typically 25 : 1–30 : 1) ensures a balanced supply of energy and nutrients for microbial metabolism, while moisture content in the range of 50–60% supports microbial activity without inducing anaerobic conditions. Temperature control is critical; thermophilic conditions (55–65 °C) accelerate organic matter degradation and ensure pathogen inactivation, but excessive heat (>70 °C) can inhibit beneficial microbes. Adequate aeration maintains aerobic conditions, preventing odor formation and facilitating efficient decomposition, while excessive aeration can lead to moisture loss and reduced microbial efficiency. Particle size affects surface area and porosity; smaller particles enhance decomposition but may impede airflow if too fine. The pH should be maintained between 6.0 and 8.0 for optimal microbial function. Other factors, such as the presence of bulking agents, initial microbial inoculum, and the nature of feedstock (e.g., lignin content, presence of inhibitors), further influence composting dynamics. Collectively, these factors determine not only the rate of composting but also the maturity, stability, and agronomic value of the final compost product [39].

8.6.1 Process Design

8.6.1.1 Feedstock Blending

Feedstock blending is a foundational strategy in composting process optimization, directly influencing the rate of decomposition, microbial activity, and the quality of the final compost product. The primary objective is to achieve a balanced C : N ratio, typically in the range of 25 : 1 to 30 : 1, which is ideal for microbial metabolism and efficient organic matter breakdown. High-carbon materials such as cereal straw, sugarcane bagasse, and sawdust provide the energy source for microbes but decompose slowly due to their lignocellulosic structure and low nitrogen content. Conversely, high-nitrogen materials like animal manures, green vegetable waste, and legume residues supply the necessary nitrogen for microbial protein synthesis but can lead to NH_3 volatilization and odor issues if used in excess [15]. Blending these materials in appropriate proportions not only achieves the target C : N ratio but also helps regulate moisture content, another critical parameter for microbial activity (optimal range: 50–60%). For example, mixing dry straw (low moisture, high C) with wet vegetable waste (high moisture, high N) can balance both C : N and water content. Additionally, the inclusion of bulking agents, such as wood chips or rice husk, improves pile porosity and aeration, preventing compaction and the development of anaerobic zones. Advanced composting operations often use mathematical models or software to predict and optimize blend compositions based on feedstock analysis. Proper feedstock blending also mitigates the risk of heavy metal contamination or phytotoxicity by diluting problematic inputs. Overall, strategic blending is a chemical engineering approach that transforms heterogeneous agro-waste streams into a homogenous, processable substrate, ensuring consistent process kinetics, minimizing emissions, and producing a mature, nutrient-rich compost suitable for agricultural application as shown in Figure 8.4 [40].

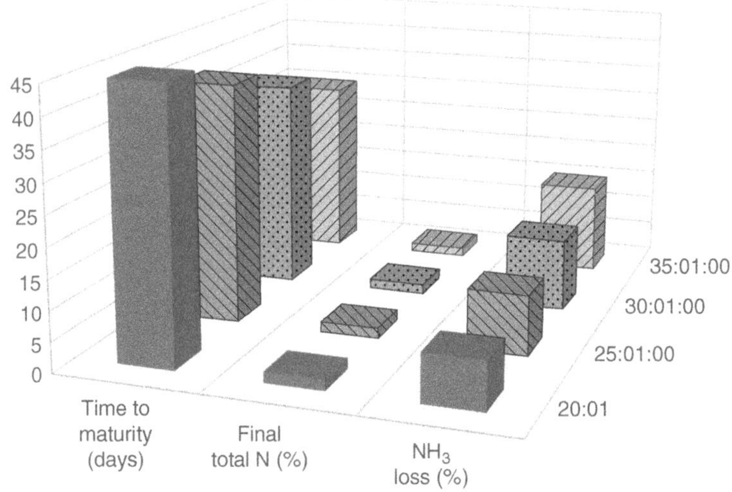

Figure 8.4 Effect of C : N ratio on composting rate and final nitrogen content.

8.6.1.2 Particle Size Reduction

Particle size reduction is a critical pretreatment step in the composting process, significantly impacting the rate and extent of organic matter degradation. Shredding or grinding agro-waste to particle sizes less than 20 mm increases the specific surface area available for microbial colonization and enzymatic hydrolysis, thereby accelerating the breakdown of complex polymers such as cellulose, hemicellulose, and lignin. This enhanced surface area facilitates more effective moisture distribution and oxygen penetration throughout the compost matrix, both of which are essential for sustaining aerobic microbial activity. Research has demonstrated that reducing particle size can decrease composting time by up to 30%, improve temperature profiles, and increase the uniformity of decomposition. However, excessive size reduction (<5 mm) can lead to compaction, reduced porosity, and restricted airflow, potentially resulting in anaerobic microenvironments that inhibit aerobic microbes and promote the formation of malodorous compounds. Therefore, an optimal particle size distribution – typically between 10 and 30 mm, depending on the feedstock – is recommended to balance surface area enhancement with adequate pile structure and aeration. In industrial-scale composting, mechanical shredders, hammer mills, or rotary choppers are commonly employed for this purpose. Particle size reduction also aids in the homogenization of diverse agro-waste inputs, facilitating more consistent process control and product quality [2]. Moreover, smaller particles are more amenable to subsequent process steps such as mixing, turning, and screening. From a chemical engineering perspective, optimizing particle size is a cost-effective intervention that enhances process kinetics, reduces energy requirements for aeration and turning, and contributes to the production of a stable, mature compost with improved agronomic value shown in Figure 8.5.

8.6.2 Aeration Control

Aeration control is a cornerstone of efficient composting, as it governs the oxygen supply necessary for aerobic microbial metabolism, heat generation, and the

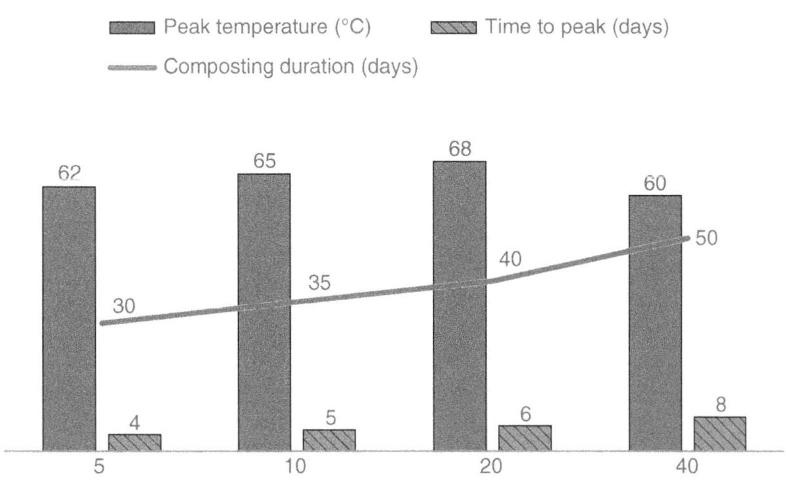

Figure 8.5 Impact of particle size on compost temperature profile.

prevention of anaerobic conditions that can lead to odor and greenhouse gas emissions. Forced aeration systems, such as aerated static piles or in-vessel reactors, utilize blowers or perforated pipes to deliver a controlled flow of air throughout the compost mass, maintaining oxygen concentrations above the critical threshold of 10%. This approach enables precise regulation of the composting environment, supporting high microbial activity and rapid organic matter degradation. Adequate aeration also facilitates the removal of excess heat and water vapor, helping to maintain optimal temperature (55–65 °C) and moisture (50–60%) conditions. Insufficient aeration can result in oxygen depletion, leading to the formation of anaerobic zones, accumulation of volatile fatty acids, and the emission of malodorous gases such as NH_3, H_2S, and CH_4 (methane). Conversely, excessive aeration may cause excessive moisture loss, cooling of the pile, and increased nitrogen loss through NH_3 volatilization. Therefore, aeration rates must be carefully calibrated, typically ranging from 0.2 to 0.5 L air/kg waste/minutes, depending on pile size, feedstock properties, and process phase. Modern composting facilities often employ automated aeration systems with feedback controls based on real-time monitoring of temperature, oxygen, and moisture levels as shown in Figure 8.6. This integration of process control and chemical engineering principles not only enhances composting efficiency and product consistency but also minimizes environmental impacts, making aeration control a pivotal factor in the sustainable management of agro-waste composting operations [33, 41].

8.6.3 Temperature Monitoring

Temperature monitoring is a vital aspect of compost process management, directly influencing microbial succession, pathogen reduction, and the overall rate of organic matter transformation. The composting process is characterized by distinct temperature phases: mesophilic (20–40 °C), thermophilic (55–65 °C), and maturation

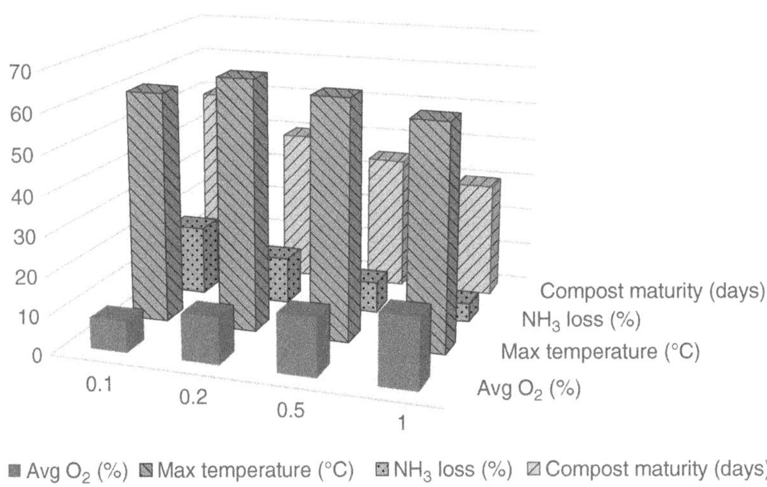

Figure 8.6 Aeration rate and its effect on oxygen, temperature, and nitrogen loss.

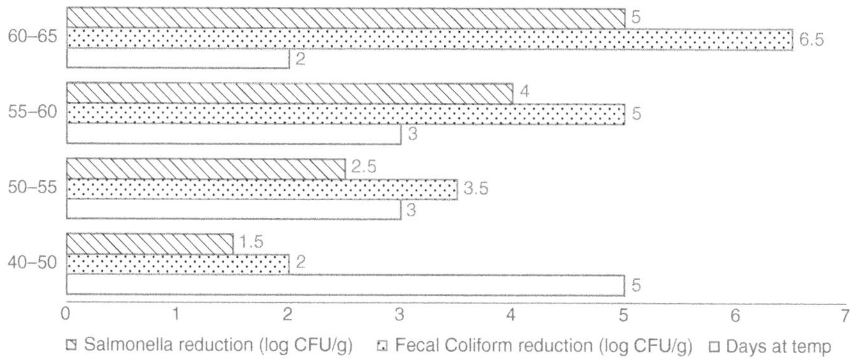

Figure 8.7 Temperature monitoring and pathogen reduction during composting.

(ambient). Maintaining the compost mass within the thermophilic range is essential for the rapid degradation of complex organic materials and the effective inactivation of pathogens and weed seeds, as stipulated by regulatory standards (e.g., sustained >55 °C for at least 3 days). Automated temperature-monitoring systems, often integrated with data loggers and wireless sensors, provide real-time feedback on the internal conditions of the compost pile or reactor as shown in Figure 8.7. This data is used to adjust aeration rates, turning frequency, and moisture addition, ensuring that optimal thermophilic conditions are sustained without overheating (>70 °C), which can inhibit microbial activity and reduce compost quality. Temperature profiles also serve as indicators of process progress and stability; a rapid rise followed by a gradual decline signifies active microbial decomposition and the transition toward compost maturity. In advanced facilities, temperature data is linked to process control software, enabling predictive adjustments and minimizing manual intervention. From a chemical engineering perspective, precise temperature control enhances process kinetics, reduces composting time, and ensures regulatory compliance for pathogen destruction. Furthermore, it contributes to the production of a mature, stable compost with high agronomic value and minimal risk of phytotoxicity or disease transmission [42].

Compost quality is regulated by standards (e.g., US EPA, EU, BIS in India) specifying limits for heavy metals, pathogens, and maturity indices ($C:N < 20$, $GI > 80\%$).

8.7 Conclusion

Composting stands as a crucial technology for sustainable agro-waste management, offering a pathway to convert waste into a valuable resource. This review highlights the evolution of composting methodologies from traditional, low-input heap and windrow systems to advanced, technology-driven vermicomposting, in-vessel, and thermal composting techniques. Advanced methods offer significant advantages in terms of process control, decomposition rate, and compost quality, but often come with higher capital and operational costs.

References

1 Batinic, B., Vaccari, M., Savvilotidou, V., Kousaiti, A., Gidarakos, E., Marinkovic, T., et al. Applied WEEE pre-treatment methods: opportunities to maximizing the recovery of critical metals. *Glob. NEST J.* vol. 20, no. 4, pp. 706–711, Sep. 2018. https://doi.org/10.30955/gnj.002589.

2 Bessi, C., Lombardi, L., Meoni, R., Canovai, A., and Corti, A. Solid recovered fuel: an experiment on classification and potential applications. *Waste Manag.* vol. 47, pp. 184–194, Jan. 2016. https://doi.org/10.1016/J.WASMAN.2015.08.012.

3 Chew, K.W., Chia, S.R., Chia, W.Y., Cheah, W.Y., Munawaroh, H.S.H., and Ong, W.J. Abatement of hazardous materials and biomass waste via pyrolysis and co-pyrolysis for environmental sustainability and circular economy. *Environ. Pollut.* 2021. https://doi.org/10.1016/j.envpol.2021.116836.

4 Obotey Ezugbe, E. and Rathilal, S. Membrane technologies in wastewater treatment: a review. *Membranes (Basel)* vol. 10, no. 5, p. 89, Apr. 2020. https://doi.org/10.3390/membranes10050089.

5 Alidoosti, Z., Sadegheih, A., Govindan, K., and Pishvaee, M.S. Sustainable network design of bioenergies generation based on municipal solid waste (MSW) management under uncertainty. *J. Renew. Sustain. Energy* vol. 15, no. 1, Jan. 2023. https://doi.org/10.1063/5.0128073.

6 Czekała, W., Drozdowski, J., and Łabiak, P. Modern technologies for waste management: a review. *Appl. Sci.* 2023. https://doi.org/10.3390/app13158847.

7 Onu, P., Pradhan, A., and Mbohwa, C. The potential of industry 4.0 for renewable energy and materials development – the case of multinational energy companies. *Heliyon* vol. 9, no. 10, Oct. 2023. https://doi.org/10.1016/j.heliyon.2023.e20547.

8 ur Rehman, K., Schwennen, C., Visscher, C., Plötz, M., Grabowski, N.T., Sultana, M.U.C., et al. Closing the loop with pretreatment and black soldier fly technology for recycling lignocellulose-rich organic by-products: a progressive review. *Carbohydr. Polym. Technol. Appl.* vol. 9, p. 100630, Mar. 2025. https://doi.org/10.1016/j.carpta.2024.100630.

9 Awasthi, S., Duan, Y., Liu, T., Zhou, Y., Qin, S., Liu, H., et al. Sequential presence of heavy metal resistant fungal communities influenced by biochar amendment in the poultry manure composting process. *J. Clean. Prod.* vol. 291, pp. 125947, 2021. https://doi.org/10.1016/j.jclepro.2021.125947.

10 Zota, R.D., Cîmpeanu, I.A., Dragomir, D.A., and Lungu, M.A. Practical approach for smart and circular cities: chatbots used in waste recycling. *Appl. Sci.* vol. 14, no. 7, p. 3060, Apr. 2024. https://doi.org/10.3390/app14073060.

11 Savon, D.Y., Kolotyri, K.P., and Romanov, A.V. Improving the ecological efficiency of the processing industry of agricultural security on the basis of economic instruments. *Russ. J. Ind. Econ.* 2019. https://doi.org/10.17073/2072-1633-2019-2-305-315.

12 Tian, Z., Niu, Y., Fan, D., Sun, L., Ficsher, G., Zhong, H., et al. Maintaining rice production while mitigating methane and nitrous oxide emissions from paddy fields in China: evaluating tradeoffs by using coupled agricultural systems models. *Agric. Syst.* vol. 159, pp. 175–186, Jan. 2018. https://doi.org/10.1016/j.agsy.2017.04.006.

13 Wasiak, A.L. Effect of biofuel production on sustainability of agriculture. *Procedia Eng.* 2017. https://doi.org/10.1016/j.proeng.2017.03.192.

14 Singhania, R.R., Guo, W., de Souza Vendenberghe, L.P., Mannina, G., and Kim, S.H. Bioresource technology for bioenergy, bioproducts & environmental sustainability. *Bioresour. Technol.* vol. 347, Mar. 2022. https://doi.org/10.1016/j.biortech.2022.126736.

15 Toldrá-Reig, F., Mora, L., and Toldrá, F. Trends in biodiesel production from animal fat waste. *Appl. Sci.* vol. 10, no. 10, p. 3644, May. 2020. https://doi.org/10.3390/app10103644.

16 Yıldız, Ü.Y., Keçili, R., and Hussain, C.M. Green and sustainable chemistry. In: *Green Imprinted Materials.* Elsevier, 2024, pp. 3–25. https://doi.org/10.1016/B978-0-443-15431-7.00014-3.

17 Ab Halim, M.H., Nor Anuar, A., Abdul Jamal, N.S., Azmi, S.I., Ujang, Z., and Bob, M.M. Influence of high temperature on the performance of aerobic granular sludge in biological treatment of wastewater. *J. Environ. Manage.* vol. 184, pp. 271–280, Dec. 2016. https://doi.org/10.1016/j.jenvman.2016.09.079.

18 Onu, P., Mbohwa, C., and Pradhan, A. An analysis of the application of machine learning techniques in anaerobic digestion. In: *2023 International Conference on Control, Automation and Diagnosis (ICCAD 2023).* 2023. https://doi.org/10.1109/ICCAD57653.2023.10152335.

19 Okoro, O.V., Sun, Z., and Birch, J. Techno-economic assessment of a scaled-up meat waste biorefinery system: a simulation study. *Materials (Basel)* vol. 12, no. 7, p. 1030, Mar. 2019. https://doi.org/10.3390/ma12071030.

20 Nanda, S., Azargohar, R., Dalai, A.K., and Kozinski, J.A. An assessment on the sustainability of lignocellulosic biomass for biorefining. *Renew. Sustain. Energy Rev.* vol. 50, pp. 925–941, Oct. 2015. https://doi.org/10.1016/j.rser.2015.05.058.

21 Dey, T., Bhattacharjee, T., Nag, P., Ritika, Ghati, A., and Kuila, A. Valorization of agro-waste into value added products for sustainable development. *Bioresour. Technol. Reports* vol. 16, Dec. 2021. https://doi.org/10.1016/j.biteb.2021.100834.

22 Bhatia, S.K., Jagtap, S.S., Bedekar, A.A., Bhatia, R.K., Patel, A.K., Pant, D., et al. Recent developments in pretreatment technologies on lignocellulosic biomass: effect of key parameters, technological improvements, and challenges. *Bioresour. Technol.* vol. 300, p. 122724, Mar. 2020. https://doi.org/10.1016/J.BIORTECH.2019.122724.

23 Khan, T.F., Ullah, M.W., and Huq, S.M.I. Heavy metal contents of different wastes used for compost. *J. Miner. Mater. Charact. Eng.* vol. 4, pp. 241–249, 2016.

24 Behnia, M., Ghahderijani, M., Kaab, A., and Behnia, M. Evaluation of sustainable energy use in sugarcane production: a holistic model from planting to harvest and life cycle assessment. *Environ. Sustain. Indic.* vol. 26, Jun. 2025. https://doi.org/10.1016/j.indic.2025.100617.

25 Borchard, R., Zeiss, R., and Recker, J. Digitalization of waste management: insights from German private and public waste management firms. *Waste Manag. Res. J. A Sustain. Circ. Econ.* vol. 40, no. 6, pp. 775–792, Jun. 2022. https://doi.org/10.1177/0734242X211029173.

26 Lu, J., Brigham, C.J., Li, S., and Sinskey, A.J. Ralstonia eutropha H16 as a platform for the production of biofuels, biodegradable plastics, and fine chemicals from diverse carbon resources. In: *Biotechnology for Biofuel Production and Optimization*. Jan. 2016, pp. 325–351. https://doi.org/10.1016/B978-0-444-63475-7.00012-1.

27 Onu, P. and Mbohwa, C. Front matter. In: *Agricultural Waste Diversity and Sustainability Issues Sub-Saharan Africa as a Case Study*. Jan. 2021, p. iii. https://doi.org/10.1016/B978-0-323-85402-3.01001-9.

28 Sujauddin, M., Huda, S.M.S., and Hoque, A.T.M.R. Household solid waste characteristics and management in Chittagong, Bangladesh. *Waste Manag.* vol. 28, no. 9, pp. 1688–1695, Jan. 2008. https://doi.org/10.1016/j.wasman.2007.06.013.

29 Brouwer, M.T., Thoden van Velzen, E.U., Augustinus, A., Soethoudt, H., De Meester, S., and Ragaert, K. Predictive model for the Dutch post-consumer plastic packaging recycling system and implications for the circular economy. *Waste Manag.* vol. 71, pp. 62–85, Jan. 2018. https://doi.org/10.1016/j.wasman.2017.10.034.

30 Agrawal, P.S., Belkhode, P.N., and Rokhum, S.L. Agro-waste-produced biofuels. In: *Sustainability in Biofuel Production Technology*. Wiley-Blackwell, 2022. https://doi.org/10.1002/9781119888864.ch7.

31 Waqas, M., Hashim, S., Humphries, U.W., Ahmad, S., Noor, R., Shoaib, M., et al. Composting processes for agricultural waste management: a comprehensive review. *Processes* 2023. https://doi.org/10.3390/pr11030731.

32 F.E., Boldrin, A., Andersen, J.K., Møller, J., Christensen, T.H. Composting and compost utilisation: accounting of greenhouse gas emissions and global warming potentials. *Waste Manag. Res. J. a Sustain. Circ. Econ.* vol. 27, pp. 65–79, no. 9, 2009.

33 Iloms, E., Ololade, O.O., Ogola, H.J.O., and Selvarajan, R. Investigating industrial effluent impact on municipal wastewater treatment plant in Vaal, South Africa. *Int. J. Environ. Res. Public Health* vol. 17, no. 3, p. 1096, Feb. 2020. https://doi.org/10.3390/ijerph17031096.

34 Bhatia, S.K., Kim, S.-H., Yoon, J.-J., and Yang, Y.-H. Current status and strategies for second generation biofuel production using microbial systems. *Energy Convers. Manag.* vol. 148, pp. 1142–1156, Sep. 2017. https://doi.org/10.1016/j.enconman.2017.06.073.

35 Kurniasih, E. Penggunaan katalis heterogen untuk produksi biodisel. *J. Sains dan Teknol. Reaksi*. 2018. https://doi.org/10.30811/jstr.v15i1.522.

36 Pecorini, I. and Iannelli, R. Characterization of excavated waste of different ages in view of multiple resource recovery in landfill mining. *Sustainability* vol. 12, no. 5, p. 1780, Feb. 2020. https://doi.org/10.3390/su12051780.

37 Hajam, Y.A., Kumar, R., and Kumar, A. Environmental waste management strategies and vermi transformation for sustainable development. *Environ. Challenges* vol. 13, p. 100747, Dec. 2023. https://doi.org/10.1016/j.envc.2023.100747.

38 Hamad, T.A., Agll, A.A., Hamad, Y.M., and Sheffield, J.W. Solid waste as renewable source of energy: current and future possibility in Libya. *Case Stud.*

Therm. Eng. vol. 4, pp. 144–152, Nov. 2014. https://doi.org/10.1016/j.csite.2014.09.004.

39 Giannothanasis, E., Spanoudaki, E., Kinnas, S., Ntatsi, G., Voogt, W., and Savvas, D. Development and validation of an innovative algorithm for sodium accumulation management in closed-loop soilless culture systems. *Agric. Water Manag*. vol. 301, Aug. 2024. https://doi.org/10.1016/j.agwat.2024.108968.

40 Sarno, M. and Iuliano, M. Biodiesel production from waste cooking oil. *Green Process. Synth*. vol. 8, no. 1, pp. 828–836, Jan. 2019. https://doi.org/10.1515/gps-2019-0053.

41 W.T., Mittal, S., Khan, M.A., and Romero, D. A knowledge-based framework for implementing industrial robotics and automation. *Prod. Plan. Control*. vol. 30, pp. 179–196, no. 16, 2019.

42 Scarlat, N., Dallemand, J.F., and Banja, M. Possible impact of 2020 bioenergy targets on European union land use. A scenario-based assessment from national renewable energy action plans proposals. *Renew. Sustain. Energy Rev*. vol. 18, pp. 595–606, 2013. https://doi.org/10.1016/j.rser.2012.10.040.

9

Glycerol: From Abundant Byproduct to Valuable Bio-oil

9.1 Introduction

Glycerol, systematically designated as 1,2,3-propanetriol and commonly known as glycerin, is a trihydric alcohol (a polyol containing three hydroxyl groups) with the molecular formula $C_3H_8O_3$. In its pure state, it is an odorless, colorless, and viscous liquid. Historically, glycerol has been a valuable commodity with extensive applications across numerous industrial sectors, including its role as a humectant, solvent, and sweetener in food and beverages; a moisturizer and emollient in cosmetics and personal care products; and as a solvent, excipient, or intermediate in the pharmaceutical industry. The molecular structure of glycerol, with its three highly reactive hydroxyl groups attached to a short propane backbone, bestows upon its exceptional chemical versatility and unique physical properties such as high solubility in water (H_2O), hygroscopicity, and the capacity for extensive hydrogen (H_2) bonding. The contemporary industrial landscape is increasingly characterized by a strong emphasis on sustainability and the utilization of renewable resources. This trend is driven by both environmental concerns and economic incentives, fostering the development of innovative processes for producing chemicals and materials from bio-based feedstocks, thereby aligning with the principles of a sustainable industrial ecosystem.

A significant factor contributing to the increased availability and economic importance of glycerol is the rapid expansion of the global biodiesel industry. Biodiesel, a renewable and biodegradable fuel produced from the transesterification of triglycerides found in sources like vegetable oils, animal fats, and recycled greases, serves as a key alternative to conventional petroleum diesel. However, the fundamental stoichiometry of the transesterification reaction dictates the co-production of glycerol. Specifically, for every approximately 100 units of biodiesel produced by mass, about 10 units of crude glycerol are generated as a significant byproduct stream [1]. The substantial and ever-increasing volume of crude glycerol produced has saturated traditional markets, leading to a marked decrease in its market value. This economic shift has created a compelling need and strong incentive for the development and implementation of innovative, economically viable, and environmentally sustainable

Agro-Waste Management and Valorization, First Edition. Pratibha S. Agrawal, et al.

methods for the high-value utilization (valorization) of this abundant crude glycerol surplus.

Glycerol, as a versatile polyol, offers numerous opportunities for chemical transformation due to its three reactive hydroxyl groups. These functional groups provide multiple potential reaction sites, enabling a wide range of chemical conversions including dehydration (removal of H_2O), oxidation (introduction of oxygen or increase in oxidation state), etherification (formation of C—O—C bonds), esterification (reaction with carboxylic acids), and hydrogenolysis (cleavage of C—C or C—O bonds with H_2) [2]. These intrinsic chemical properties, combined with its non-toxicity and biodegradability, position glycerol as a promising platform molecule for the synthesis of a diverse portfolio of value-added chemicals and biofuels.

The production of glycerol in large volumes as a co-product from biodiesel transesterification presents both challenges and opportunities [3]. While biodiesel itself contributes to sustainability by displacing fossil fuels, the large-scale accumulation of glycerol raises concerns regarding waste management and environmental impact if not properly utilized [4, 5]. Therefore, the conversion of this abundant byproduct into higher-value products, often termed glycerol valorization, holds significant importance. This strategy directly reduces the volume of waste generated, minimizing the environmental burden associated with disposal. It also transforms a waste stream into a renewable resource, enabling the production of valuable chemicals and fuels from a bio-based source, thereby contributing to resource efficiency and the development of a circular economy [6]. Furthermore, successful valorization pathways create new economic opportunities, potentially turning a disposal cost into a revenue stream, thus improving the overall economic sustainability of biodiesel production. These benefits underscore the strategic importance of developing efficient glycerol conversion technologies (Figure 9.1).

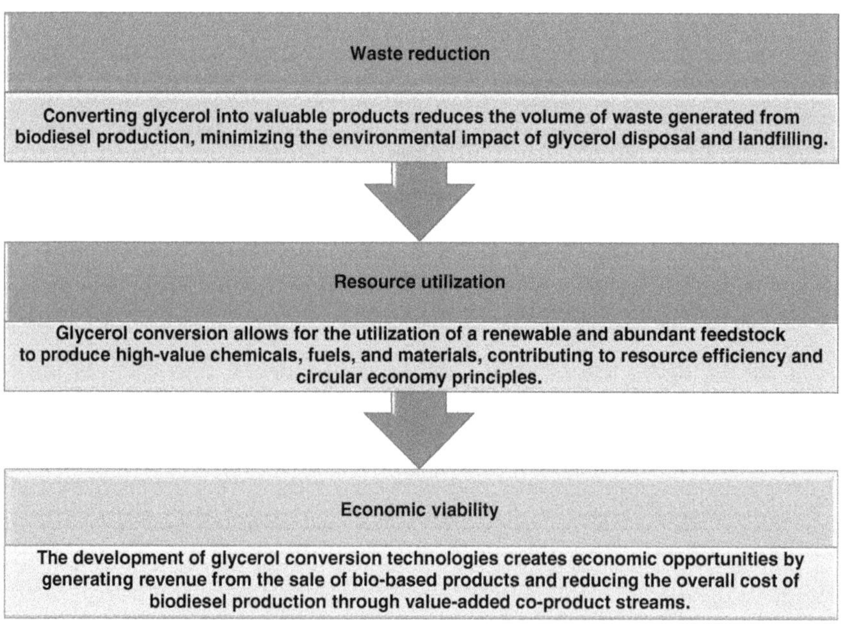

Figure 9.1 Importance of glycerol conversion.

Among the various valorization pathways explored for glycerol, its conversion into bio-oil has attracted significant attention from both researchers and industry [7]. Bio-oil, a complex organic liquid, is typically produced from the thermochemical decomposition of biomass or biomass-derived feedstocks. From glycerol, bio-oil consists predominantly of oxygenated organic compounds, including aldehydes, ketones, carboxylic acids, alcohols, and esters. This product stream is seen as a promising renewable energy source and a potential feedstock for subsequent upgrading into transportation fuels or for the recovery of valuable chemical intermediates. Glycerol can be converted to bio-oil through several thermochemical processes, notably pyrolysis, hydrothermal liquefaction (HTL), and catalytic pyrolysis [8].

The objectives articulated for this chapter are designed to provide a comprehensive and technically detailed review of the subject:

1. To provide an in-depth description of the physical and chemical properties of glycerol, including a detailed examination of its molecular structure, reactivity, and how these characteristics influence its conversion potential.
2. To elucidate the biochemical process of transesterification in detail, focusing on its role in the industrial production of biodiesel and explicitly explaining how this process results in the co-generation of glycerol as a major byproduct.
3. To critically assess the importance of glycerol valorization through conversion to bio-oil and to thoroughly explore the various conversion technologies, including their mechanisms and performance characteristics, and the potential applications for the resulting bio-based products in energy, chemicals, and materials sectors.
4. To conduct a detailed discussion of the current technical, economic, and logistical challenges associated with glycerol conversion, highlight promising opportunities for advancement, and examine future prospects, considering technological innovations, market dynamics, and the impact of policy frameworks.

By addressing these objectives with rigorous technical detail and critical analysis, this chapter aims to serve as a valuable resource for researchers, engineers, and policymakers interested in the sustainable utilization of glycerol and the development of bio-based industries.

9.2 Production of Glycerol in Biodiesel Production

The pervasive availability of glycerol as a potential feedstock for valorization is directly attributable to its integral role as a major co-product in the industrial-scale synthesis of biodiesel [9]. Biodiesel is predominantly manufactured through the transesterification reaction, which involves the chemical reaction between triglycerides (the primary components of vegetable oils, animal fats, and other lipid feedstocks) and a short-chain alcohol, typically methanol (methanolysis) or ethanol (ethanolysis). This reaction is catalyzed by an acid, base, or enzyme, yielding fatty acid alkyl esters (FAME or FAEE), which constitute biodiesel, and glycerol as a distinct phase [10–12]. The core reaction involves a series of reversible steps where the alcohol molecule displaces the glycerol moiety from each ester linkage

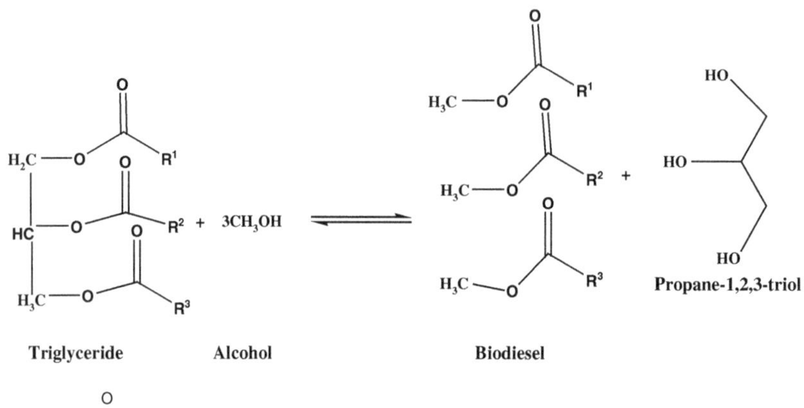

Figure 9.2 Chemical reaction of glycerol formation. [15]/Elsevier.

in the triglyceride. The process effectively cleaves the ester bonds, resulting in the formation of three molecules of fatty acid alkyl ester and one molecule of glycerol. The overall stoichiometry is generally represented as shown in Figure 9.2 [13, 14].

In large-scale industrial operations, base-catalyzed transesterification is the most widely adopted method due to its high reaction rates under relatively mild conditions (typically 50–65 °C and atmospheric pressure). Homogeneous base catalysts, such as sodium hydroxide (NaOH) or potassium hydroxide (KOH), are commonly used [16]. The catalytic mechanism involves the abstraction of a proton from the alcohol by the strong base, generating a highly reactive alkoxide anion. This anion then initiates a nucleophilic attack on the carbonyl carbon of the ester group in the triglyceride. A tetrahedral intermediate form and subsequently collapses, releasing a fatty acid alkyl ester molecule and a diglyceride anion [17, 18]. This sequence is repeated successively for the remaining two ester linkages on the glycerol backbone, ultimately liberating a molecule of glycerol. Heterogeneous catalysts, including solid bases like calcium oxide (CaO) or magnesium oxide (MgO), solid acids such as sulfated metal oxides, or immobilized enzymes, are also employed to facilitate transesterification. These catalysts offer advantages in terms of ease of separation from the reaction products and potential for reuse, which simplifies downstream processing and reduces waste, although they may require different reaction conditions or exhibit lower intrinsic activity compared to homogeneous bases, and can be susceptible to deactivation by impurities in the feedstock [19]. Irrespective of the specific catalyst system or alcohol employed, the chemical pathway of transesterification inherently leads to the generation of a substantial stream of crude glycerol alongside the desired biodiesel product.

9.2.1 Quantity and Quality of Glycerol Generated

The quantitative output of crude glycerol from biodiesel production is directly linked to the scale of the operation and the fundamental stoichiometry of the transesterification reaction [11, 20]. As a general guideline, the production of approximately 100 units of biodiesel by weight typically results in the co-production of about

10 units of crude glycerol [21, 22]. With the global biodiesel market experiencing continuous growth driven by renewable energy mandates and sustainability initiatives, the total volume of glycerol produced annually has reached millions of tons. This large-scale generation has created a significant surplus on the global market, fundamentally altering the supply–demand balance and causing a sharp decline in the market price of crude glycerol, making its valorization economically crucial.

However, the quality of this crude glycerol varies considerably depending on numerous factors, including the quality of the initial triglyceride feedstock (e.g., presence of free fatty acids, H_2O), the specific biodiesel production process parameters (e.g., temperature, pressure, reactant ratios), the type and concentration of catalyst utilized, and the efficiency of subsequent separation and purification steps [23, 24]. Crude glycerol obtained from base-catalyzed transesterification, the most common industrial process, is a complex mixture that contains, in addition to glycerol, various impurities such as unreacted alcohol (methanol or ethanol), residual catalyst (alkali soaps and salts), unreacted triglycerides and partial glycerides (mono- and diglycerides), free fatty acids, soap (formed by the saponification of triglycerides or neutralization of free fatty acids by the alkaline catalyst), H_2O (present in the feedstock or formed during saponification), and other organic non-glycerol materials (ONGM) [25]. The concentration and nature of these impurities significantly influence the physical and chemical properties of the crude glycerol, such as its viscosity, pH, and corrosivity. High levels of H_2O, salts, and soap make the crude glycerol unsuitable for many traditional high-purity applications (pharmaceuticals, cosmetics) without extensive and costly purification processes. Variability in impurity content between different batches or producers poses challenges for downstream processing technologies that require a consistent feedstock quality.

9.2.2 Challenges in Glycerol Management

The characteristics of crude glycerol – its large volume, variable quality, and low market price – present significant management challenges for biodiesel producers and the broader bio-based chemicals and fuels industries:

1. **Surplus and Depressed Market Value:** The massive increase in glycerol production has overwhelmed traditional markets, which have relatively limited demand for technical and pharmaceutical grade glycerol compared to the volume of crude glycerol available. This surplus has led to a significant depression in the price of crude glycerol, often below the cost of purification to technical grade [26]. For biodiesel producers, this transforms what was once a valuable co-product revenue stream into a cost for disposal or requires investment in purification facilities, impacting the overall economics of biodiesel production. Finding new, large-volume markets that can absorb the surplus is a critical challenge.
2. **Impurity Management and Quality Control:** The presence of diverse impurities in crude glycerol, such as salts, soaps, H_2O, and organic residues, creates significant technical challenges for downstream valorization processes. Many

catalysts used in thermochemical or biochemical conversion processes are sensitive to these impurities, which can cause deactivation through poisoning, fouling, or sintering [27]. Biological conversion processes can be inhibited by high salt concentrations, pH variations, or specific organic inhibitors. The variability in the composition of crude glycerol from different sources or batches requires either robust and costly pretreatment steps to achieve consistent feedstock quality or the development of conversion technologies that are tolerant to a range of impurities, which remains a significant research challenge.

3. **Diversification of Applications and Market Development:** The traditional markets for glycerol are insufficient to absorb the current production volume. There is a strong need to develop and commercialize new, large-scale applications for glycerol beyond its established uses [28]. This involves identifying promising conversion pathways (e.g., to bio-oil, specific platform chemicals, polymers) and developing the necessary processing technologies to produce marketable products [8]. Establishing supply chains, gaining regulatory approvals, and building consumer acceptance for these new bio-based products are also essential aspects of market development.
4. **Environmental Risks of Disposal:** Improper disposal of crude glycerol can have significant negative environmental consequences. Crude glycerol is characterized by high chemical oxygen demand (COD) and biological oxygen demand (BOD), making direct discharge into H_2O bodies environmentally unacceptable and subject to strict regulations. While options like anaerobic digestion or incineration exist, these methods have associated costs and potential environmental impacts (e.g., air emissions). Developing sustainable valorization pathways is crucial for avoiding these disposal challenges and transforming a potential environmental liability into a valuable resource.

Successfully addressing these interconnected challenges through integrated technological development, market strategy, and supportive policy frameworks is essential for unlocking the full potential of glycerol as a sustainable and economically viable resource and enhancing the long-term sustainability of the bio-based economy [29].

9.3 Conversion Technologies of Glycerol to Bio-oil

The conversion of glycerol into bio-oil can be achieved through a variety of technological approaches, broadly categorized as thermochemical and biological processes. Each method utilizes different fundamental principles and operating conditions to break down or rearrange the glycerol molecule, yielding liquid products with varying compositions and properties. These technologies represent distinct engineering and chemical challenges and opportunities, and selecting the most appropriate route depends on factors such as the quality of the glycerol feedstock, the desired properties of the bio-oil product, and economic considerations. Ongoing research is focused on improving the efficiency, selectivity, and economic viability of these conversion pathways.

9.3.1 Pyrolysis of Glycerol: Thermally Induced Decomposition

Pyrolysis is a thermochemical process involving the decomposition of organic materials through heating in an inert atmosphere (typically nitrogen [N]) or under vacuum [30]. For glycerol, this process typically occurs at elevated temperatures, commonly ranging from 350 °C to over 600 °C. The high temperature provides the energy required to break the chemical bonds within the glycerol molecule (C—C and C—O bonds) and initiate fragmentation and rearrangement reactions. Unlike the pyrolysis of complex lignocellulosic biomass, glycerol pyrolysis involves the decomposition of a relatively simple C_3 oxygenate molecule. The primary reaction pathways involve initial dehydration and cracking steps, leading to the formation of volatile intermediate species. These intermediates then undergo a complex cascade of secondary reactions in the vapor phase, including further cracking, polymerization, condensation, cyclization, and char formation. The products of glycerol pyrolysis typically comprise a non-condensable gas phase (e.g., carbon monoxide (CO), carbon dioxide (CO_2), H_2, light hydrocarbons like methane and ethylene), a liquid bio-oil fraction, and a solid char residue. The relative yields and the chemical composition of these product fractions are highly sensitive to the pyrolysis parameters, particularly the temperature, heating rate, and vapor residence time.

Pyrolysis processes are often categorized based on the heating rate and residence time:

- **Slow Pyrolysis:** This regime is characterized by relatively slow heating rates (typically less than 10 °C/minutes) and long vapor residence times (ranging from minutes to hours) [31]. These conditions favor solid-state reactions and secondary reactions in the vapor phase, leading to higher yields of char and gases and a lower yield of the liquid bio-oil fraction [32]. While useful for producing char (biochar), it is generally not preferred for maximizing liquid bio-oil production from glycerol.
- **Fast Pyrolysis:** This process employs very high heating rates (typically hundreds to thousands of °C/minutes) and very short vapor residence times (typically less than 2 seconds) [6, 33]. The objective is to rapidly heat the feedstock, quickly volatilize the decomposition products, and rapidly quench the resulting vapors to condense the maximum amount of liquid bio-oil while minimizing secondary reactions that lead to char and non-condensable gases. This rapid heating and quenching are critical for capturing reactive intermediates in the liquid phase. Fluidized-bed reactors (bubbling or circulating) are commonly used for fast pyrolysis due to their excellent heat and mass transfer properties, which enable rapid heating of the feedstock. Fast pyrolysis bio-oil from glycerol is a complex mixture predominantly composed of oxygenated compounds such as acrolein, acetol, carboxylic acids, light aldehydes, ketones, and furan derivatives.
- **Flash Pyrolysis:** Represents an extreme case of fast pyrolysis, characterized by extremely high heating rates and ultra-short vapor residence times [34]. This aims to maximize the yield of primary pyrolysis products by minimizing both primary char formation and secondary vapor-phase reactions. Specialized

reactor designs like ablative reactors or entrained flow reactors can achieve flash pyrolysis conditions. The product spectrum is similar to fast pyrolysis but with potentially higher yields of the initial reactive intermediates.

The liquid fraction obtained from glycerol pyrolysis is referred to as bio-oil [35]. As a consequence of glycerol's high oxygen content, this bio-oil is typically highly oxygenated (ranging from 35 to over 50% oxygen by weight for non-catalytic pyrolysis bio-oil). This high oxygen content imparts several undesirable properties, including high acidity (due to carboxylic acids), corrosivity, low heating value compared to fossil fuels, and chemical instability over time (aging, polymerization) [36], making it challenging to use directly in conventional engines or pipelines. Significant research focuses on optimizing pyrolysis parameters, reactor design, and integrating catalytic approaches (catalytic pyrolysis) to improve bio-oil yield, reduce its oxygen content, and enhance its properties for subsequent upgrading or direct use.

9.3.2 HTL of Glycerol: Conversion in Hot Compressed H_2O

HTL is a thermochemical conversion process that utilizes subcritical or supercritical H_2O as the reaction medium [37, 38]. This process operates at elevated temperatures, typically ranging from 250 °C to 370 °C, and high pressures, commonly from 5 MPa up to 25 MPa or even higher, pushing into the supercritical regime (above 374 °C and 22 MPa for H_2O) [39]. Under these conditions, H_2O behaves as a reactive solvent with unique properties. Its dielectric constant decreases significantly at high temperatures, making it a better solvent for organic compounds, while its ionic product increases, allowing it to act as an acid–base catalyst and facilitating hydrolysis reactions. Recent advances carried on catalysts for improving hydrocarbon compounds in bio-oil of biomass catalytic pyrolysis [40]. The high pressure keeps the H_2O and other components in a liquid or dense phase, suppressing gas-phase reactions and favoring the formation of a liquid bio-oil product.

For glycerol, HTL in hot compressed H_2O initiates a complex series of reactions. H_2O acts as a reactant, promoting hydrolysis, and as a solvent facilitating mass transfer. Key reactions include dehydration, cracking of C—C bonds, decarboxylation (removal of CO_2), and decarbonylation (removal of CO). The presence of H_2O can also promote in situ H_2 generation through steam-reforming reactions, providing H_2 for hydrodeoxygenation (HDO), a crucial step for reducing oxygen content. HTL is particularly advantageous for processing wet feedstocks like crude glycerol solutions, as it avoids the energy-intensive drying step required for pyrolysis. Bio-oil produced from glycerol via HTL typically exhibits a lower oxygen content, reduced acidity, and a higher heating value compared to bio-oil from conventional pyrolysis, making it a more suitable feedstock for fuel applications [41]. However, HTL requires expensive, high-pressure reactors and equipment, which increases capital costs. Furthermore, a significant amount of aqueous phase byproduct is produced, containing dissolved organic and inorganic compounds, which requires extensive and potentially costly treatment or valorization before disposal [42]. Research efforts in HTL of glycerol focus on optimizing reaction temperature, pressure, residence time, and water-to-glycerol ratio; exploring the use of catalysts (homogeneous or heterogeneous) to enhance reaction rates and selectivity; and developing efficient strategies for managing the aqueous phase.

9.3.3 Catalytic Conversion of Glycerol: Tailored Transformations

Catalytic conversion plays a crucial and transformative role in glycerol valorization, enabling the selective production of desired products, including bio-oil components or specific platform chemicals, often under milder conditions or with higher efficiency than purely thermal processes. Catalysts are substances that increase the rate of specific chemical reactions without being consumed in the process. Critically, in complex systems like glycerol conversion, catalysts also influence the reaction pathway by providing specific active sites that lower the activation energy for particular transformations, thus controlling reaction kinetics and product distributions. Given glycerol's structure with three reactive hydroxyl groups, catalytic processes can facilitate a wide spectrum of reactions, including dehydration (removal of H_2O), hydrogenolysis (cleavage of C—O or C—C bonds with H_2), HDO (removal of oxygen as H_2O), esterification (reaction with carboxylic acids), etherification (formation of C—O—C bonds), oxidation, cracking (cleavage of the carbon backbone), oligomerization (formation of larger molecules), and aromatization (formation of aromatic hydrocarbons) [43].

The design and selection of the catalyst are paramount, as different types of active sites promote different reaction mechanisms. For example, acid catalysts (Brønsted or Lewis) are highly effective for catalyzing dehydration reactions, leading to the formation of reactive intermediates like acrolein and acetol. Metal catalysts, particularly noble metals (e.g., Pt, Pd, Ru) or non-noble metals (e.g., Ni, Cu, Co), often supported on high-surface-area materials, are essential for hydrogenolysis and HDO reactions, which are crucial for reducing the oxygen content of glycerol and its derivatives and producing hydrocarbons or less oxygenated compounds suitable for biofuels [44]. Bifunctional catalysts, incorporating multiple types of active sites (e.g., acid–metal sites, or different types of acid sites) on the same material, are particularly powerful as they can catalyze sequential or tandem reactions within a single reactor, enhancing overall efficiency and selectivity. For example, a catalyst with both acid sites for dehydration and metal sites for hydrogenolysis can directly convert glycerol to propane or specific diols in a single step [45]. Catalytic conversion offers the potential to precisely tailor the composition of the bio-oil product, reducing undesirable characteristics like high oxygen content, acidity, and viscosity, while increasing its heating value and stability, thereby minimizing the need for severe and energy-intensive post-processing upgrading. However, a significant challenge in catalytic glycerol conversion is catalyst deactivation, which can occur due to coke formation, metal sintering, poisoning by impurities (e.g., alkali metals, sulfur [S]) in crude glycerol, or leaching of active sites [46]. Research is intensely focused on developing more robust and stable catalyst formulations, understanding deactivation mechanisms, and developing effective regeneration strategies.

While thermochemical routes are the primary focus for bulk bio-oil production, other methods contribute to glycerol valorization and can yield bio-oil components or related products. Supercritical Fluid Extraction (SFE), typically utilizing CO_2

in its supercritical state, is fundamentally a separation technique [47]. It can be employed to extract specific valuable components from crude glycerol or from a produced bio-oil, serving as a purification or fractionation step. Supercritical fluids can also function as reactive solvents or co-solvents in catalytic conversion processes, enhancing solubility and mass transfer in demanding reactions like catalytic deoxygenation. Biological conversion processes, which employ microorganisms (e.g., bacteria, yeast) or enzymes, operate under milder conditions than thermochemical routes. While fermentation of glycerol is primarily used to produce specific chemicals such as 1,3-propanediol, ethanol, or organic acids, certain metabolic pathways or enzymatic reactions could potentially lead to the synthesis of complex lipid-like molecules or other products that might be considered as components or precursors of bio-oil [48]. However, biological processes often yield specific, pure chemicals rather than a broad bio-oil mixture and can face challenges related to fermentation rates, product recovery, and sensitivity to impurities present in crude glycerol.

A portfolio of conversion technologies exists for transforming glycerol into bio-oil or related products, each with distinct advantages and limitations concerning operating conditions, energy requirements, capital and operating costs, and the resulting product yield and composition [49]. The optimal choice of technology is a complex decision influenced by factors such as the availability and quality of the crude glycerol feedstock, the desired properties and intended applications of the bio-oil or derived products, and the overall economic and environmental context (Table 9.1). Effectively leveraging these technologies is essential for valorizing glycerol as a renewable feedstock, contributing to the production of biofuels, chemicals, and materials, and supporting the transition toward a more sustainable and circular bioeconomy.

9.4 Bio-oil Properties and Applications

Bio-oil, produced from the thermochemical conversion of biomass-derived feedstocks including glycerol, is a complex organic liquid that exhibits properties significantly different from conventional petroleum crude oil [50]. Its effective utilization across various potential applications is contingent upon a thorough understanding of its intrinsic chemical composition, its resulting physical properties, and the implications these properties have for its handling, storage, upgrading, and ultimate end-use. Bio-oil is not a single chemical entity but rather a highly heterogeneous mixture of numerous compounds, whose precise makeup is profoundly influenced by the specific feedstock used (e.g., glycerol vs. lignocellulosic biomass), the conversion technology employed (e.g., pyrolysis vs. HTL), the catalyst utilized (if any), and the specific operating parameters (temperature, pressure, heating rate, residence time) [51].

9.4.1 Chemical Composition of Bio-oil: Molecular Diversity

The chemical composition of glycerol-derived bio-oil is intrinsically complex and highly variable, a direct consequence of the numerous parallel and

Table 9.1 Comparative analysis of different technologies for glycerol conversion to bio-oil.

Technology	Temperature (°C)	Pressure (MPa)	Advantages	Disadvantages	Key products	Applications
Pyrolysis	400–600	Atmospheric	Simple, relatively inexpensive	High oxygen content in bio-oil, char formation	Acrolein, hydroxy-acetone, aldehydes, ketones	Fuel, chemical feedstock
Hydrothermal liquefaction (HTL)	250–350	5–20	Can process wet glycerol, lower oxygen content in bio-oil	High pressure equipment needed; wastewater treatment required	Alcohols, carboxylic acids, esters	Fuel, chemical feedstock
Catalytic pyrolysis	300–500	Atmospheric	Improved bio-oil quality, enhanced product selectivity	Catalyst cost, catalyst deactivation	Depends on catalyst: aromatics, olefins, alcohols	Fuel, chemical feedstock

sequential reactions (dehydration, cracking, oligomerization, etc.) that occur during the conversion process. As glycerol is a C_3 molecule with three oxygen atoms, the initial products and intermediates are typically C_3 oxygenates and their derivatives. The primary organic constituents of glycerol bio-oil are predominantly oxygenated compounds. These include a wide range of functional groups and molecular structures [52, 53]:

- **Oxygenated Compounds:** These typically form the largest fraction of the organic content:
  - *Aldehydes:* Most notably acrolein ($CH_2{=}CH{-}CHO$), a key product of 1,2-dehydration [54]. Due to its α,β-unsaturated structure, acrolein is highly reactive and prone to polymerization and other secondary reactions, contributing to bio-oil instability and char formation. Other shorter-chain aldehydes (e.g., acetaldehyde, propionaldehyde) can be formed via cracking.
  - *Ketones:* Principally acetol (hydroxyacetone, CH_3COCH_2OH), formed via 1,3-dehydration [55]. Acetol is a relatively stable oxygenate and a valuable chemical intermediate [56]. Acetone can also be produced through cracking.
  - *Carboxylic Acids:* Formed from the oxidation or cracking of other oxygenates (e.g., formic acid, acetic acid, propionic acid) [57]. These compounds are responsible for the high acidity and corrosive nature of bio-oil.
  - *Alcohols:* May include unreacted glycerol, as well as products of hydrogenolysis like 1,2-propanediol and 1,3-propanediol [58], and other alcohols formed during conversion.
  - *Furans and Pyranones:* Cyclic oxygenates formed through dehydration and cyclization reactions of intermediates. Furan derivatives can contribute to bio-oil viscosity and reactivity [59].
  - *Phenols:* Can be formed, particularly at higher temperatures or with certain catalysts, through condensation and cyclization reactions of oxygenated intermediates. Phenols can act as antioxidants but also contribute to viscosity and potential stability issues [59].
  - *Ethers:* Formed via etherification reactions, such as polyglycerols (from glycerol self-condensation) or cyclic ethers.
- **Hydrocarbons:** While typically a smaller fraction in bio-oil from non-catalytic glycerol conversion compared to biomass bio-oil, hydrocarbons (alkanes, alkenes, cycloalkanes, aromatics) can be formed through more extensive deoxygenation and cracking pathways, particularly with the use of appropriate catalysts (e.g., metal catalysts, zeolites) and H_2. The formation of light olefins (e.g., propene) and aromatic hydrocarbons (e.g., BTX – benzene, toluene, xylenes) is particularly desirable for fuel applications.
- **Water:** A significant component of bio-oil, either introduced with the feedstock (especially in HTL processes utilizing aqueous glycerol solutions) or produced as a byproduct of dehydration, decarboxylation, and HDO reactions [60]. High H_2O content reduces the energy density of the bio-oil and can lead to phase separation if conditions change.

- **Heteroatoms and Inorganics:** Depending on the quality of the crude glycerol feedstock and the conversion process (e.g., catalyst residues), heteroatoms such as N and S can be present in trace amounts in the bio-oil as organic or inorganic compounds. Alkali metals (from the base catalyst in biodiesel production) and other inorganic impurities can also carry over into the bio-oil. These heteroatoms and inorganics can act as poisons for downstream upgrading catalysts and potentially contribute to undesirable emissions during combustion.

The defining characteristic of bio-oil, particularly from pyrolysis, is its high oxygen content, which can be as high as 50% by weight. This high oxygen content is directly responsible for many of the challenging properties discussed below, differentiating it significantly from conventional hydrocarbon fuels [61].

9.4.2 Physical Properties of Bio-oil

The complex and variable chemical composition of glycerol-derived bio-oil results in a range of distinct physical properties that are critical considerations for its practical handling, storage, transport, and utilization. These properties significantly influence the design requirements for equipment and processes.

- **Viscosity:** Bio-oil is typically a relatively viscous liquid, although its viscosity can vary widely depending on the specific conversion technology, the H_2O content, the concentration of heavy oxygenates and oligomers, and the presence of any solid char particles [61]. For lighter fractions with high H_2O content, the viscosity might be closer to H_2O, but for heavier or aged bio-oil, it can be comparable to molasses. High viscosity poses challenges for pumping, atomization in combustion systems, and filtration, often requiring preheating or blending with less viscous liquids to facilitate flow and processing in conventional equipment. Viscosity can also increase significantly during storage due to polymerization reactions (aging).
- **Density:** Bio-oil is generally denser than petroleum-based fuels, with typical values ranging from 1.0 to 1.3 g/cm^3 [26]. This higher density is primarily attributed to the presence of polar, oxygenated compounds and H_2O, which have higher densities than non-polar hydrocarbons.
- **Heating Value (Lower Heating Value – LHV):** A direct consequence of its high oxygen and H_2O content, bio-oil possesses a significantly **LHV** per unit mass or volume compared to conventional hydrocarbon fuels. While petroleum diesel has an LHV of around 43 MJ/kg, typical LHVs for biomass bio-oil range from 15 to 20 MJ/kg [62]. Glycerol-derived bio-oil falls within or near this range, with the specific value dependent on the degree of deoxygenation achieved during conversion. This lower energy density means that a larger mass or volume of bio-oil is required to provide the same amount of energy as a comparable amount of fossil fuel, impacting storage requirements, transportation costs, and combustion system design. Upgrading processes are essential to reduce oxygen and H_2O content, thereby increasing the heating value to fuel-grade specifications.

- **pH and Corrosivity:** Bio-oil is typically acidic, with pH values often ranging from 2 to 4. This acidity is primarily due to the presence of low-molecular-weight carboxylic acids (e.g., acetic acid, formic acid) and other acidic oxygenates formed during the conversion process. The low pH makes bio-oil highly corrosive to many common construction materials, particularly carbon steel [63]. Specialized, corrosion-resistant materials such as stainless steel or certain polymers are often required for equipment used in the handling, storage, and processing of bio-oil to prevent equipment degradation and ensure safety and reliability.
- **Stability:** Crude bio-oil is generally chemically unstable and prone to degradation during storage, a phenomenon often referred to as "aging" [64]. Aging involves a complex series of reactions, including polymerization, condensation, and oxidation, driven by the presence of reactive oxygenated compounds such as aldehydes (especially acrolein), ketones, and unsaturated compounds. These reactions lead to undesirable changes in bio-oil properties, including a significant increase in viscosity, potential phase separation (especially if H_2O content changes), and the formation of solid precipitates. These changes can make the bio-oil difficult to pump, filter, and atomize, and can clog fuel injectors and other equipment. Stabilization strategies, such as catalytic upgrading (e.g., hydrotreating), blending with stabilizers, or storage under controlled conditions (e.g., low temperature, inert atmosphere), are necessary for enabling long-term storage and transport.

9.4.3 Potential Applications of Bio-oil: A Versatile Renewable Resource

Despite the challenges presented by its properties, glycerol-derived bio-oil holds significant promise as a versatile renewable resource with a wide range of potential applications across the energy, chemicals, and materials sectors [63]. Its utility stems from its origin as a renewable, bio-based feedstock and its complex chemical composition, which can be modified or tailored through specific conversion pathways and subsequent upgrading processes.

- **Biofuels:** One of the most significant potential applications is as a feedstock for the production of renewable transportation fuels that can potentially serve as "drop-in" replacements for fossil fuels [65]. Crude bio-oil is generally not suitable for direct use in conventional engines due to its unfavorable properties (high oxygen content, acidity, viscosity, instability). However, it can be catalytically upgraded through processes such as hydrotreating (hydrogenation and HDO), catalytic cracking, or steam reforming [66]. These upgrading steps reduce the oxygen content, remove impurities, improve stability, and increase the heating value, yielding hydrocarbon-rich liquids (alkanes, cycloalkanes, aromatics) that are compatible with existing fuel infrastructure and meet fuel specifications for gasoline, diesel, or even jet fuel. Glycerol's relatively simple C_3 structure makes it a potentially interesting feedstock for producing light hydrocarbons like propane or specific C_3 oxygenates (propanediols) that can then be converted or blended into fuels [67].

- **Chemical Feedstock:** Bio-oil is a rich source of oxygenated molecules that can serve as platform chemicals or precursors for the synthesis of a wide variety of industrial chemicals and materials [68]. The specific composition of bio-oil, which can be influenced by the conversion technology and catalyst, determines its potential as a chemical feedstock. For example, key intermediates like acrolein (used in acrylic acid production) and acetol (used as a solvent or precursor for propylene glycol) are direct dehydration products of glycerol. Bio-oil fractions or derivatives can be utilized as building blocks or intermediates in the synthesis of polymers, resins, adhesives, solvents, surfactants, and other specialty chemicals. Selective conversion pathways aimed at maximizing the yield of specific valuable oxygenates are particularly relevant for chemical applications.
- **Heat and Power Generation:** Crude bio-oil can be directly combusted in industrial boilers, furnaces, or modified gas turbines to generate heat and electricity [69]. While this requires addressing the challenges related to its high viscosity, acidity, and potential for incomplete combustion (leading to higher CO and particulate emissions compared to fossil fuels), it offers a direct energy utilization pathway, particularly at production sites or in industrial facilities with existing combustion infrastructure. Modifications to combustion systems, such as using corrosion-resistant materials and specialized burners, may be necessary. Co-firing bio-oil with conventional fuels (e.g., natural gas, coal) can also help mitigate some of these issues.
- **Biorefinery Integration:** Integrating glycerol-to-bio-oil conversion technologies into advanced biorefinery concepts represents a key strategy for maximizing resource efficiency and creating synergistic value streams. Glycerol can be co-processed alongside other biomass feedstocks (e.g., lignocellulosic residues, algae) using technologies like co-pyrolysis or co-HTL. This can potentially improve bio-oil yield or quality, or allow for the production of tailored mixtures of biofuels and biochemicals within an integrated facility. Glycerol valorization thus enhances the flexibility and economic robustness of the biorefinery by utilizing a concentrated, readily available byproduct stream and contributing to a portfolio of bio-based products.
- **Soil Amendment:** The solid char co-produced during the pyrolysis of glycerol (especially slow pyrolysis) can be utilized as biochar [70]. Biochar applied to soil can improve soil fertility, enhance H_2O retention, and act as a stable form of carbon, contributing to carbon sequestration. While not a direct application of the liquid bio-oil itself, it is a valuable co-product stream from the conversion process.
- **Specialty Applications:** Specific components isolated or derived from bio-oil may find niche applications in higher-value markets, such as components in lubricants, wood preservatives, or even fine chemicals and flavor compounds, depending on the ability to purify them to high standards.

Effectively realizing the full potential of glycerol-derived bio-oil across these diverse applications requires continued technological advancements in conversion efficiency, product tailoring, cost-effective upgrading, and the development of robust supply chains and market outlets.

9.5 Catalytic Processes for Glycerol Conversion to Bio-oil

Catalytic processes are fundamental to many advanced glycerol valorization strategies aimed at producing bio-oil or specific bio-oil components with high yield and desired properties. Unlike purely thermal decomposition, catalysis allows for the selective promotion of specific reaction pathways, enabling transformations to occur at lower temperatures and pressures, with faster reaction rates, and most importantly, with improved selectivity toward valuable products while minimizing undesired side reactions and byproduct formation. Catalysts achieve this by providing active sites that lower the activation energy for particular chemical reactions, thereby influencing the reaction kinetics and guiding the complex network of potential transformations that glycerol can undergo.

The central role of catalysts in glycerol conversion are as follows:

- Increasing reaction rates to achieve desired conversion levels efficiently.
- Enhancing selectivity toward specific products or classes of products (e.g., hydrocarbons, acrolein, acetol, propanediols).
- Enabling reactions that are kinetically or thermodynamically unfavorable under purely thermal conditions.
- Potentially allowing for milder operating conditions, reducing energy consumption and equipment costs.
- Influencing the overall composition and properties of the resulting bio-oil (e.g., reducing oxygen content, acidity, viscosity).

A wide array of catalyst types has been investigated for glycerol conversion, each offering unique advantages and promoting different reaction mechanisms (Table 9.2):

- **Solid Acid Catalysts:** These materials are particularly effective in catalyzing dehydration, cracking, isomerization, and esterification reactions [49]. They possess Brønsted acid sites (proton donors) and/or Lewis acid sites (electron pair acceptors) that interact with the hydroxyl groups of glycerol and its

Table 9.2 Various types of catalyst utilized in bio-oil production.

Solid acids	Solid acids, such as zeolites, sulfated ZrO_2, and alumina, are effective catalysts for glycerol dehydration and cracking. They promote the formation of acrolein, hydroxyacetone, and other unsaturated compounds
Solid bases	Solid bases, such as MgO, CaO, and hydrotalcites, are effective catalysts for glycerol transesterification and aldol condensation. They promote the formation of esters and higher molecular weight compounds
Zeolites	Zeolites are crystalline aluminosilicates with well-defined pore structures that can be tailored to specific reactions. Zeolites can be used for glycerol cracking, aromatization, and isomerization
Metal oxides	Metal oxides, such as TiO_2, ZrO_2, and Nb_2O_5, can be used as catalysts for glycerol oxidation and dehydration

derivatives. Examples include zeolites (e.g., H-ZSM-5, H-Beta, H-Y, mordenite), sulfated metal oxides (e.g., sulfated zirconia [ZrO_2]), modified clays, amorphous silica-alumina, and certain metal oxides (e.g., γ-Al_2O_3, TiO_2). The strength, density, and distribution of these acid sites significantly influence the catalytic activity and product selectivity. For instance, stronger acid sites tend to favor more extensive cracking or polymerization reactions, while milder acidity can promote selective dehydration. Zeolites, with their ordered pore structures, can also exert shape selectivity, influencing which molecules can access the active sites or exit the pores based on size and shape. Solid acid catalysts are crucial for producing light oxygenates like acrolein and acetol via dehydration. However, they are often susceptible to deactivation by coke formation, especially those with strong acidity or narrow pores, and can be poisoned by impurities like alkali metals in crude glycerol.

- **Solid Base Catalysts:** These catalysts possess basic sites (electron pair donors or hydroxide donors) and are effective in catalyzing reactions such as aldol condensation, esterification, transesterification, and etherification. Examples include metal oxides (e.g., CaO, MgO, SrO), layered double hydroxides (hydrotalcites), and alkali-promoted support materials (e.g., Cs-promoted silica). While less commonly used alone for producing the bulk bio-oil fraction compared to acid catalysts, they can promote reactions that contribute to the formation of higher-molecular-weight oxygenates, esters, or polyols, influencing bio-oil composition. Solid bases can also be effective for transesterification reactions if residual free fatty acids or triglycerides are present in the crude glycerol feedstock, helping to reduce acidity.
- **Zeolites:** Crystalline aluminosilicate materials with precisely defined, ordered pore structures, zeolites are highly versatile catalysts due to their tunable acidity, high surface area, and shape selectivity. As solid acids, their framework aluminum atoms create Brønsted acidity, while extraframework species can contribute to Lewis acidity. Zeolites are extensively used for catalyzing dehydration, cracking, isomerization, and importantly, aromatization reactions that convert oxygenates into hydrocarbons [71]. The pore size and connectivity (e.g., 8-ring pores in ZSM-5, 12-ring pores in Beta and Y) can dictate which molecules can enter and react, and which products can exit, thereby controlling selectivity. Zeolites like H-ZSM-5 are particularly effective in converting oxygenates into aromatic hydrocarbons, which are valuable components of fuels. However, coke deposition within the pores is a major deactivation mechanism, especially during reactions with highly reactive intermediates like acrolein or olefins. Research focuses on developing hierarchical zeolites (combining micropores and mesopores) or modifying their surface properties to improve coke resistance and stability. Examples of zeolites used include H-ZSM-5, H-Beta, and H-Y.
- **Metal Oxides:** Oxides of various metals (e.g., Al_2O_3, SiO_2, TiO_2, ZrO_2, Nb_2O_5, V_2O_5, MoO_3) can act as catalysts or catalyst supports for glycerol conversion [72]. Their catalytic activity can arise from intrinsic acidity or basicity, redox properties, or oxygen vacancies. For instance, alumina (γ-Al_2O_3) and silica-alumina are widely used acid catalysts for dehydration. ZrO_2 can exhibit both acidic and basic properties depending on its preparation, acting as a versatile catalyst.

Supported metal oxides or mixed metal oxides are also explored to enhance activity, tune acidity/basicity, or improve stability.

- **Supported Metal Catalysts:** Transition metals, particularly noble metals (e.g., Platinum [Pt], Palladium [Pd], Ruthenium [Ru], Rhodium [Rh]) and less expensive non-noble metals (e.g., Nickel [Ni], Copper [Cu], Cobalt [Co]), are frequently supported on high-surface-area materials (e.g., carbon, metal oxides, zeolites, silica, alumina) and are essential for catalyzing hydrogenolysis and HDO reactions in the presence of H_2 [73]. These metal sites activate H_2 and facilitate the cleavage of C—O and C—C bonds, enabling the removal of oxygen as H_2O and the formation of hydrocarbons or less oxygenated molecules (e.g., converting glycerol to propanediols or propane). Noble metals are highly active but expensive. Non-noble metals are more economical but may require higher temperatures and pressures and can be more susceptible to poisoning and sintering. The choice of metal and support, as well as metal dispersion, critically impacts the activity, selectivity, and stability of these catalysts [14].

The selection of the optimal catalyst for a given glycerol conversion process, such as catalytic pyrolysis or catalytic HTL, depends critically on the desired product distribution and the specific reaction conditions. For instance, if the goal is primarily light oxygenates like acrolein, strong solid acid catalysts are often preferred. If the objective is to produce deoxygenated hydrocarbons for fuel blending, supported metal catalysts operating under H_2 pressure are necessary. Catalytic pyrolysis strategically integrates catalysts within a pyrolysis reactor to influence the thermal decomposition pathways and improve bio-oil quality, as conceptually illustrated in Figure 9.3 (schematic of a pyrolysis system, which can be adapted for catalytic

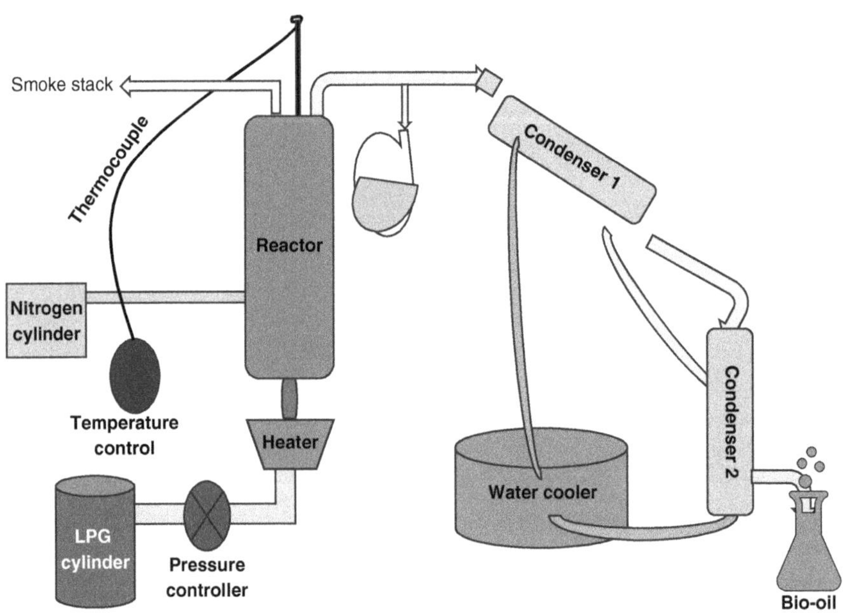

Figure 9.3 Schematic flow diagram of pyrolysis system for bio-oil production.

variants). Achieving high efficiency and selectivity requires a careful match between the catalyst properties, the reaction environment, and the targeted chemical transformations, alongside strategies to mitigate catalyst deactivation over prolonged operation.

9.6 Chemistry of Glycerol Conversion to Bio-oil

The conversion of glycerol into bio-oil is a result of a complex interplay of chemical reactions that occur simultaneously and sequentially. Understanding the fundamental reaction pathways, the intermediate species formed, and the factors influencing their rates is critical for designing catalysts and optimizing processes to control the composition and properties of the final bio-oil product. Glycerol's structure, with its three hydroxyl groups and relatively short carbon chain, makes it highly reactive and capable of undergoing a variety of transformations depending on the thermal energy and the presence of catalysts.

9.6.1 Dehydration: The Initial Oxygen Removal Step

Dehydration, the process of removing H_2O molecules, is a fundamental and often initial reaction pathway in glycerol conversion, particularly favored under acidic conditions or in the presence of catalysts possessing dehydration activity. Glycerol can undergo different dehydration routes depending on which hydroxyl groups are involved and the reaction mechanism. Two primary dehydration pathways are widely recognized:

- **1,2-Dehydration:** This involves the removal of a H_2O molecule from adjacent carbon atoms (C1 and C2 or C2 and C3), typically proceeding via a vinylic alcohol intermediate that quickly tautomerizes to an aldehyde. The primary product of 1,2-dehydration is acrolein (propenal, $CH_2{=}CH{-}CHO$). Acrolein is a highly reactive α,β-unsaturated aldehyde, which contributes to the high reactivity and instability of crude bio-oil. It is prone to polymerization and other secondary reactions, including Diels–Alder reactions and aldol condensations, that can lead to the formation of heavier compounds, oligomers, and ultimately char. Despite its reactivity challenges, acrolein is itself a key platform chemical, used in the production of acrylic acid.
- **1,3-Dehydration:** This pathway involves the removal of a H_2O molecule across a 1,3 arrangement, typically proceeding through a cyclic intermediate or a carbocation rearrangement involving the terminal and central hydroxyl groups. This reaction primarily yields acetol (hydroxyacetone, CH_3COCH_2OH) [10]. Acetol is a simpler oxygenate compared to acrolein and is generally more stable. It is also a valuable chemical intermediate, used as a solvent and a precursor for the production of propylene glycol.

The selectivity between the 1,2-dehydration (leading to acrolein) and 1,3-dehydration (leading to acetol) pathways is highly dependent on the type and properties of the catalyst used, particularly the nature and strength of the acid sites

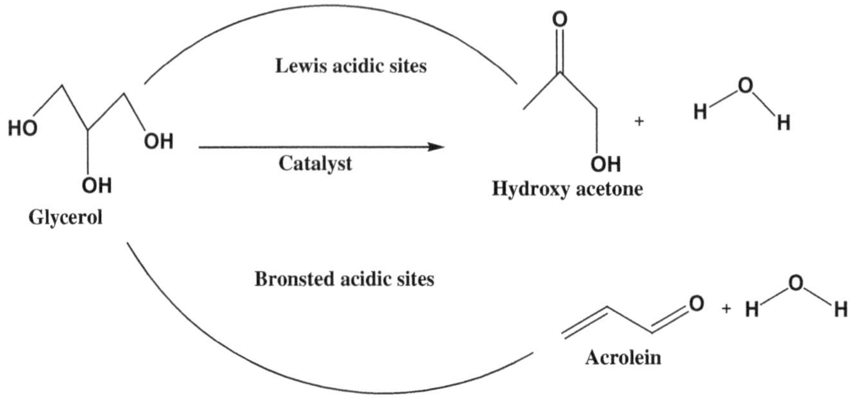

Figure 9.4 Dehydration routes of glycerol showing the formation of hydroxyacetone and acrolein.

(Brønsted vs. Lewis acidity) and the pore structure in the case of heterogeneous catalysts like zeolites. Stronger Brønsted acid sites often favor the formation of acrolein, while milder or Lewis acid sites may favor acetol formation [74]. The dehydration products, acrolein and acetol, are crucial intermediates that can then undergo further reactions such as cracking, oligomerization, or reactions leading to furan derivatives, significantly contributing to the complexity of the final bio-oil composition. Figure 9.4 schematically illustrates these two primary dehydration routes to form hydroxyacetone (acetol) and acrolein [54].

9.6.2 Cracking: Breaking the Carbon Backbone

Cracking reactions involve the breaking of carbon–carbon (C—C) and carbon–oxygen (C—O) bonds within the glycerol molecule itself or, more commonly, its intermediate products such as acrolein, acetol, and other oxygenates. This leads to the formation of smaller molecular fragments, typically ranging from C_1 to C_2 oxygenates, light hydrocarbons, and permanent gases (such as CO, CO_2, H_2, methane, ethylene). Cracking can be induced by thermal energy at high temperatures (thermal cracking) or catalyzed by acidic or metallic sites at lower temperatures.

In catalytic cracking, acid sites (Brønsted or Lewis) catalyze bond cleavage via the formation and subsequent fragmentation of carbocation intermediates. For example, acetol can undergo acid-catalyzed cracking to produce acetaldehyde and formaldehyde. Metal catalysts can also promote cracking, often in conjunction with hydrogenation or deoxygenation reactions. The extent of cracking significantly influences the molecular weight distribution and volatility of the bio-oil product [75]. More severe cracking, typically occurring at higher temperatures or with highly active cracking catalysts, leads to increased production of light ends and gases, which reduces the overall yield of the liquid bio-oil fraction but can result in a bio-oil with a lower average molecular weight and potentially less oxygenated compounds. Controlled cracking is essential to reduce the size of the initial oxygenated fragments and produce precursors for hydrocarbon formation through subsequent deoxygenation steps. The specific products formed during cracking are highly sensitive to the reaction temperature, pressure, residence time, and, critically, the type and activity of the catalyst employed. Catalytic cracking pathways are vital for transforming the initial oxygen-rich C_3 intermediates into smaller oxygenates or hydrocarbons, contributing to the reduction of bio-oil acidity and the improvement of its heating value.

9.6.3 Oligomerization: Formation of Larger Molecules

Oligomerization refers to reactions where two or more smaller molecules combine to form larger molecules (dimers, trimers, and higher oligomers). In glycerol conversion, this can involve the self-condensation of glycerol or the reaction of reactive intermediates such as acrolein, acetol, furans, and other oxygenates. These reactions typically proceed via condensation mechanisms, often catalyzed by acidic or basic sites on the catalyst or via non-catalyzed thermal polymerization. Glycerol itself can undergo intermolecular dehydration to form diglycerols, triglycerols, and polyglycerols, increasing the molecular weight and viscosity. Reactive unsaturated compounds like acrolein are particularly prone to polymerization, forming heavy, often intractable compounds that contribute to increased viscosity, phase separation, and ultimately char formation. Acetol and other oxygenates can also undergo aldol condensation or other polymerization pathways. While some controlled oligomerization might contribute to the formation of heavier, less volatile components of the liquid bio-oil, excessive or uncontrolled oligomerization is detrimental as it leads to the formation of high-molecular-weight, insoluble compounds and char. Char reduces the yield of the desired liquid product and can cause significant catalyst deactivation and equipment fouling. Controlling oligomerization is crucial for obtaining a stable, pumpable bio-oil product. The presence of highly reactive intermediates, catalyst properties (acidity, basicity, pore structure), reactant concentration, and reaction temperature significantly influence the extent of oligomerization. Acidic catalysts, while promoting desirable dehydration and cracking, can also catalyze undesired condensation and polymerization reactions, highlighting the delicate balance required in catalyst design and process optimization.

9.6.4 Aromatization: Synthesis of Cyclic Hydrocarbons

Aromatization is a desirable reaction pathway for converting aliphatic oxygenates and hydrocarbons into aromatic compounds, such as benzene, toluene, and xylenes (collectively referred to as BTX), which are valuable components of gasoline and key petrochemical feedstocks [76]. This complex transformation typically requires specific catalysts, most effectively shape-selective zeolites (such as H-ZSM-5) often promoted by metals like gallium (Ga) or zinc (Zn), and generally occurs at higher temperatures (often above 400 °C). The mechanism involves a complex sequence of reactions: initial dehydration and cracking of glycerol and its intermediates to produce light oxygenates and olefins (C_2—C_4), followed by oligomerization of these olefins, cyclization of the oligomers, and subsequent dehydrogenation to form stable aromatic rings. Intermediates like acrolein, acetol, and light olefins formed from the cracking of glycerol and its derivatives serve as precursors for aromatization. While producing highly desirable BTX products, aromatization over zeolite catalysts is highly susceptible to catalyst deactivation due to rapid coke formation on the strongly acidic sites and within the pore structure. Coke deposition blocks active sites and pore entrances, severely reducing catalyst activity and lifespan. Achieving high selectivity toward specific aromatics while minimizing coke formation is a major challenge in catalytic aromatization of glycerol derivatives. Research focuses on designing catalysts with optimal acidity, metal dispersion, and pore morphology, and optimizing operating conditions to balance reaction rates with deactivation rates.

9.6.5 Other Reactions: Diverse Transformation Routes

In addition to the primary pathways described above, several other significant reactions contribute to the complex product distribution and influence the final properties of glycerol-derived bio-oil, and some offer specific routes to valuable chemicals:

- **Esterification:** Glycerol, being a triol, can react with carboxylic acids (either present as impurities in crude glycerol, such as free fatty acids from incomplete transesterification, or formed as byproducts during the conversion process) to form mono-, di-, or triglycerides (esters). This reaction is typically catalyzed by acids. Esterification of free fatty acids with glycerol is beneficial as it reduces the overall acidity of the bio-oil and consumes undesirable impurities. A well-known example is the reaction of glycerol with acetic acid (which can be a cracking product or added ex situ) to form mono-, di-, or triacetin (glycerol acetate esters). Triacetin is a valuable compound used as a fuel additive, solvent, and plasticizer. Figure 9.5a illustrates the formation of triacetin [77].
- **Etherification:** Glycerol can undergo etherification reactions, forming C—O—C bonds by reacting with alcohol groups (either its own hydroxyl groups, leading to polyglycerols like diglycerol, or with external alcohols like methanol or ethanol). These reactions are generally catalyzed by acids and involve the elimination of a H_2O molecule between two hydroxyl groups. Intramolecular etherification can also occur, leading to cyclic ethers. Etherification increases the molecular weight and viscosity of the bio-oil and can potentially reduce the reactivity of hydroxyl groups by converting them into ether linkages. Specific glycerol ethers can have applications as fuel additives or chemical intermediates. Figure 9.5b shows the etherification of glycerol with ethanol to form ethoxypropanediols. Decarboxylation involves the removal of CO_2 from carboxylic acids. These reactions can influence the composition and properties of the bio-oil product.
- **Hydrogenolysis and HDO:** These are crucial reactions for reducing the oxygen content of glycerol and its oxygenated derivatives, essential for producing fuel-grade hydrocarbons [78]. These reactions occur in the presence of H_2 gas and require appropriate metal catalysts (e.g., supported noble metals like Pt, Pd, Ru, or non-noble metals like Ni, Cu, Co). Hydrogenolysis involves

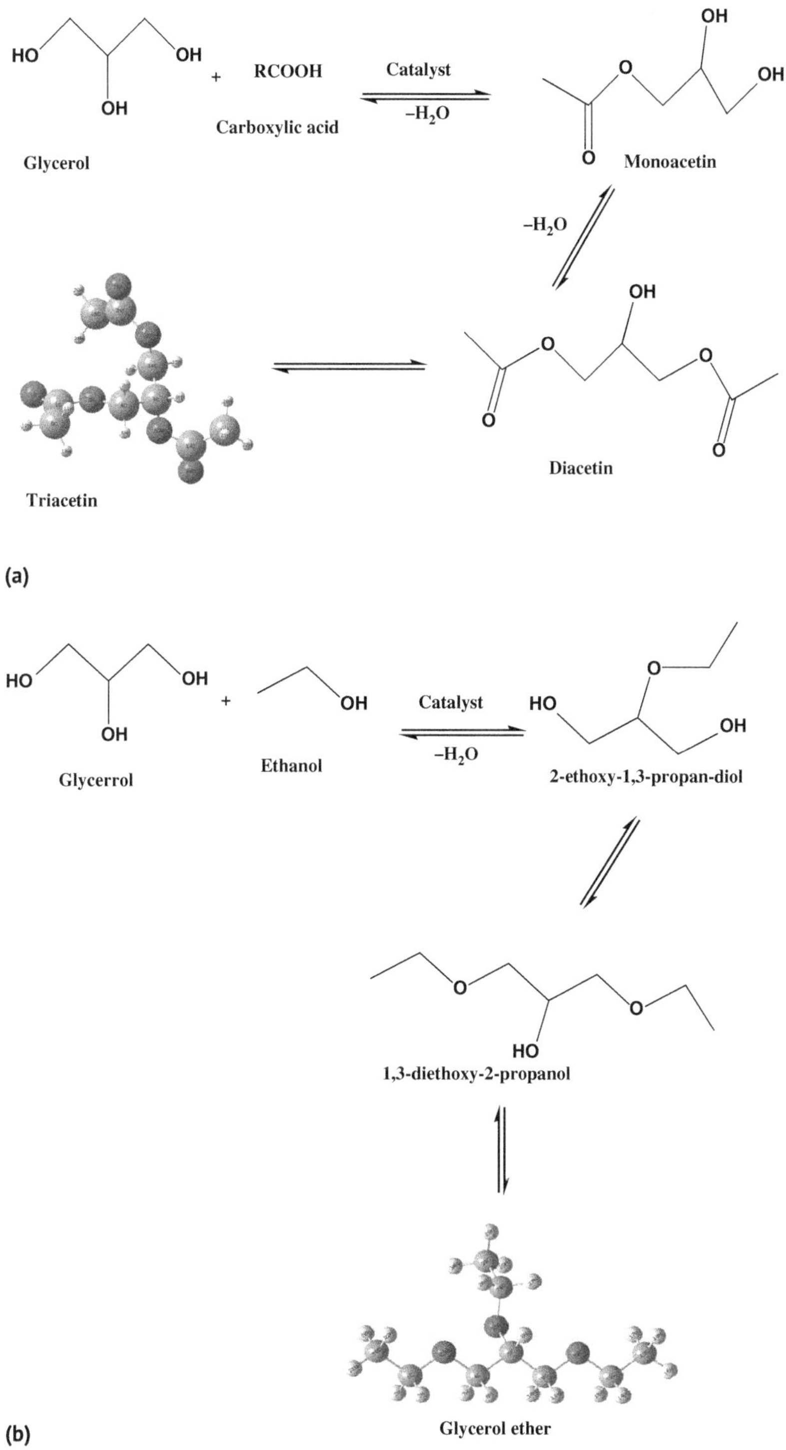

Figure 9.5 Formation reaction of (a) triacetin and (b) glycerol ether.

the cleavage of C—C or C—O bonds accompanied by hydrogenation, leading to molecules with fewer or no oxygen atoms (e.g., converting glycerol to 1,2-propanediol, 1,3-propanediol, or propane). HDO specifically targets the removal of oxygen atoms as H_2O from organic molecules. For example, HDO of acetol can produce propane. These reactions are vital for upgrading oxygen-rich bio-oil into stable, non-polar hydrocarbon mixtures suitable for transportation fuels, increasing their heating value and reducing corrosivity.

- **Decarboxylation and Decarbonylation:** Carboxylic acids formed during glycerol conversion can undergo decarboxylation, losing a carbon atom as CO_2. Carbonyl compounds (aldehydes, ketones) can undergo decarbonylation, losing a carbon atom as CO [79]. These reactions further reduce the oxygen content of the bio-oil and can lead to the formation of hydrocarbons or shorter-chain oxygenates. They are often catalyzed by metals or metal oxides and are important deoxygenation pathways in processes like HTL or catalytic upgrading.

The complex composition of the final bio-oil product from glycerol is the result of the intricate interplay, relative rates, and extents of these diverse reaction pathways. Factors such as reaction temperature and pressure, reactant concentration (including H_2O), residence time, and most critically, the type and design of the catalyst employed, all influence which pathways are favored and the resulting product distribution. Controlling this complex chemical network through precise parameter optimization and sophisticated catalyst engineering is the central scientific and engineering challenge and opportunity in efficient and selective glycerol valorization. The fundamental bonding structure and specific geometrical parameters (bond lengths, bond angles, dihedral angles) of glycerol, its intermediates, and products are key inputs for computational chemistry and kinetic modeling studies aimed at elucidating reaction mechanisms at the molecular level and predicting catalytic performance.

9.7 Reactor Design and Process Optimization for Glycerol Conversion to Bio-oil

Beyond the intrinsic chemical reactions and catalyst performance, the successful technical and economic implementation of glycerol-to-bio-oil conversion is significantly dependent on the engineering aspects of the process, particularly the design of the chemical reactor system and the rigorous optimization of operating parameters. The chemical reactor provides the controlled environment where the desired transformations occur, facilitating effective heat and mass transfer, managing the distribution of different phases (gas, liquid, solid catalyst), and enabling efficient contact between reactants and catalysts. Different reactor configurations are suited to specific reaction conditions, catalyst types, and desired product profiles, each presenting unique engineering challenges and offering particular advantages for glycerol conversion. Process optimization, involving the systematic adjustment of key operating variables, is essential for maximizing desired outcomes such as bio-oil yield, selectivity toward valuable components, overall energy efficiency, and catalyst longevity.

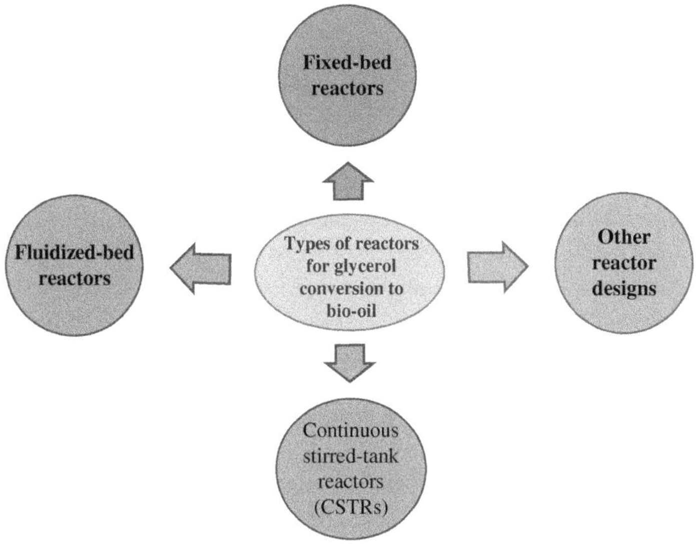

Figure 9.6 Various types of reactors utilized in bio-oil formation.

The selection of an appropriate reactor design is therefore a critical early step in process development. The choice depends on factors such as whether the reaction is homogeneous or heterogeneous, the phases of the reactants and products at reaction conditions, the required temperature and pressure, the reaction kinetics, and the propensity for catalyst deactivation or solid formation. Figure 9.6 illustrates various types of reactors commonly used or investigated in processes related to bio-oil formation, many of which are applicable to glycerol conversion [80].

- **Fixed-bed Reactors:** These reactors feature a stationary bed of solid catalyst particles packed into a tube or vessel. Reactants (liquid, vapor, or mixed phase) flow through the packed bed, interacting with the catalyst surface. Advantages include their relative simplicity in design and construction, ease of scaling up (to a degree, by using multiple tubes), and lower capital costs compared to more complex designs. They are well-suited for heterogeneous catalytic reactions, particularly those that are predominantly gas-phase or involve trickle-bed flow, and can provide a relatively long residence time profile, which can be beneficial for achieving high conversion or sequential reactions. However, fixed beds can suffer from poor heat transfer characteristics, potentially leading to radial and axial temperature gradients (hotspots or coldspots) that negatively impact selectivity and catalyst stability. Pressure drop across the packed bed can build up over time, especially if solids (like char or coke) are formed or if catalyst particles are small, which increases pumping costs and can lead to flow channeling. Catalyst deactivation, particularly by coke deposition in processes like catalytic pyrolysis, is a major challenge in fixed beds, as it reduces catalyst activity and requires periodic process shutdown for regeneration. Managing multiphase flow and ensuring uniform distribution of reactants across the bed can also be complex.

- **Fluidized-bed Reactors:** In these reactors, solid catalyst particles are suspended and vigorously mixed in a fluid stream (gas or liquid), exhibiting fluid-like behavior. This configuration offers significant advantages, particularly for highly exothermic or endothermic reactions where precise temperature control is essential, due to excellent heat transfer rates and uniform temperature distribution throughout the bed. The intense mixing also improves mass transfer between the fluid phase and the catalyst surface, enhancing reaction rates. Fluidized beds are well-suited for processes involving fine catalyst particles and are particularly effective at handling feedstocks or reactions that produce solid byproducts like char, as the fluidization action helps prevent agglomeration and facilitates continuous catalyst withdrawal and addition. For fast pyrolysis and catalytic pyrolysis of glycerol, fluidized beds (e.g., bubbling or circulating fluidized beds) provide rapid heating rates for the feedstock and efficient quenching of vapor products, which helps maximize liquid bio-oil yield and minimize undesirable secondary reactions [81].

  However, fluidized beds are more complex and expensive to design, build, and operate compared to fixed beds, requiring careful control of fluidization velocity to prevent defluidization or excessive catalyst elutriation. Catalyst attrition (mechanical breakdown) and entrainment of fine particles requiring downstream separation are also considerations.
- **Continuous Stirred-tank Reactors (CSTRs):** CSTRs are typically employed for liquid-phase reactions and are characterized by intensive mixing, aiming for uniform concentration and temperature throughout the reactor volume [82]. Reactants are continuously fed into the tank, and the reaction mixture is continuously withdrawn. CSTRs offer excellent control over reaction conditions and are well-suited for reactions with relatively slow kinetics where a long average residence time is required, or for achieving high conversions in multiple reactors arranged in series (a CSTR cascade). They are applicable for homogeneous or heterogeneous (slurry) catalytic reactions where the catalyst can be effectively suspended and mixed in the liquid phase. For liquid-phase glycerol reactions such as catalytic esterification, etherification, or hydrogenolysis under specific conditions, CSTRs can be utilized. However, achieving the same conversion for first-order reactions typically requires a larger reactor volume in a single CSTR compared to an ideal plug flow reactor (like a fixed bed). Managing solid catalysts in a slurry CSTR presents challenges related to ensuring uniform suspension, separating catalyst from the liquid effluent, and minimizing catalyst attrition.
- **Other and Emerging Reactor Designs:** Research and development efforts are continuously exploring novel and intensified reactor concepts to improve the efficiency, selectivity, and sustainability of glycerol conversion processes. Microwave reactors utilize microwave radiation to rapidly and selectively heat-specific materials (e.g., polar reactants like glycerol or specific catalyst materials), potentially leading to different reaction pathways, faster kinetics, or reduced energy consumption compared to conventional heating [83]. Microreactors, characterized by very small channel dimensions (typically tens to hundreds of micrometers), provide exceptional surface area-to-volume

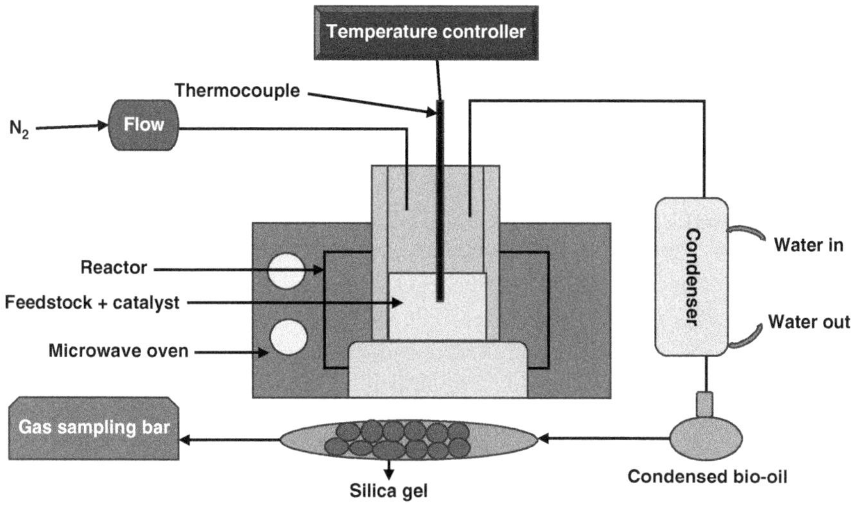

Figure 9.7 Experimental setup for pyrolysis of crude glycerol into bio-oil. Adapted from [85].

ratios, enabling extremely rapid and efficient heat and mass transfer and precise temperature control [84]. These attributes are beneficial for managing highly exothermic or fast reactions and can potentially improve selectivity by quickly removing products from the reaction zone. Supercritical fluid reactors utilize fluids such as H_2O or CO_2 in their supercritical state as reactive solvents and transport media. The unique properties of supercritical fluids (high density, low viscosity, high diffusivity, tunable solvent power) can enhance solubility and mass transfer, facilitating reactions involving polar and non-polar components and enabling operation at high reactant concentrations. Figure 9.7 illustrates a schematic diagram of an experimental setup for microwave-assisted pyrolysis of crude glycerol, indicating one such advanced reactor concept.

Rigorous process optimization is crucial for maximizing the performance of the chosen reactor system and achieving the desired outcomes, such as maximizing bio-oil yield, optimizing selectivity toward specific valuable compounds, minimizing undesired byproduct formation (e.g., char, gases), enhancing energy efficiency, and extending catalyst lifespan. This involves systematically studying and adjusting key operating parameters (Figure 9.6). Key operating parameters critically influence the conversion of glycerol to bio-oil, fundamentally dictating reaction rates, favoring specific competing pathways, and impacting product yield, selectivity, and catalyst performance. Reaction temperature is a primary factor, governing the kinetics of transformations like dehydration, cracking, oligomerization, and aromatization, while also significantly influencing catalyst stability and deactivation. Pressure plays a critical role, particularly in processes involving liquid or supercritical phases such as HTL where it affects fluid properties and can suppress gas formation; in catalytic processes utilizing gaseous reactants (e.g., H_2 in HDO), it impacts reaction equilibria, rates, and catalyst performance. The specific catalyst type and amount fundamentally determine the catalyzed reaction pathways and their relative rates, thereby dictating product selectivity and influencing the overall conversion rate and potential for catalyst deactivation. The glycerol concentration in the feed stream affects reaction kinetics and can influence undesirable intermolecular side reactions like oligomerization and char formation, while the concentration of other feed components, such as H_2O, is particularly crucial in HTL. Residence time, the duration reactants spend in the reactor, controls the overall extent of conversion; insufficient time limits product yield, but excessively long times can lead to over-reaction, increased char formation, gas production, and significant catalyst deactivation. Finally, the heating rate, especially critical in pyrolysis, influences the initial decomposition pathways and the subsequent fate of volatile intermediates; high rates characteristic of fast pyrolysis promote rapid volatilization and quenching, essential for maximizing liquid bio-oil yield by minimizing char formation. Optimizing this complex interplay of parameters is essential for designing efficient, selective, and economically viable glycerol conversion processes.

Optimization of these numerous process parameters is a complex task that often involves systematic experimental design methodologies (e.g., Design of Experiments [DoE], Response Surface Methodology [RSM]) or computational approaches such as kinetic modeling, process simulation, and optimization algorithms (e.g., genetic algorithms, particle swarm optimization) [86]. Effective reactor design, coupled with rigorous process optimization informed by a deep understanding of the underlying reaction chemistry and catalyst behavior, is essential for translating laboratory-scale successes into scalable, efficient, economically viable, and environmentally sound glycerol-to-bio-oil production processes.

9.8 Challenges and Limitations

Despite the significant progress made in research and development, the conversion of glycerol to bio-oil faces several persistent technical, economic, and logistical challenges and limitations that must be addressed to facilitate widespread commercialization and deployment. These hurdles are inherent to the nature of the feedstock, the complexity of the conversion processes, and the characteristics of the bio-oil product.

Scaling up glycerol conversion technologies from laboratory-scale experiments, which typically handle grams or kilograms of feedstock, to commercial-scale production facilities capable of processing tons per hour presents a complex array of engineering, logistical, and economic challenges [87]. Laboratory studies are essential for elucidating reaction mechanisms, evaluating catalyst performance, and optimizing parameters under controlled conditions, but translating these findings to larger scales is non-trivial and requires addressing issues inherent to increased size and throughput.

Engineering challenges include designing and building reactors and associated equipment that can operate reliably and safely at high temperatures and pressures while maintaining desired temperature and concentration profiles across larger volumes. Maintaining uniform heat transfer rates needed for fast pyrolysis or controlling multiphase flow dynamics in large fluidized beds becomes significantly more complex. Managing catalyst handling, activation, deactivation, and regeneration for large catalyst volumes in continuous processes is also a major engineering challenge. Ensuring the mechanical integrity and safety of equipment operating under demanding conditions is paramount.

Logistical challenges involve establishing robust and consistent supply chains for large volumes of crude glycerol feedstock, which can have variable quality and may require centralized collection and pretreatment facilities. Handling, storing, and transporting large volumes of the produced bio-oil, which can be corrosive, viscous, and unstable, requires specialized infrastructure and logistics. Managing and disposing of or valorizing large quantities of process byproducts, such as the aqueous phase from HTL or char from pyrolysis, are also significant logistical considerations.

Economic challenges at commercial scale include high capital expenditures (CAPEX) for building large-scale reactors, specialized equipment (e.g., high-pressure vessels for HTL), and downstream processing units (separation, upgrading). Operating expenses (OPEX) include the cost of feedstock (though crude glycerol price is low, purification costs may apply), energy, catalyst purchase and regeneration, labor, and maintenance. The economic viability of a commercial plant is highly sensitive to these costs and the market price obtainable for the produced bio-oil or its upgraded products. Demonstrating the technical reliability and economic feasibility at larger scales often necessitates pilot-scale or demonstration plants to validate performance, gather operational data, and de-risk the technology before full commercial deployment.

Overcoming these scale-up challenges requires a multidisciplinary effort involving chemical engineers, mechanical engineers, material scientists, chemists, and economists. It necessitates significant investment in process design, simulation, pilot testing, and optimization to ensure that the technology is not only chemically viable but also technically scalable, economically competitive, and environmentally sound.

9.9 Commercial Glycerol-to-bio-oil Plant

The transition from pilot and demonstration scale to fully commercial-scale plants dedicated specifically to the production of bio-oil from glycerol is an ongoing process. While dedicated, standalone commercial glycerol-to-bio-oil facilities are not yet widespread compared to, for example, biodiesel plants, several companies and initiatives are actively pursuing commercial implementation, often integrating glycerol conversion into broader bio-based production or waste-processing facilities.

Fulcrum BioEnergy (United States): Fulcrum BioEnergy is a company focused on converting municipal solid waste (MSW) into low-carbon transportation fuels, primarily utilizing a process involving gasification of the waste followed by Fischer–Tropsch synthesis. While their primary feedstock is MSW, Fulcrum has expressed interest in incorporating other relevant waste and byproduct streams into their conversion process to diversify their feedstock base and enhance the sustainability of their operations [88–90]. This includes potentially utilizing glycerol as a carbon source. Converting glycerol into a suitable intermediate (e.g., synthesis gas via reforming, or potentially a liquid bio-oil stream) could fit within their broader waste-to-fuels platform. This indicates a potential pathway for large-scale glycerol valorization within centralized waste-processing and energy recovery facilities.

The trend appears to favor integrating glycerol conversion into larger bio-based production complexes or waste-to-energy facilities to leverage existing infrastructure and create synergistic value streams. The continued success of these and future initiatives will be crucial for demonstrating the technical and economic viability of large-scale glycerol-to-bio-oil production. As the global impetus toward sustainable energy and chemicals intensifies, the field of glycerol valorization, and specifically its conversion to bio-oil and related products, is poised for significant advancements and innovations. Ongoing research and emerging technological trends are shaping the future trajectory of this field and its role within the expanding bioeconomy. Continued investment in fundamental research, applied development, pilot testing, and commercial demonstration, coupled with supportive policies and market development, will be crucial for realizing the full potential of glycerol valorization as a cornerstone of the future bioeconomy.

9.10 Conclusion

Glycerol, generated in significant quantities as a necessary co-product of the rapidly expanding biodiesel industry, has evolved from a low-value byproduct into a highly promising, abundant, and versatile renewable feedstock. Its effective valorization is not merely an economic opportunity but a critical imperative for enhancing the overall sustainability of biodiesel production, alleviating waste management burdens, and contributing significantly to the global transition toward a more circular and bio-based economy.

Throughout this comprehensive review, several key findings and insights have been highlighted:

1. **Glycerol's High Valorization Potential:** Glycerol's inherent molecular structure, featuring three reactive hydroxyl groups on a short carbon chain, makes it a highly versatile platform molecule. It can be effectively converted into a diverse spectrum of valuable products, ranging from renewable biofuels and fuel additives to a wide variety of biochemicals, platform chemicals, and even biomaterials, through a range of conversion pathways, including thermochemical processes (such as pyrolysis and HTL), biological routes (like fermentation and enzymatic conversions), and various catalytic transformations.
2. **Diversity of Conversion Technologies:** A portfolio of conversion technologies exists for transforming glycerol into liquid bio-oil or specific oxygenated intermediates. These technologies, including pyrolysis, HTL, and various catalytic processes (e.g., catalytic pyrolysis, catalytic hydrogenolysis/HDO), each offer distinct approaches to breaking down or reforming the glycerol molecule. They differ in their required operating conditions (temperature, pressure), energy requirements, capital and operating costs, and the characteristics (yield, composition, quality) of the resulting bio-oil product. Catalytic methods are particularly important for enhancing reaction rates, improving process efficiency, and

selectively tailoring bio-oil composition toward desired components by guiding the complex reaction network.

3. **Bio-oil Characteristics and Utility:** The bio-oil produced from glycerol conversion is a complex liquid mixture, predominantly composed of oxygenated hydrocarbons, with properties such as high viscosity, acidity, and instability that distinguish it from conventional petroleum fuels. While these characteristics present challenges for direct use in existing infrastructure and necessitate upgrading, bio-oil serves as a valuable and versatile intermediate. It holds promise as a feedstock for subsequent catalytic upgrading into renewable transportation fuels (gasoline, diesel, jet fuel); it can be a source for the extraction or synthesis of valuable platform chemicals; or it can be utilized directly for heat and power generation in modified systems.
4. **Key Challenges and Research Imperatives:** Despite significant progress, major challenges remain in optimizing glycerol-to-bio-oil conversion for widespread commercialization. These include overcoming catalyst deactivation (particularly by coke formation and impurities), achieving high and stable product selectivity toward desired value streams within the complex reaction network, improving the overall energy efficiency of conversion processes, and successfully scaling up laboratory and pilot-scale technologies to commercial production volumes. Addressing these challenges necessitates continued, focused research and development efforts in catalyst design, reactor engineering, process integration, and rigorous techno-economic and lifecycle assessments.

By strategically investing in these interconnected areas of research and development, the technical and economic barriers to glycerol valorization can be overcome, enabling its full potential as a key renewable feedstock to be realized, thereby contributing significantly to a more sustainable, resource-efficient, and prosperous future.

References

1 Zulqarnain, M., Yusoff, M.H.M., Ayoub, M., Jusoh, N., and Abdullah, A.Z. The challenges of a biodiesel implementation program in Malaysia. *Processes*. 2020. https://doi.org/10.3390/pr8101244.

2 Umpierre, A.P. and Machado, F. Glycerochemistry and glycerol valorization. *Rev. Virtual. Quim*. 2013. https://doi.org/10.5935/1984-6835.20130010.

3 Gujar, J.P. and Modhera, B. A review on catalytic conversion of biodiesel derivative glycerol to bio-olefins. *Mater. Today Proc*. 2023. https://doi.org/10.1016/j.matpr.2022.09.418.

4 Okoye, P.U., Arias, D.M., Hameed, B.H., Sebastian, P.J., Li, S., and Song, W. Thermocatalytic routes and reactor strategies for valorization of biodiesel-derived glycerol to fuels. *Appl. Therm. Eng*. 2022. https://doi.org/10.1016/j.applthermaleng.2022.118901.

5 Pant, D. Polycarbonate waste management using glycerol. *Process Saf. Environ. Prot*. 2016. https://doi.org/10.1016/j.psep.2015.12.012.

6 Onwudili, J.A. and Scaldaferri, C.A. Catalytic upgrading of intermediate pyrolysis bio-oil to hydrocarbon-rich liquid biofuel via a novel two-stage solvent-assisted process. *Fuel*. 2023. https://doi.org/10.1016/j.fuel.2023.129015.

7 Chilakamarry, C.R., Khilji, I.A., Sirohi, R., Pandey, A., Baskar, G., and Satyavolu, J. Maximizing the value of biodiesel industry waste: exploring recover, recycle, and reuse for sustainable environment. *Environ. Technol. Innov*. 2023. https://doi.org/10.1016/j.eti.2023.103447.

8 Chen, Y., Wu, Y., Hua, D., Li, C., Harold, M.P., Wang, J., et al. Thermochemical conversion of low-lipid microalgae for the production of liquid fuels: challenges and opportunities. *RSC Adv*. 2015. https://doi.org/10.1039/c4ra13359e.

9 Jana, A., Pati, S., and Chowdhury, R. Comparative technoeconomic analysis of algal biorefineries with and without methanol recycle in Aspen Plus. *Sadhana – Acad. Proc. Eng. Sci*. 2024. https://doi.org/10.1007/s12046-023-02372-0.

10 Kishor, N. and Agrawal, P.S. Nitration of glycerol-a byproduct of biodiesel production. *Mater. Today Proc*. 2020. https://doi.org/10.1016/j.matpr.2020.03.065.

11 Changmai, B., Laskar, I.B., and Rokhum, S.L. Microwave-assisted synthesis of glycerol carbonate by the transesterification of glycerol with dimethyl carbonate using Musa acuminata peel ash catalyst. *J. Taiwan Inst. Chem. Eng*. 2019. https://doi.org/10.1016/j.jtice.2019.06.014.

12 Ao, S., Alghamdi, L.A., Kress, T., Selvaraj, M., Halder, G., Wheatley, A.E.H., et al. Microwave-assisted valorization of glycerol to solketal using biomass-derived heterogeneous catalyst. *SSRN. Electron. J*. 2023. https://doi.org/10.2139/ssrn.4398152.

13 Tiwari, R., Agrawal, P.S., Belkhode, P.N., Brijpuriya, D., Devase, A., and Samarth, P. Utilization of waste transformer oil for the synthesis of biodiesel and glyptal. *Lect. Notes Civ. Eng*. 2024. https://doi.org/10.1007/978-981-99-4186-5_7.

14 Agrawal, P. and Rokhum, S.L. Supported metal/metal oxide catalysts in biodiesel production: an overview. In: *Biodiesel Production: Feedstocks, Catalysts, and Technologies*. Wiley, 2022, pp. 125–146. https://doi.org/10.1002/9781119771364.ch7.

15 Bashir, M.A., Wu, S., Zhu, J., Krosuri, A., Khan, M.U., and Aka, R.J.N. Recent development of advanced processing technologies for biodiesel production: a critical review. *Fuel Process. Technol*. 2022. https://doi.org/10.1016/j.fuproc.2021.107120.

16 Wang, S. High-efficiency separation of Bio-Oil. In: *Biomass Now - Sustainable Growth and Use*. IntechOpen, 2013, pp. 225–238. https://doi.org/10.5772/51423.

17 Agrawal, P.S., Belkhode, P.N., and Rokhum, S.L. Chemically produced biofuels. *Sustain. Biofuel Prod. Technol*. 2022. https://doi.org/10.1002/9781119888864.ch5.

18 Agrawal, P.S., Belkhode, P.N., and Rokhum, S.L. *Sustainability in Biofuel Production Technology*. John Wiley & Sons Ltd, 2022. https://doi.org/10.1002/9781119888864.

19 Sihombing, J.L., Herlinawati, H., Pulungan, A.N., Simatupang, L., Rahayu, R., and Wibowo, A.A. Effective hydrodeoxygenation bio-oil via natural zeolite supported transition metal oxide catalyst. *Arab. J. Chem*. 2023. https://doi.org/10.1016/j.arabjc.2023.104707.

20 Das, A., Li, H., Kataki, R., Agrawal, P.S., Moyon, N.S., Gurunathan, B., et al. Terminalia arjuna bark – A highly efficient renewable heterogeneous base catalyst for biodiesel production. *Renew. Energy*. 2023. https://doi.org/10.1016/j.renene.2023.05.066.

21 Kong, P.S., Cognet, P., Pérès, Y., Esvan, J., Daud, W.M.A.W., Aroua, M.K. Development of a novel hydrophobic ZrO2-SiO2 based acid catalyst for catalytic esterification of glycerol with oleic acid. *Ind. Eng. Chem. Res*. 2018. https://doi.org/10.1021/acs.iecr.8b01609.

22 Kong, P.S., Pérès, Y., Cognet, P., Senocq, F., Daud, W.M.A.W., Aroua, M.K., et al. Structure–selectivity relationship of a zirconia-based heterogeneous acid catalyst in the production of green mono- and dioleate product. *Clean Technol. Environ. Policy*. 2021. https://doi.org/10.1007/s10098-020-01830-1.

23 Khosravanipour Mostafazadeh, A., Solomatnikova, O., Drogui, P., and Tyagi, R.D. A review of recent research and developments in fast pyrolysis and bio-oil upgrading. *Biomass Convers. Biorefinery*. 2018. https://doi.org/10.1007/s13399-018-0320-z.

24 Olson, A.L., Tunér, M., and Verhelst, S.A. Concise review of glycerol derivatives for use as fuel additives. *Heliyon*. 2023. https://doi.org/10.1016/j.heliyon.2023.e13041.

25 Attarbachi, T., Kingsley, M.D., and Spallina, V. New trends on crude glycerol purification: a review. *Fuel*. 2023. https://doi.org/10.1016/j.fuel.2023.127485.

26 He, R., Ye, X.P., English, B.C., and Satrio, J.A. Influence of pyrolysis condition on switchgrass bio-oil yield and physicochemical properties. *Bioresour. Technol*. 2009. https://doi.org/10.1016/j.biortech.2009.02.069.

27 Quispe, C.A.G., Coronado, C.J.R., and Carvalho, J.A. Glycerol: production, consumption, prices, characterization and new trends in combustion. *Renew. Sustain. Energy Rev*. 2013. https://doi.org/10.1016/j.rser.2013.06.017.

28 Ilyin, S.O. and Makarova, V.V. Bio-oil: production, modification, and application. *Chem. Technol. Fuels Oils*. 2022. https://doi.org/10.1007/s10553-022-01348-w.

29 Sheel, A. and Pant, D. Epoxy paint from waste polycarbonate (PC). *Int. J. Environ. Sci. Technol*. 2023. https://doi.org/10.1007/s13762-022-04365-8.

30 Almazrouei, M. and Janajreh, I. Model-fitting approach to kinetic analysis of non-isothermal pyrolysis of pure and crude glycerol. *Renew. Energy*. 2020. https://doi.org/10.1016/j.renene.2019.07.095.

31 Salema, A.A. and Ani, F.N. Microwave induced pyrolysis of oil palm biomass. *Bioresour. Technol*. 2011. https://doi.org/10.1016/j.biortech.2010.09.115.

32 Mohan, K., Pai, S.D.K.R., Reghunath, B.S., and Pinheiro, D. Unveiling cutting edge innovations in the catalytic valorization of biodiesel byproduct glycerol into value added products. *Chemistry Select*. 2023. https://doi.org/10.1002/slct.202204501.

33 Lødeng, R. and Bergem, H. Stabilisation of pyrolysis oils. *Direct Thermochem. Liq. Energy Appl*. 2017. https://doi.org/10.1016/B978-0-08-101029-7.00006-0.

34 Zhang, Z., Fang, Y., Yang, J., and Li, X. A comprehensive review of bio-oil, bio-binder and bio-asphalt materials: their source, composition, preparation and performance. *J. Traffic Transp. Eng*. 2022. https://doi.org/10.1016/j.jtte.2022.01.003.

35 Chen, H., Xu, H., Zhu, H., Yan, S., Zhang, S., Zhang, H., et al. A review on bioslurry fuels derived from bio-oil and biochar: preparation, fuel properties and application. *Fuel*. 2024. https://doi.org/10.1016/j.fuel.2023.129283.

36 Schmitt, N., Apfelbacher, A., Jäger, N., Daschner, R., Stenzel, F., and Hornung, A. Thermo-chemical conversion of biomass and upgrading to biofuel: the thermo-catalytic reforming process – a review. *Biofuels Bioprod. Biorefining*. 2019. https://doi.org/10.1002/bbb.1980.

37 Singh, N., Kalbande, P.N., and Sudarsanam, P. Heterogeneous catalytic routes for bio-glycerol-based acrylic acid synthesis. *Biodiesel Prod. Feed. Catal. Technol.* 2022. https://doi.org/10.1002/9781119771364.ch18.

38 Smith, J., Balana, B.B., Black, H., von Blottnitz, H., Casson, E., Glenk, K., et al. The potential of small-scale biogas digesters to alleviate poverty and improve long term sustainability of ecosystem services in Sub-Saharan Africa. *1st World Sustain. Forum*. 2013.

39 Huber, G.W., Iborra, S., and Corma, A. Synthesis of transportation fuels from biomass: chemistry, catalysts, and engineering. *Chem. Rev.* 2006. https://doi.org/10.1021/cr068360d.

40 Bhoi, P.R., Ouedraogo, A.S., Soloiu, V., and Quirino, R. Recent advances on catalysts for improving hydrocarbon compounds in bio-oil of biomass catalytic pyrolysis. *Renew. Sustain. Energy Rev.* 2020. https://doi.org/10.1016/j.rser.2019.109676.

41 Tymchyshyn, M., Yuan, Z., and Xu, C. Direct conversion of glycerol into bio-oil via hydrotreatment using supported metal catalysts. *Fuel*. 2013. https://doi.org/10.1016/j.fuel.2013.04.069.

42 Knez, Z., Markočič, E., Hrnčič, M.K., Ravber, M., and Škerget, M. High pressure water reforming of biomass for energy and chemicals: a short review. *J. Supercrit. Fluids*. 2015. https://doi.org/10.1016/j.supflu.2014.06.008.

43 Martínez-Klimov, M.E., Mäki-Arvela, P., and Murzin, D.Y. Catalysis for production of jet fuel from renewable sources by hydrodeoxygenation and hydrocracking. *Catalysis*. 2021. https://doi.org/10.1039/9781839163128-00181.

44 Ravindran, H., Thangalazhy-Gopakumar, S., Adhikari, S., Fasina, O., Tu, M., Via, B., et al. Production of bio-oil from underutilized forest biomass using an auger reactor. *Energy Sources, Part A Recover Util Environ Eff.* 2015. https://doi.org/10.1080/15567036.2011.613894.

45 Robinson, A.M., Hensley, J.E., and Medlin, J.W. Bifunctional catalysts for upgrading of biomass-derived oxygenates: a review. *ACS Catal.* 2016. https://doi.org/10.1021/acscatal.6b00923.

46 Lei, N., Miao, Z., Liu, F., Wang, H., Pan, X., Wang, A., et al. Understanding the deactivation behavior of Pt/WO3/Al2O3 catalyst in the glycerol hydrogenolysis reaction. *Chinese J. Catal.* 2020. https://doi.org/10.1016/S1872-2067(20)63549-5.

47 Hueso, J.L., Mallada, R., and Santamaria, J. Gas-solid contactors and catalytic reactors with direct microwave heating: current status and perspectives. *Catal. Today* 2023. https://doi.org/10.1016/j.cattod.2022.10.009.

48 Jędrzejczyk, M., Podlaska, A., Cieluch, K., Ryczkowski, R., Goscianska, J., and Grams, J. La-doped Ni/ZrO2 catalyst for the production of H2-rich gas by upgrading vapors coming from pyrolysis of biomass and co-pyrolysis of biomass with plastic. *Int. J. Hydrogen Energy*. 2024. https://doi.org/10.1016/j.ijhydene.2023.09.231.

49 Hu, X. and Gholizadeh, M. Progress of the applications of bio-oil. *Renew. Sustain. Energy Rev*. 2020. https://doi.org/10.1016/j.rser.2020.110124.

50 Sekar, M., Mathimani, T., Alagumalai, A., Chi, N.T.L., Duc, P.A., Bhatia, S.K., et al. A review on the pyrolysis of algal biomass for biochar and bio-oil – Bottlenecks and scope. *Fuel*.2021. https://doi.org/10.1016/j.fuel.2020.119190.

51 Haverly, M.R., Whitmer, L.E., and Brown, R.C. Evaluation of polymer compatibility with Bio-oil produced from thermochemical conversion of biomass. *Energy Fuels*. 2015. https://doi.org/10.1021/acs.energyfuels.5b01943.

52 Wu, Q., Yang, Q., Zeng, X., Deng, J., Zhang, L., and Qiu, J. Catalytic conversion of cellulose-based biomass to diols. *Prog. Chem*. 2022. https://doi.org/10.7536/PC220236.

53 Qiyue, Y., Qiaomei, W., Jiarong, Q., Xianhai, Z., Xing, T., and Liangqing, Z. Catalytic conversion of bio-based platform compounds to fufuryl alcohol. *Prog. Chem*. 2022. https://doi.org/10.7536/PC211007.

54 Talebian-Kiakalaieh, A., Amin, N.A.S., and Hezaveh, H. Glycerol for renewable acrolein production by catalytic dehydration. *Renew. Sustain. Energy Rev*. 2014. https://doi.org/10.1016/j.rser.2014.07.168.

55 Liu, L., Yu, F., Wang, S., and Ye, X.P. Glycerol dehydration to acrolein catalyzed by silicotungstic acid: effect of mesoporous support. *Eng*. 2023. https://doi.org/10.3390/eng4010012.

56 Salehi, E., Abedi, J., and Harding, T. Bio-oil from sawdust: effect of operating parameters on the yield and quality of pyrolysis products. *Energy Fuels*. 2011. https://doi.org/10.1021/ef200688y.

57 Mahadevan, R., Shakya, R., Neupane, S., and Adhikari, S. Physical and chemical properties and accelerated aging test of bio-oil produced from in situ catalytic pyrolysis in a bench-scale fluidized-bed reactor. *Energy Fuels*. 2015. https://doi.org/10.1021/ef502353m.

58 Chun Minh, A.L., Samudrala, S.P., and Bhattacharya, S. Valorisation of glycerol through catalytic hydrogenolysis routes for sustainable production of value-added C3chemicals: current and future trends. *Sustain. Energy Fuels*. 2022. https://doi.org/10.1039/d1se01333e.

59 Gracia-Vitoria, J., Gándara, S.C., Feghali, E., Ortiz, P., Eevers, W., Triantafyllidis, K.S., et al. The chemical and physical properties of lignin bio-oils, facts and needs. *Curr. Opin. Green Sustain. Chem*. 2023. https://doi.org/10.1016/j.cogsc.2023.100781.

60 Nomanbhay, S., Hussein, R., and Ong, M.Y. Sustainability of biodiesel production in Malaysia by production of bio-oil from crude glycerol using microwave pyrolysis: a review. *Green Chem. Lett. Rev*. 2018. https://doi.org/10.1080/17518253.2018.1444795.

61 Cai, J., Rahman, M.M., Zhang, S., Sarker, M., Zhang, X., Zhang, Y., et al. Review on aging of bio-oil from biomass pyrolysis and strategy to slowing aging. *Energy Fuels*. 2021. https://doi.org/10.1021/acs.energyfuels.1c01214.

62 Sandid, A., Spallina, V., and Esteban, J. Glycerol to value-added chemicals: state of the art and advances in reaction engineering and kinetic modelling. *Fuel Process. Technol.* 2024. https://doi.org/10.1016/j.fuproc.2023.108008.

63 Wang, B., Liu, F., Guan, W., Wang, A., and Zhang, T. Promoting the effect of Au on the selective hydrogenolysis of glycerol to 1,3-Propanediol over the Pt/WOx/Al2O3Catalyst. *ACS Sustain. Chem. Eng.* 2021. https://doi.org/10.1021/acssuschemeng.1c00880.

64 Hu, H., Luo, Y., Zou, J., Zhang, S., Yellezuome, D., Rahman, M.M., et al. Exploring aging kinetic mechanisms of bio-oil from biomass pyrolysis based on change in carbonyl content. *Renew. Energy*. 2022. https://doi.org/10.1016/j.renene.2022.09.049.

65 Neto, C.J.D., Sydney, E.B., de Souza Vandenberghe, L.P., and Soccol, C.R. Green fuels technology. *Green Energy Technol.* 2016.

66 Phusuwan, U., Atong, D., Nisamaneenate, J., and Sricharoenchaikul, V. Sustainable fuel production from steam reforming of waste motor oil over olivine-supported Fe catalyst. *Energy Rep.* 2021. https://doi.org/10.1016/j.egyr.2021.07.095.

67 Medel, A., Lugo, F., and Meas, Y. Application of electrochemical processes for treating effluents from hydrocarbon industries. *Electrochem. Water Wastewater Treat.* 2018. https://doi.org/10.1016/B978-0-12-813160-2.00014-6.

68 Lachos-Perez, D., Martins-Vieira, J.C., Missau, J., Anshu, K., Siakpebru, O.K., Thengane, S.K., et al. Review on biomass pyrolysis with a focus on Bio-oil upgrading techniques. *Analytica*. 2023. https://doi.org/10.3390/analytica4020015.

69 No, S.Y. Application of bio-oils from lignocellulosic biomass to transportation, heat and power generation – a review. *Renew. Sustain. Energy Rev.* 2014. https://doi.org/10.1016/j.rser.2014.07.127.

70 Khare, P. and Goyal, D.K. Effect of high and low rank char on soil quality and carbon sequestration. *Ecol. Eng.* 2013. https://doi.org/10.1016/j.ecoleng.2012.12.101.

71 Gomes, G.J., Zalazar, M.F., and Arroyo, P.A. New insights into the effect of the zeolites framework topology on the esterification reactions: a comparative study from experiments and theoretical calculations. *Top. Catal.* 2022. https://doi.org/10.1007/s11244-022-01606-5.

72 Fernandes Barbosa F. and Pinheiro Braga, T. Catalytic conversion of glycerol to acetol and acrolein using metal oxides: surface reactions, prospects and challenges. *Chem.Cat.Chem.* 2023. https://doi.org/10.1002/cctc.202200950.

73 Paredes-Quevedo, L.C., Carriazo, J.G., and Velasquez, M. CuCo2O4 spinel supported on dealuminized metakaolinite for partial glycerol oxidation. *Catal. Commun.* 2023. https://doi.org/10.1016/j.catcom.2023.106676.

74 Cho, E.K., Kim, Y.E., and Lee, S.H. Review on the synthesis and biological activities of acrolein and its acetal form. *Yakhak Hoeji*. 2024. https://doi.org/10.17480/psk.2024.68.1.1.

75 Zhang, S., Yin, H., Wang, J., Zhu, S., and Xiong, Y. Catalytic cracking of biomass tar using Ni nanoparticles embedded carbon nanofiber/porous carbon catalysts. *Energy*. 2021. https://doi.org/10.1016/j.energy.2020.119285.

76 Baluna, A.S., Galan, A., Leigh, D.A., Smith, G.D., Spence, J.T.J., Tetlow, D.J., et al. In search of wasserman's catenane. *J. Am. Chem. Soc.* 2023. https://doi.org/10.1021/jacs.3c01939.

77 Sakthivel, R. and Ramesh, K. Influence of temperature on yield, composition and properties of the sub-fractions derived from slow pyrolysis of *Calophyllum inophyllum* de-oiled cake. *J. Anal. Appl. Pyrolysis.* 2017. https://doi.org/10.1016/j.jaap.2017.08.012.

78 Chaudhari, R.V., Torres, A., Jin, X., and Subramaniam, B. Multiphase catalytic hydrogenolysis/hydrodeoxygenation processes for chemicals from renewable feedstocks: kinetics, mechanism, and reaction engineering. *Ind. Eng. Chem. Res.* 2013. https://doi.org/10.1021/ie400709d.

79 Wan Khalit, W.N.A., Asikin-Mijan, N., Marliza, T.S., Safa-Gamal, M., Shamsuddin, M.R., Azreena, I.N., et al. One-pot decarboxylation and decarbonylation reaction of waste cooking oil over activated carbon supported nickel-zinc catalyst into diesel-like fuels. *J. Anal. Appl. Pyrolysis.* 2022. https://doi.org/10.1016/j.jaap.2022.105505.

80 Hooshdaran, B., Haghshenasfard, M., Hosseini, S.H., Esfahany, M.N., Lopez, G., and Olazar, M. CFD modeling and experimental validation of biomass fast pyrolysis in a conical spouted bed reactor. *J. Anal. Appl. Pyrolysis.* 2021. https://doi.org/10.1016/j.jaap.2020.105011.

81 Gouws, S.M., Carrier, M., Bunt, J.R., and Neomagus, H.W.J.P. Co-pyrolysis of coal and raw/torrefied biomass: a review on chemistry, kinetics and implementation. *Renew. Sustain. Energy Rev.* 2021. https://doi.org/10.1016/j.rser.2020.110189.

82 Novoselov, K.S., Geim, A.K., Morozov, S.V., Jiang, D., Zhang, Y., Dubonos, S.V., et al. Electric field effect in atomically thin carbon films supplementary. *Science.* 2004. https://doi.org/10.1126/science.aab1343.

83 Dabros, T.M.H., Stummann, M.Z., Høj, M., Jensen, P.A., Grunwaldt, J.D., Gabrielsen, J., et al. Transportation fuels from biomass fast pyrolysis, catalytic hydrodeoxygenation, and catalytic fast hydropyrolysis. *Prog. Energy Combust. Sci.* 2018. https://doi.org/10.1016/j.pecs.2018.05.002.

84 Abou Rjeily, M., Gennequin, C., Pron, H., Abi-Aad, E., and Randrianalisoa, J.H. Pyrolysis-catalytic upgrading of bio-oil and pyrolysis-catalytic steam reforming of biogas: a review. *Environ. Chem. Lett.* 2021. https://doi.org/10.1007/s10311-021-01190-2.

85 Ng, J.H., Leong, S.K., Lam, S.S., Ani, F.N., and Chong, C.T. Microwave-assisted and carbonaceous catalytic pyrolysis of crude glycerol from biodiesel waste for energy production. *Energy Convers. Manag.* 2017. https://doi.org/10.1016/j.enconman.2017.04.024.

86 Peirovi Minaee, R., Afsharnia, M., Moghaddam, A., Ebrahimi, A.A., Askarishahi, M., and Mokhtari, M. Calibration of water quality model for distribution networks using genetic algorithm, particle swarm optimization, and hybrid methods. *MethodsX.* 2019. https://doi.org/10.1016/j.mex.2019.03.008.

87 Lamers, P., Searcy, E., Hess, J.R., and Stichnothe, H. (Editors) *Developing the Global Bioeconomy: Technical, Market, and Environmental Lessons from Bioenergy.* 2016. https://doi.org/10.1016/c2015-0-02444-3.

88 B.P. Johnson, M. License innovative waste-to-fuels technology to biofuels producer Fulcrum BioEnergy. *Focus Catal.* 2019. https://doi.org/10.1016/j.focat.2019.04.043.

89 Fulcrum BioEnergy produces low-carbon fuel from landfill waste. *MSW Manag.* 2023.

90 Bettenhausen, C. Fulcrum BioEnergy starts making syngas from trash near Reno, Nevada. *Chem. Eng. News.* 2022. https://doi.org/10.47287/cen-10020-buscon10.

10

Production of Biochar

10.1 Introduction

The agricultural sector is fundamental to global food security and economic development, yet its activities generate prodigious quantities of organic residues and byproducts, collectively termed agro-waste. Estimates by the Food and Agriculture Organization (FAO) indicate global food losses and waste reach approximately 1.3 billion tons annually, with a substantial portion originating from agricultural production and processing [FAOSTAT, 1997]. This vast biomass stream, ranging from cereal straws and fruit peels to livestock manure and forestry residues, poses significant environmental challenges when improperly managed. Common practices such as open burning release substantial amounts of greenhouse gases (GHGs) – including CO_2, CH_4, and N_2O – and particulate matter, contributing to climate change and air pollution. Decomposition in landfills also generates potent GHGs, particularly methane. Furthermore, runoff from agro-waste stockpiles can lead to nutrient leaching (e.g., nitrates and phosphates), causing eutrophication and water body contamination, while potential heavy metal content can impact soil health and food safety [1]. Addressing these environmental burdens necessitates the development and implementation of sustainable agro-waste management strategies that prioritize resource recovery and circular economy principles. Converting agro-waste into biochar through thermochemical processes represents a compelling solution that simultaneously manages waste and produces a stable, carbon-rich material with diverse environmental and agricultural benefits.

10.1.1 Biochar: Properties, Applications, and Significance

Biochar is a solid carbonaceous material obtained from the thermochemical conversion of organic biomass in an oxygen-limited environment, typically through pyrolysis. Unlike charcoal, which is primarily used for fuel, biochar is primarily intended for non-fuel applications, particularly soil amendment and environmental remediation, though their production methods overlap significantly. Activated carbon, another related material, undergoes further physical or chemical activation to develop extremely high surface areas and porosity for specialized adsorption

Agro-Waste Management and Valorization, First Edition. Pratibha S. Agrawal, et al.

applications [2–4]. The valuable properties of biochar are directly linked to its production process and feedstock, which include:

- **High Carbon Content:** Typically ranging from 50 to >80% (dry weight basis), depending on pyrolysis temperature and feedstock, contributing to its stability and carbon sequestration potential [5].
- **Porous Structure:** Possessing a complex pore network (macro-, meso-, and micropores) resulting from the biomass cellular structure breakdown during thermochemical conversion. This leads to a high specific surface area, often ranging from 10 to over 500 m^2/g (for non-activated biochar), facilitating adsorption and microbial habitat formation [6].
- **Surface Chemistry:** Featuring various functional groups (e.g., carboxyl, hydroxyl, phenolic) that impart surface charge and reactivity, influencing nutrient and contaminant adsorption, pH buffering, and interaction with soil components and microorganisms [7]. The abundance and type of functional groups are highly dependent on feedstock and process conditions, especially lower temperatures, which preserve more oxygen-containing groups.
- **Alkalinity:** Agro-waste biochar is often alkaline (pH > 7), particularly when derived from feedstocks with high ash content rich in alkali and alkaline earth metals (K, Ca, Mg) [8]. This property makes it effective for ameliorating acidic soils. However, biochar pH can vary depending on feedstock and production parameters, occasionally being neutral or slightly acidic at very high pyrolysis temperatures or from low-ash feedstocks.
- **Nutrient Content:** Biochar can contain essential plant nutrients (P, K, Ca, Mg) concentrated from the feedstock's inorganic components, making it a potential slow-release fertilizer. Nitrogen, however, is largely lost as volatile compounds during high-temperature conversion processes but can be retained more effectively in hydrochar produced at lower temperatures [9].

Properties of biochars vary with the charring temperatures [10]. With increasing charring temperature, yields declined rapidly at 300 °C and remained relatively stable above 600 °C for rice straw-derived biochars (Figure 10.1).

The versatility stemming from these properties enables biochar's use in numerous applications:

- **Soil Amendment:** Improving soil structure, water retention, nutrient availability (cation exchange capacity), reducing bulk density, and enhancing microbial activity, leading to increased crop yields and resilience [11].
- **Carbon Sequestration:** Offering a long-term sink for atmospheric carbon when incorporated into soil, as its highly stable carbon structure resists microbial decomposition over centuries to millennia [12]. This directly contributes to climate change mitigation.
- **Environmental Remediation:** Adsorbing inorganic and organic pollutants from contaminated water and soil, including heavy metals, pesticides, and emerging contaminants, due to its high surface area and surface chemistry [13].
- **Catalysis Support:** Serving as a low-cost, carbon-neutral support material for catalysts in various chemical processes, often enhanced by its porous structure and tunable surface properties [14].

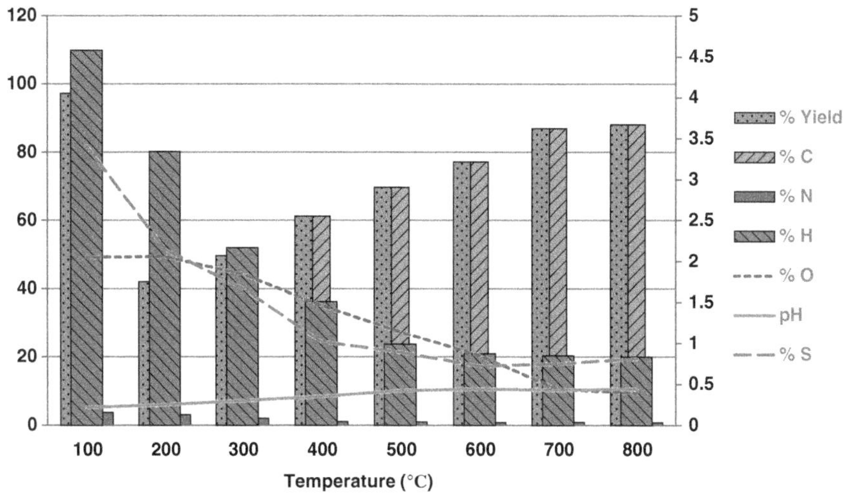

Figure 10.1 The trend of rice straw biochar yield and chemical composition vs. pyrolysis temperature. Adapted from [10].

- **Energy Storage:** Modified biochar materials [15] can be explored as electrodes for supercapacitors and batteries, leveraging their electrical conductivity and porous structure.

10.1.2 Agro-waste-to-biochar Conversion: A Sustainable and Circular Approach

The conversion of agro-waste to biochar aligns fundamentally with circular economy principles by transforming a low-value waste stream into a valuable product, reducing the need for traditional disposal methods (like burning or landfilling) and their associated environmental costs. This approach fosters a closed-loop system within agriculture and beyond, where nutrients and carbon are recycled. The advantages of this conversion pathway are multifold, which are well depicted in Table 10.1.

Thermochemical conversion technologies are central to transforming agro-waste into biochar since these processes involve subjecting the biomass to high temperatures in controlled atmospheres. The primary technologies are pyrolysis, gasification, and hydrothermal carbonization (HTC). Each technology operates under distinct conditions and yields different product distributions, influencing the properties and yield of the resulting biochar. The choice of conversion technology depends on the characteristics of the agro-waste feedstock, the desired biochar properties, and economic considerations, as summarized in Table 10.2.

10.1.3 Scope and Objectives

This chapter provides an in-depth chemical engineering perspective on biochar production from agro-waste. The scope is focused on the technical aspects of utilizing

Table 10.1 Advantages of agro-waste-to-biochar conversion.
[16]/Elsevier.

Benefit	Description
Waste valorization	Converts agricultural residues, manures, and byproducts from a disposal challenge into a commercially viable product, lowering waste management costs and creating new economic opportunities
Environmental mitigation	Reduces greenhouse gas emissions (CO_2, CH_4, N_2O) from decomposition/burning, minimizes nutrient leaching, and sequesters atmospheric carbon. Decreased risk of water and soil pollution from uncontrolled stockpiles
Soil improvement	Enhances physical, chemical, and biological properties of soil, potentially reducing reliance on synthetic fertilizers and irrigation and leading to increased water retention, nutrient availability, and crop productivity
Resource recovery	Concentrates and stabilizes valuable nutrients (P, K, Ca, Mg) within the biochar matrix, facilitating their beneficial recycling to land
Economic opportunities	Creates new value chains, potentially stimulates rural economies, and diversifies income streams for agricultural operations
Promotion of circular economy	Closes nutrient and carbon loops, reducing reliance on external inputs (fertilizers, soil conditioners) and promoting sustainable resource use

diverse agro-waste feedstocks in various thermochemical conversion processes. The specific objectives include:

- To provide a detailed characterization of the physical and chemical properties of diverse agro-waste feedstocks and their suitability for biochar production, including proximate/ultimate analysis and their impact on conversion kinetics and product properties.
- To conduct a comparative analysis of key thermochemical conversion technologies (pyrolysis, gasification, HTC) from a chemical engineering standpoint, detailing process principles, typical reactor designs, mass and energy flows, and factors influencing product yields and distribution.
- To analyze in detail the influence of critical process parameters (temperature, pressure, heating rate, residence time, gasifying agent, water-to-biomass ratio) on the chemical reactions, char formation mechanisms, biochar yield, and resulting physiochemical properties, referencing specific kinetic models and characterization techniques.
- To evaluate the environmental implications (via LCA insights on GHG balance, energy consumption, water usage, pollutant emissions) and economic feasibility (via TEA insights on capital costs, operating costs, product value) of agro-waste-to-biochar pathways.
- To identify current technological challenges (e.g., scale-up, continuous processing, energy integration, ash management) and outline future research and development directions necessary to optimize conversion efficiency, tailor biochar properties for advanced applications, and ensure the widespread, sustainable implementation of agro-waste-to-biochar technologies within a biorefinery context.

Table 10.2 Comparison of thermochemical conversion technologies for biochar production. Adapted from [17].

Technology	Operating conditions (temperature, pressure)	Product distribution (biochar, bio-oil, gas)	Advantages	Disadvantages	Suitable feedstocks
Pyrolysis (slow)	300–700 °C, ambient-slight positive, inert (N_2)	High biochar yield (30–50%), lower bio-oil/syngas	Relatively simple, versatile, can handle diverse feedstocks, produces bio-oil as a valuable byproduct. High biochar yield, simple process, relatively low capital cost, stable char	Long residence time, lower energy products (bio-oil quantity), requires drier feedstock	Dry crop residues, wood waste, nut shells, manures
Pyrolysis (fast)	400–650 °C, ambient-slight positive, inert (N_2)	High bio-oil yield (50–75%), moderate syngas, low biochar (10–20%)	Maximizes liquid bio-oil production, shorter residence time, compact reactor systems	Lower biochar yield, bio-oil requires upgrading, sensitive to feedstock moisture, requires rapid quenching	Finer dry biomass particles, wood, crop residues
Pyrolysis (flash)	500–1000 °C, ambient-slight positive, inert (N_2)	Very high volatiles/bio-oil (~>75%), minimal biochar (<10%)	Maximizes specific volatile products/chemicals, extremely short residence time	Very low biochar yield, highly sensitive to parameters, requires sophisticated equipment, difficult to scale	Fine dry biomass particles
Gasification	700–1000 °C, ambient-elevated, oxidizing (air, O_2, steam, CO_2)	High syngas yield (60–80%), moderate ash, lower biochar (5–15%)	Produces syngas for energy/chemical synthesis, can handle wetter feedstocks than dry pyrolysis, high efficiency	Lower biochar yield (often higher ash), syngas cleaning required, higher operating temperatures, more complex	Wood waste, energy crops, some agricultural residues
HTC	180–250 °C, elevated pressure (2–6 MPa), aqueous	High hydrochar yield (40–70%), aqueous phase, gases	Handles wet feedstocks efficiently (no pre-drying), nutrient retention in hydrochar, lower temperature	Hydrochar often lower carbon/surface area than pyrolysis char, requires pressurized reactors, produces wastewater	Wet biomass, sewage sludge, food waste, wet manures

10.2 Agro-waste Feedstocks for Biochar Production

The characteristics of the feedstock are paramount in determining the suitability for specific conversion technologies and the properties of the resultant biochar or hydrochar. This section delves into the classification, composition, and pretreatment requirements of common agro-waste streams.

10.2.1 Classification and Characteristics of Agro-waste Feedstocks

Agro-waste streams can be broadly categorized based on their origin:

1. **Crop Residues:** Post-harvest plant biomass (straws, stalks, husks, shells, leaves). These are typically lignocellulosic, composed primarily of cellulose, hemicellulose, and lignin, with varying ash content. Availability is often seasonal and geographically specific. Global production is immense; for example, rice straw alone accounts for hundreds of millions of tons annually, often leading to problematic disposal methods like field burning [18]. Table 10.3 summarizes the approximate global production of some common crop residues. Examples include rice straw, wheat straw, corn stover, sugarcane bagasse, coconut shells, and nut shells. High lignocellulose content makes them suitable for pyrolysis and gasification. Low moisture content favors dry thermochemical processes. Variability in composition (especially ash) impacts reactor design and biochar properties (e.g., alkali metals in ash can catalyze cracking, affecting bio-oil composition and char structure) [19].
2. **Animal Manure:** Excreta from livestock (cattle, poultry, swine). High in moisture, nutrients (N, P, K), and often ash. Pathogen content and odor are significant management challenges. High moisture content (>60%) makes it less suitable for dry pyrolysis/gasification without pre-drying, but ideal for HTC. High ash content and nutrient content (N, P) significantly impact biochar properties (alkalinity, nutrient load) and potential reactor fouling. Nitrogen loss as NH_3 is a concern at higher temperatures. Typical examples include cattle manure, poultry manure, and swine manure. Table 10.4 provides a comparative overview of the characteristics of different animal manures.
3. **Agro-industrial Byproducts:** Waste streams from processing agricultural products. Composition varies widely based on the process. Can be wet or dry, fibrous or pulpy. It offers centralized, high-volume sources. Moisture content

Table 10.3 Approximate annual global production of selected crop residues.

Crop residue	Estimated annual production (million tons)	References
Rice straw	731	[20]
Wheat straw	864	[21]
Corn stover	734	[22]
Sugarcane bagasse	434	[23]

Table 10.4 Typical characteristics of different animal manures.

Manure type	Total solids (%)	Volatile solids (%)	Total nitrogen (N) (%)	Total phosphorus (P) (%)	Total potassium (K) (%)	References
Cattle manure	10–20	70–80	0.3–0.6	0.1–0.3	0.2–0.4	[24]
Poultry manure	20–30	75–85	1.5–4.0	1.0–3.0	0.8–2.0	[25]
Swine manure	3–8	75–90	0.2–0.8	0.1–0.5	0.1–0.6	[26]

Table 10.5 Examples of agro-industrial byproducts and their relevance to biochar production. Adapted from [27].

Agro-industrial byproduct	Source	Relevance to biochar production
Olive pomace	Olive oil production	High lignin content, moderate moisture, high calorific value. Suitable for high-carbon biochar production via pyrolysis
Fruit peels	Juice production	Rich in organic matter, can be used to produce biochar with enhanced nutrient content
Coffee husks	Coffee processing	Fibrous, moderate moisture, residual caffeine/lignin. Suitable for pyrolysis. Resulting char can be used as adsorbent
Cocoa shells	Chocolate production	Moderate lignin/cellulose, residual fats/sugars. Potential for nutrient-rich or adsorbent biochar
Brewery waste	Beer production	High moisture, protein, spent grains. Rich in nitrogen and phosphorus, can produce nutrient-rich biochar for agricultural uses

determines technology choice. Composition (e.g., high lignin in olive pomace, high nutrients in brewery waste) influences biochar yield and properties. Potential contaminants (oils, processing chemicals) may require consideration. The typical examples include olive pomace (olive oil), fruit peels/pomace (juice/wine), coffee husks/pulp (coffee), cocoa shells (chocolate), brewery waste (beer), and bagasse (sugar – also crop residue). Table 10.5 lists common agro-industrial byproducts and their potential applications in biochar production.

The global waste generation was about 2.02 billion tons in 2016 and has been estimated to become 2.59 billion tons in 2030 and 3.4 billion tons in 2050 (Figure 10.2). The global waste management market value has been estimated to increase from US$1.61 trillion in 2020 to US$2.5 trillion in 2030 [28].

10.2.2 Influence of Feedstock Composition on Biochar Production and Properties

The chemical makeup of agro-waste feedstocks profoundly dictates their thermal decomposition behavior and the characteristics of the resulting biochar. From a chemical engineering perspective, understanding the feedstock composition is

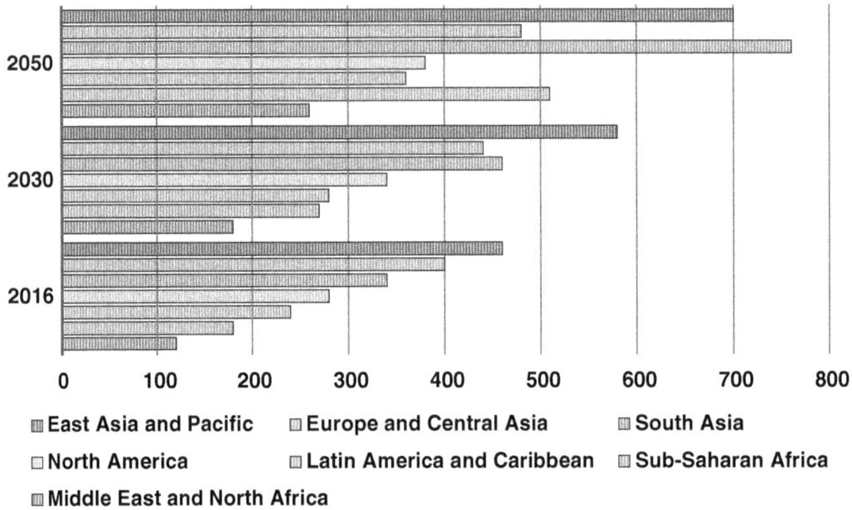

Figure 10.2 Approximate global distribution of major agro-waste types. Adapted from [28].

critical for process design and optimization. Table 10.6 summarizes the key characteristics of various agro-waste feedstocks relevant to biochar production. Major components are as discussed below:

a. **Lignocellulosic Polymers (Cellulose, Hemicellulose, Lignin):** These structural polymers undergo distinct decomposition pathways and kinetics during thermochemical conversion.
   - **Hemicellulose:** Less crystalline, amorphous structure. Decomposes at lower temperatures (200–300 °C) via depolymerization and degradation into furans, aldehydes, and organic acids. Higher hemicellulose content tends to yield more volatiles and less char at moderate temperatures.
   - **Cellulose:** Crystalline structure. Decomposes at higher temperatures (300–400 °C) primarily via depolymerization (levoglucosan formation) and subsequent cracking/carbonization. Higher cellulose content favors bio-oil production at fast heating rates but contributes significantly to char yield at slow rates [29].

b. **Lignin:** Complex, highly cross-linked aromatic polymer. Decomposes over a broad temperature range (280–500 °C and above) via cleavage of ether and carbon–carbon linkages, yielding phenolic compounds and contributing substantially to the stable carbon structure of biochar [30]. Feedstocks with higher lignin content typically produce biochar with higher fixed carbon content, greater aromaticity, and potentially higher structural integrity and surface area at high temperatures. Wood-based feedstocks generally have higher lignin content (20–30%) than herbaceous residues like straw (10–20%) [31].

c. **Ash Content:** The inorganic residue after combustion. It does not decompose at typical biochar production temperatures but becomes concentrated in the solid product. High ash content in the feedstock (e.g., rice husk >15%, some

Table 10.6 Key characteristics of common agro-waste feedstocks for biochar production (dry weight basis).

Feedstock	Cellulose (%)	Hemi-cellulose (%)	Lignin (%)	Ash (%)	Moisture content (%)	N (%)	P (%)	K (%)	References
Rice straw	32–47	19–27	5–24	8–20	8–15	0.5–1	0.1–0.2	0.8–1.5	[32]
Wheat straw	30–48	20–35	15–25	4–10	8–15	0.4–0.8	0.1–0.2	0.7–1.2	[33]
Corn stover	32–45	20–30	10–20	5–10	15–25	0.6–1.2	0.1–0.3	1.0–1.8	[34]
Sugarcane bagasse	32–48	21–32	17–24	1–5	40–60	0.2–0.5	0.1–0.2	0.3–0.6	[35]
Cattle manure	15–25	8–15	5–12	15–30	60–80	1.5–3.0	0.5–1.5	0.4–1.0	[36]
Poultry manure	18–28	10–18	8–15	10–20	50–70	3.0–6.0	2.0–4.0	1.5–3.0	[37]
Olive pomace	20–35	15–25	25–40	5–10	50–70	0.5–1.5	0.1–0.3	0.5–1.0	[38]
Coffee husks	40–50	15–25	25–35	5–10	10–15	2.0–3.0	0.2–0.4	1.5–2.5	[39]

manures >20%) leads to high ash content in biochar, which can dilute the carbon content and affect porosity, surface area, and catalytic activity. The composition of ash (rich in Si, K, Ca, Mg, P) influences biochar's alkalinity, nutrient profile, and potential catalytic effects on char formation reactions.

d. **Moisture Content:** The amount of water in the feedstock. High moisture (>20–30%) significantly reduces the energy efficiency of dry thermochemical processes (pyrolysis, gasification) because energy is consumed for water vaporization (latent heat of vaporization is ~2.26 MJ/kg at 100 °C. This increases operating costs and can lower reactor throughput. HTC is specifically designed to handle wet feedstocks.

10.2.3 Pretreatment of Agro-waste Feedstocks

Pretreatment steps are often indispensable in preparing raw agro-waste feedstocks for efficient and optimized thermochemical conversion into biochar and other products. These initial processing stages significantly influence not only the energy balance and operational feasibility of the conversion technology but also play a critical role in dictating the yield, physical characteristics, and chemical properties of the final biochar product. Cueva et al. [40] proposed that the selection and design of pretreatment are guided by the specific type and characteristics of the raw agro-waste, the chosen conversion technology (e.g., dry pyrolysis/gasification vs. wet HTC), and the desired quality and application of the biochar. Implementing appropriate pretreatment can transform a variable, challenging feedstock into a consistent, process-ready material, minimizing operational issues, improving product quality, and ultimately enhancing the overall techno-economic and environmental sustainability of the waste-to-value pathway.

10.2.3.1 Drying: Essential Moisture Management

The primary purpose of drying agro-waste prior to dry thermochemical conversion processes like pyrolysis and gasification is the reduction of moisture content. Raw agro-wastes, especially fresh crop residues, manures, and certain agro-industrial byproducts, often contain high levels of moisture, ranging from 15 to over 80% wet basis. Water within the biomass exists in various forms, including free surface water, capillary water held within pores, and bound water chemically adsorbed to the biomass structure. Removal of this water is a critical step because its presence significantly impacts the energy efficiency and operational performance of high-temperature dry conversion processes. From an energy balance perspective, vaporizing water is an endothermic process, requiring approximately 2.26 MJ of energy per kilogram of water at 100 °C (latent heat of vaporization). This energy must be supplied by the conversion system, either through internal combustion of a portion of the feedstock or process products (like syngas) or via external energy sources. High moisture content directly increases the total energy input required per unit of *dry* biomass processed, leading to lower net energy output from the system or necessitating larger, more expensive reactors to handle the increased thermal load. For instance, drying 1 ton of biomass from 60 to 10% moisture removes 555 kg of water, requiring over 1250 MJ of energy just for vaporization, excluding sensible heating and drying inefficiency. This energy penalty can substantially reduce the overall process efficiency and increase operating costs, particularly for large-scale operations [41].

Various methods are employed for drying agro-waste, each with its own trade-offs in terms of cost, speed, and energy source. Simplest methods include passive processes like solar drying or open-air drying, which are low-cost and utilize renewable energy (sun/wind) but are slow, require large land areas, are weather-dependent, and can lead to biomass degradation or nutrient losses due to biological activity. As reported by Saharudin et al. [42] in continuous industrial processes, thermal dryers are commonly used for faster and controlled drying processes. These utilize heat transfer (convection, conduction, radiation) to drive water evaporation. Common types include rotary dryers, belt dryers, flash dryers, and fluidized bed dryers. Thermal dryers can use hot air, flue gases from the thermochemical process (an excellent strategy for energy integration and heat recovery), or steam as the heating medium. Designing an efficient thermal dryer involves optimizing parameters like air temperature and humidity, airflow rate, feedstock residence time, and contact method to minimize energy consumption and prevent overheating or charring of the biomass. From a chemical engineering standpoint, thermal drying involves complex simultaneous heat and mass transfer phenomena within the porous biomass structure, often modeled using diffusion equations and heat transfer coefficients [43].

Beyond energy efficiency, drying impacts the conversion process kinetics and reactor performance. Introducing large amounts of water vapor into a pyrolysis or gasification reactor can alter the partial pressures of reactive gases, potentially influencing secondary reactions. For instance, steam can participate in gasification reactions ($C + H_2O \rightarrow CO + H_2$), which might be undesirable if biochar yield is prioritized in pyrolysis, or can affect syngas composition in gasification. Furthermore, excessive moisture can cause particle agglomeration or bridging in certain reactor types, disrupting flow and heat transfer. Efficient drying ensures a more stable and predictable feedstock moisture content, leading to more consistent reactor performance, higher heating rates (as less energy is diverted to vaporization), and ultimately, a more uniform biochar product. For fast and flash pyrolysis, achieving the necessary rapid heating rates (<1 second) is only feasible with very dry feedstock (<10% moisture) and small particle sizes, as internal heat transfer resistance and the energy sink of vaporization become limiting factors at higher moisture levels. While drying is not strictly necessary for HTC – which is designed for wet feedstocks – managing moisture is still important for slurry pumpability and energy input optimization in HTC systems. In summary, drying is a fundamental pretreatment that significantly impacts the feasibility and performance of dry thermochemical agro-waste conversion, playing a vital role in the overall energy balance and operational reliability [44].

10.2.3.2 Size Reduction: Optimizing Heat and Mass Transfer

Size reduction, also known as grinding or milling, is a mechanical pretreatment step that decreases the average particle size of agro-waste feedstock. This process increases the total external surface area of the biomass particles relative to their volume. The significance of this increased surface area is rooted in the principles of heat and mass transfer, which are fundamental to thermochemical conversion processes. In pyrolysis and gasification, heat must be transferred from the reactor environment (hot walls, gas, or sand in fluidized beds) to the biomass particles to initiate and sustain the endothermic decomposition reactions. This heat transfer occurs via convection, radiation, and conduction. For larger particles, heat conduction through the particle's interior becomes the rate-limiting step, creating significant internal temperature gradients. By reducing particle size, the path length for heat conduction is shortened, allowing the particles to heat up more rapidly and uniformly. This is particularly crucial in fast and flash pyrolysis, where heating rates of hundreds or thousands of degrees Celsius per second are required for rapid devolatilization; such rates are unachievable with large, slow-heating particles. Grinding agro-waste to particle sizes typically ranging from <1 mm to a few millimeters, depending on the reactor type, ensures that the entire particle reaches the reaction temperature quickly, leading to more synchronized decomposition kinetics across the feedstock batch or flow [45].

Simultaneously, size reduction enhances mass transfer. During thermochemical conversion, volatile organic compounds (VOCs) are released from the biomass structure and must diffuse out of the particle and away from the reaction zone. Inversely, in gasification, the gasifying agent (e.g., steam, CO_2) must diffuse into the solid char particle to react with the carbon matrix. Diffusion rates are inversely proportional to the distance the diffusing species must travel. Smaller particles reduce the internal diffusion resistance for volatile escape and the external diffusion resistance for gas–solid reactions. Rapid removal of primary volatiles is essential in fast pyrolysis to minimize secondary reactions (cracking, repolymerization) that occur if vapors reside too long at high temperatures, thus maximizing bio-oil yield. For char formation, as in slow pyrolysis, faster heat and mass transfer still contribute to more uniform conversion throughout the feedstock bed and potentially more consistent char structure and properties. The particle size distribution (PSD) resulting from grinding is also important; a consistent PSD is desirable for stable operation of reactors like fluidized beds, where particle size affects fluidization quality, bed expansion, and potential for elutriation (fines being carried out with the gas flow) or agglomeration. In fixed or moving beds, PSD impacts pressure drop, which affects gas flow distribution and contact efficiency.

Size reduction is typically accomplished using mechanical comminution equipment. Initial coarse reduction of large pieces (like wood branches or fibrous stalks) might use chippers or shredders. Finer grinding employs mills such as hammer mills (impact crushing), ball mills (impact and attrition), or knife mills (cutting/shearing). The specific equipment choice depends on the feedstock characteristics (hardness, moisture, fiber content) and the desired final particle size. While essential for process performance, grinding is an energy-intensive operation, especially for achieving very fine particle sizes [46]. The energy required for grinding biomass can be significant, sometimes accounting for a substantial portion of the total process energy consumption. Equipment wear, particularly when processing abrasive feedstocks like rice husk (high silica content), is another operational challenge and cost factor. Furthermore, dry grinding of biomass can generate significant amounts of fine dust, which poses explosion hazards (dust deflagration/explosion) and occupational health risks, requiring strict safety measures and dust collection systems. Despite these challenges, appropriate size reduction is a fundamental pretreatment enabling efficient heat and mass transfer, crucial for achieving optimal reaction kinetics and consistent product quality in thermochemical conversion processes.

10.2.3.3 Washing: Cleaning and Purification of Feedstock

Washing agro-waste involves leaching out soluble inorganic impurities and salts using a liquid medium, typically water or dilute acid/alkaline solutions. The primary motivations for washing are multifaceted, aiming to improve the quality and safety of the resulting biochar, prevent operational issues within the thermochemical reactor, and potentially recover valuable elements. Agro-wastes, particularly crop residues grown in saline or fertilized soils, animal manures, and certain agro-industrial byproducts (like fruit peels that may contain residual pesticides or cleaning agents), can contain varying amounts of inorganic minerals, alkali and alkaline earth metals (Na, K, Ca, Mg), chlorides, sulfates, and trace amounts of heavy metals. These components do not convert to biochar and become concentrated in the solid product during pyrolysis or gasification.

Removing these impurities through washing has several beneficial impacts on the process and product. From a reactor performance perspective, alkali metals (especially K and Na) and chlorides can significantly lower the melting point of ash, leading to undesirable phenomena like slagging (melting and solidification of ash) and fouling (deposit formation on reactor surfaces) at the high temperatures encountered in pyrolysis and gasification (>700 °C for significant issues, but can start lower). These deposits can block gas flow, reduce heat transfer efficiency, corrode reactor materials, and ultimately lead to unplanned shutdowns and increased maintenance costs. Washing out a portion of these inorganic species mitigates these risks, improving operational uptime and equipment lifespan [47]. Furthermore, alkali metals can act as catalysts for both the primary decomposition of biomass polymers and secondary reactions of volatile compounds and char. While sometimes leveraged (e.g., in catalytic gasification), uncontrolled catalytic activity from feedstock ash can lead to undesirable product distributions (e.g., increased gas yield, altered tar composition) or affect the resulting char structure and reactivity in ways that are not optimal for specific biochar applications. Washing can help reduce or control this inherent catalytic effect, allowing for more predictable product characteristics.

On the product side, washing directly impacts the biochar's composition and properties. Reducing the ash content increases the relative proportion of carbon in the biochar. By removing soluble salts, particularly alkali chlorides, washing can also potentially increase the specific surface area of the biochar by clearing mineral deposits that might block or occlude pores formed during carbonization. The removal of alkali metals also slightly reduces the pH of the resulting biochar, although it generally remains alkaline if significant amounts of Ca and Mg remain in the ash. Critically, washing is a primary method for reducing the concentration of heavy metals (e.g., Cd, Pb, Cr, As) and other potentially toxic elements that may be present in the feedstock, particularly in contaminated manures or sludges. Concentrating these elements in biochar intended for soil application poses environmental and health risks. Washing, especially with dilute acid solutions (e.g., HCl, H_2SO_4, organic acids), can effectively leach out many problematic heavy metals, making the biochar safer and compliant with regulations for land application [48].

The major challenge associated with washing is the generation of a significant volume of wastewater. This wash water is typically contaminated with dissolved organic compounds, removed inorganic salts, and potentially heavy metals and pathogens. Managing and treating this wastewater before discharge or reuse is essential to avoid transferring the environmental burden from solid waste to water pollution. Wastewater treatment adds significant capital investment and operating costs to the overall conversion process. Treatment options can range from simple settling and filtration to more complex processes like biological treatment (anaerobic/aerobic digestion), adsorption, membrane filtration, or evaporation, depending on the level and type of contamination and regulatory requirements [49]. Despite the added complexity and cost of wastewater management, washing remains a vital pretreatment step for specific agro-waste feedstocks, particularly those with high ash, problematic ash composition, or significant levels of contaminants, ensuring the production of higher quality, safer biochar and improving the reliability of the conversion process. Table 10.7 summarizes the effects of different pretreatment methods on biochar yield and properties.

10.2.3.4 Torrefaction: Thermal Upgrading for Enhanced Feedstock Properties

Distinct from the physical pretreatments of drying, size reduction, and washing, torrefaction is a mild thermochemical conversion process that serves as a thermal

Table 10.7 Effects of pretreatment methods on biochar yield and properties.

Pretreatment methods	Effect on biochar yield	Effect on biochar properties	Explanation
Drying	Increased energy efficiency, higher throughput, potential for slightly increased yield (relative to wet feedstock)	Increased energy density of feedstock, potentially higher carbon content (if volatiles retained)	Removes water, reducing energy required for vaporization (endothermic process), allowing reactor energy to focus on biomass decomposition; higher solid content fed
Size reduction	Improved heat transfer kinetics, more uniform product, potential for slightly higher yield (at optimal size)	Improved porosity (potentially), more uniform particle size distribution	Increases external surface area, facilitating faster heat penetration into particles; reduces internal diffusion limitations for volatile release
Washing	Reduces reactor fouling/slagging, improves consistency	Reduced ash content, higher carbon content (relative basis), altered surface chemistry, reduced heavy metal concentration	Removes inorganic components (minerals, salts, heavy metals) and potentially catalytic alkali metals. Modifies surface functional groups depending on washing agent

upgrading step rather than primarily removing impurities. Kumar et al. [50] proposed that torrefaction process, conducted at lower temperatures than typical pyrolysis, generally ranging from 200 to 300 °C, and under an inert or reducing atmosphere (such as nitrogen, carbon dioxide, or recirculated process gas) to prevent combustion, torrefaction intentionally alters the chemical and physical structure of the agro-waste feedstock. From a chemical engineering perspective, torrefaction involves controlled partial devolatilization and depolymerization of the biomass polymers. At these moderate temperatures, the less stable hemicellulose fraction is the most significantly affected, undergoing extensive depolymerization and fragmentation into smaller volatile compounds (e.g., furans, organic acids, aldehydes, CO_2, CO) and some char. Cellulose, with its more crystalline structure, is less reactive in this temperature range but undergoes some degree of depolymerization and reduction in crystallinity, while lignin, being the most stable polymer, degrades to a lesser extent but contributes significantly to the char residue formed during torrefaction. Simultaneously, significant dehydration reactions occur throughout the biomass matrix, cleaving C—O bonds and removing oxygen-containing functional groups (hydroxyl, carboxyl, carbonyl) as water, CO_2, and light oxygenated organics. This preferential removal of oxygen relative to carbon atoms is the key mechanism behind the increase in the carbon content and the reduction in the O : C and H : C ratios of the solid product, making it chemically closer to higher-rank fuels like coal or char. The chemical transformations during torrefaction result in profound changes to the physical properties of the agro-waste, which are highly beneficial for subsequent processing. The depolymerization and loss of functional groups weaken the internal structure of the biomass cell walls, causing the torrefied material to become significantly more brittle. This increased brittleness dramatically improves the grindability of the feedstock. Quantitative studies often show that the specific energy required for grinding torrefied biomass to a target particle size is reduced by 50% or more compared to raw biomass. This energy saving in the grinding stage can be substantial, contributing positively to the overall process economics, particularly when fine particle sizes are needed for downstream processes like fast pyrolysis or circulating fluidized bed gasification. Furthermore, the removal of polar hydroxyl groups through dehydration renders the torrefied biomass hydrophobic. This means it repels water and is less susceptible to moisture reabsorption from the atmosphere compared to hydrophilic raw biomass. Hydrophobicity is advantageous for storage and transport, preventing spoilage, biological degradation, and weight increase, reducing handling costs, and ensuring a consistent moisture content for the subsequent conversion step. Enaime et al. [51] pointed out that biochar (Figure 10.3) produced by torrefaction contains O/C ratio ~ >0.4, which results in poor quality; therefore, torrefaction technique is considered as pre-process to decrease the moisture content and hence increasing the biomass heating rate. The mass loss incurred during torrefaction (typically 20–40%, depending on severity) is primarily due to the removal of moisture and low-energy volatiles; this results in a concentrated solid product with a higher energy density (higher heating value per unit mass). The energy density of torrefied biomass can be 20–30% higher than the raw feedstock, meaning less material needs to be handled and fed to a reactor to achieve a given thermal output. Su et al. [52] pointed out that biochar has the property of improving the soil quality, and various agricultural wastes (straw and manure) were subjected to torrefaction in order to produce biochar, which was then applied to polluted soil to improve its fertility and microbial atmosphere. Torrefaction processes with different reaction conditions have been discussed in Table 10.8.

When torrefied agro-waste is subsequently subjected to high-temperature pyrolysis or gasification, its altered properties influence the reaction kinetics and product distribution of these main conversion steps. The torrefied material essentially acts as

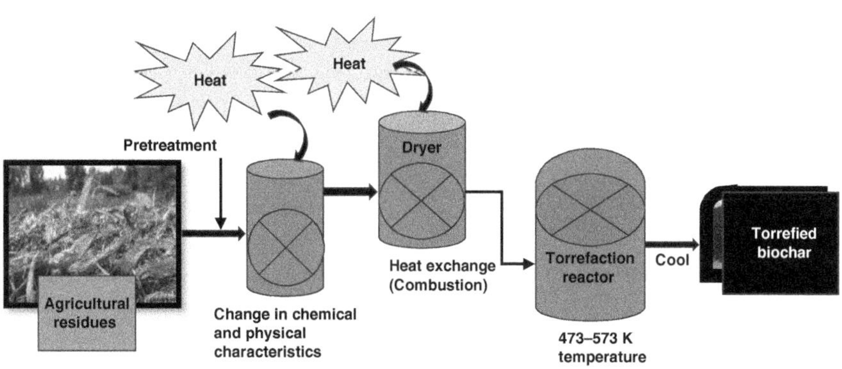

Figure 10.3 Torrefaction process for production of biochar. Adapted from [51].

Table 10.8 Reaction conditions of various torrefaction processes. Adapted from [52].

Torrefaction types	Heating rate	Reaction time	Temperature	Pressure
Catalytic torrefaction	Medium-fast	Few seconds–minutes	~700–1500	≥Atmospheric
Microwave torrefaction	Slow	Few minutes–hours	~400–900	Atmospheric
Slow torrefaction	<10 °C/minutes	Few seconds–hours	~350–800	Atmospheric
Fast torrefaction	~1000 °C/s	Few seconds	~400–600	Vacuum-atmospheric

a pre-char or upgraded feedstock. Its lower volatile content, higher carbon density, and altered structure can lead to faster heating rates and different devolatilization and char conversion behaviors compared to raw biomass. The stable carbon matrix formed during torrefaction provides a foundation upon which further carbonization occurs at higher temperatures. Consequently, the properties of the final biochar produced from torrefied feedstock can differ from biochar produced directly from raw biomass under the same conditions. For instance, torrefaction can influence the development of pore structure during subsequent high-temperature treatment, potentially affecting the final biochar's surface area and porosity. The reduced O/C ratio of the torrefied input material will also be reflected in the resulting biochar [53]. While torrefaction itself requires an energy input (primarily for heating the feedstock and providing latent heat for water removal), the energy savings achieved in subsequent grinding, drying (avoided reabsorption), and potentially more efficient high-temperature conversion can make it a net energy-positive or economically favorable pretreatment strategy within an integrated biochar production facility, enhancing the overall efficiency and viability of converting agro-waste into valuable biochar, especially for dry thermochemical pathways. Table 10.9 summarizes the impact of torrefaction pretreatment on biochar production.

Thus, agro-waste represents a diverse and readily available resource for biochar production, and selection of the appropriate feedstock depends on factors such as availability, cost, and desired biochar properties. Pretreatment methods, including drying, size reduction, washing, and torrefaction, play a vital role in enhancing the efficiency of biochar production and improving the quality of the resulting biochar. Further research is needed to optimize pretreatment methods and pyrolysis conditions for specific agro-waste feedstocks and to explore the potential of biochar for various applications, including soil amendment, carbon sequestration, and environmental remediation.

Application of different thermochemical conversion technologies – pyrolysis (slow, fast, and flash), gasification, and HTC – for biochar production from agro-waste, focusing on their operating conditions, biochar yields, biochar properties, environmental impacts, and economic considerations, provides a sustainable approach to agro-waste management.

Table 10.9 Impact of torrefaction pretreatment on biochar production. Adapted from [54].

Aspect	Effect of torrefaction	Explanation
Grindability	Significantly improved	Torrefaction embrittles the agro-waste, making it easier and less energy-intensive to grind into smaller particle sizes, facilitating more efficient pyrolysis
Energy density	Increased	By removing moisture and volatile compounds, torrefaction concentrates the energy content of the biomass, leading to a higher heating value for subsequent pyrolysis
Hydrophobicity	Increased	The hydrophobic nature of torrefied biomass reduces moisture uptake during storage and improves handling characteristics, preventing degradation and maintaining feedstock quality
Biochar yield	Can be influenced; may increase or decrease	The impact on biochar yield is complex and depends on the specific torrefaction conditions and feedstock type. Some studies show increased yields due to the removal of volatile matter, while others report a decrease due to the loss of carbon during torrefaction. Careful optimization is crucial
Biochar properties	Altered; can impact surface area, carbon content, and pore structure	The altered composition of the feedstock after torrefaction can affect the properties of the resulting biochar, potentially leading to biochar with better carbon content, improved surface area, and enhanced performance in specific applications

10.3 Pyrolysis: Thermal Decomposition for Biochar Maximization or Bio-oil Production

Pyrolysis stands as a cornerstone thermochemical process for valorizing agro-waste into valuable products, fundamentally involving the thermal decomposition of organic materials in an oxygen-limited or inert atmosphere. This process, occurring at temperatures typically ranging from 300 to 700 °C, fractures the complex polymeric structure of biomass – comprising cellulose, hemicellulose, and lignin – into solid biochar, condensable liquid bio-oil (also known as pyrolysis oil or tar), and non-condensable gases (primarily CO, CO_2, H_2, and light hydrocarbons, forming syngas). The distribution and characteristics of these products are highly sensitive to the operating conditions and feedstock properties, making pyrolysis a versatile tool that can be tuned for maximum biochar yield (slow pyrolysis) or maximum liquid yield (fast/flash pyrolysis). The journey of biomass through a pyrolysis reactor is a sequence of overlapping physical and chemical transformations, initiating with the removal of moisture below 150 °C, an endothermic step consuming thermal energy required for vaporization. Following drying, as temperature increases, primary devolatilization commences, typically between 150 and

450 °C. During this phase, the lignocellulosic polymers undergo rapid depolymerization and fragmentation through complex heterogeneous reactions, releasing the bulk of the VOCs and primary gases [55]. Hemicellulose, with its amorphous structure and weaker glycosidic bonds, decomposes first at lower temperatures (~200–300 °C) , followed by the more crystalline cellulose (~300–400 °C) via mechanisms like chain scission and transglycosylation, yielding anhydrosugars like levoglucosan. Lignin, characterized by its aromatic structure and robust ether and C—C linkages, degrades over a broader temperature range (~280–500 + °C), contributing phenolic compounds and aromatic fragments. The kinetics of these primary decomposition reactions are crucial; they are highly temperature-dependent, typically following Arrhenius-type relationships, and also influenced by feedstock composition and particle size, which govern heat transfer rates [56]. As temperatures rise above approximately 450 °C and the volatile products evolve, secondary reactions become increasingly significant. These involve further thermal cracking, repolymerization, and condensation reactions of the primary volatile compounds in the gas phase and heterogeneous reactions between the volatiles and the solid char surface. These secondary processes are pivotal in determining the final product distribution and, specifically, the properties of the solid char. Longer residence times of volatile vapors within the hot zone, especially in contact with the reactive char surface, promote these secondary reactions [57]. Cracking breaks down larger molecules into smaller ones and gases, while repolymerization and condensation reactions build more complex, stable aromatic structures, contributing to increased char yield and density while reducing bio-oil yield and altering its composition. The solid residue remaining throughout these stages constitutes the biochar, whose yield and micro/macrostructure are ultimately a product of the initial devolatilization extent and the subsequent secondary char-forming reactions. The physiochemical properties of the resulting biochar, such as fixed carbon content, surface area, porosity, pH, and stability, are thus intricately linked to the specific temperature profile, heating rate, and vapor–solid residence times experienced during the process.

Kwaku Armah et al. [58] proposed that thermochemical process pyrolysis (Figure 10.4) applied to agricultural wastes/biomass resulted in products biochar, syngas, and bio-oil. During the process chemical mechanisms such as fragmentation, depolymerization, and cross-linking occur at specific temperatures, forming subsequent product state (solid, liquid, and gas) for hemicellulose

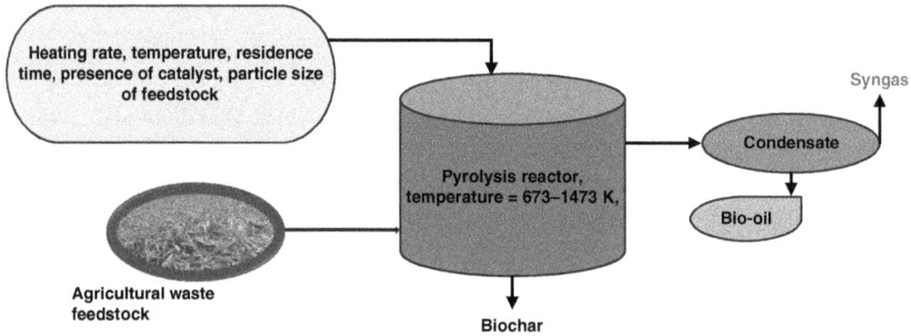

Figure 10.4 General scheme of pyrolysis technique. Adapted from [58].

and cellulose (lignocellulosic components). Based on the deliberate manipulation of these operating parameters, distinct pyrolysis modes are defined to favor specific products and biochar characteristics. **Slow pyrolysis** is characterized by relatively low heating rates (typically 0.1–1 °C/s) and long solid and vapor residence times, ranging from minutes to hours. This regime is specifically designed to maximize biochar yield, often achieving 30–50% on a dry feedstock basis. The slow heating allows thermal energy to penetrate larger particles, and the extended residence time provides ample opportunity for secondary reactions to deposit carbon onto the char surface, promoting extensive carbonization and structural rearrangement, resulting in a highly stable biochar with high fixed carbon and aromaticity, making it well-suited for soil amendment and carbon sequestration. Reactor types like fixed-bed and moving-bed (e.g., auger, rotary kiln) systems are commonly employed for slow pyrolysis due to their ability to easily accommodate longer solid residence times [59].

In stark contrast, **fast pyrolysis** employs high heating rates (typically 10–200 °C/s) and, crucially, very short vapor residence times (measured in seconds), while solid residence times are generally shorter than slow pyrolysis but sufficient for devolatilization. The core objective here is to maximize the yield of liquid bio-oil, which can reach 50–75%, with biochar being a lower-yield byproduct (typically 10–20%). This is achieved by rapidly heating fine feedstock particles (<1–2 mm) to quickly release volatiles and then rapidly quenching these vapors (e.g., to <400 °C as they exit the hot zone to prevent secondary cracking and capture them as a liquid [60]. Reactors like bubbling and circulating fluidized beds are highly effective for fast pyrolysis due to their excellent heat transfer capabilities (facilitated by fine particle mixing with a hot solid medium like sand or char) and the ability to control vapor residence time through gas flow rates. An even more extreme variant, **flash pyrolysis**, operates at ultra-high heating rates (>1000 °C/s) and extremely short vapor residence times (milliseconds). This highly kinetically controlled regime aims to maximize the yield of specific primary volatile compounds or high-quality bio-oil fractions by instantly decomposing the biomass polymers and immediately cooling the products. While achieving very low biochar yields (<10%), often with higher residual volatile matter, flash pyrolysis is explored for producing specialty chemicals. Entrained flow reactors, where very fine particles (<0.1 mm) are rapidly carried through a high-temperature zone by a carrier gas, are suited for achieving the extreme conditions required for flash pyrolysis. Thus, by precisely controlling temperature, heating rate, and residence times within appropriate reactor designs, pyrolysis offers flexible pathways to produce biochar with tailored characteristics or prioritize other valuable products from agro-waste (Table 10.10) [61].

From a chemical reaction engineering perspective, the interplay of process parameters governs the extent and pathways of biomass decomposition and, hence, affects the quality and yield of biochar:

1. **Temperature:** Higher temperatures promote more complete decomposition of organic matter into volatile compounds and gases, leading to lower biochar yield. However, higher temperatures also increase the degree of carbonization and condensation reactions in the solid phase, resulting in biochar with higher

Table 10.10 Comparison of pyrolysis techniques and their biochar yields. Adapted from [62].

Pyrolysis technique	Heating rate (°C/s)	Temperature (°C)	Residence time	Biochar yield (%)	Primary products
Slow pyrolysis	0.1–1	300–700	Hours	30–50	Biochar
Fast pyrolysis	10–200	400–650	Seconds	10–20	Bio-oil > biochar, syngas
Flash pyrolysis	>1000	500–1000	Milliseconds	<10	Bio-oil, specific chemicals, limited biochar

fixed carbon content, greater aromaticity, potentially higher surface area (due to pore development), and higher pH (due to loss of acidic functional groups and concentration of inorganic elements). The rate constants for devolatilization increase significantly with temperature, following Arrhenius kinetics.

2. **Heating Rate:** Influences the rate of temperature increase. Slower heating rates allow more time for solid-phase reactions (carbonization and repolymerization of volatiles onto the char) to occur, favoring higher biochar yields. Faster heating rates favor the rapid evolution of volatiles, minimizing char formation and maximizing bio-oil/gas yields [63].
3. **Residence Time (Solid and Vapor):**
   - **Solid Residence Time:** Longer times at temperature allow for more complete devolatilization and further carbonization of the solid char, increasing fixed carbon content and biochar stability. Can also increase ash content on a mass basis (though not necessarily concentration) if volatile inorganic elements are lost. Excessive residence time can lead to char degradation or gasification reactions, reducing yield.
   - **Vapor Residence Time:** The time volatile products spend in the hot reactor zone. Longer vapor residence times promote secondary gas-phase cracking and char-forming reactions, decreasing bio-oil yield and increasing gas and potentially char yield. Shorter times, especially with rapid quenching, preserve primary volatiles as bio-oil (as in fast pyrolysis) [64].
4. **Feedstock Composition:** As discussed in Section 2.2, the initial proportions of cellulose, hemicellulose, lignin, and ash fundamentally determine the potential product yields and the properties of the resulting biochar. High lignin content favors high char yield and stability. High ash content increases biochar ash content and alkalinity and can influence reactions catalytically [65].

10.4 Gasification: Primarily Syngas Production with Biochar as a Byproduct

Gasification is a thermochemical process that converts carbonaceous materials into a mixture of combustible gases, primarily carbon monoxide (CO) and hydrogen (H_2), known as syngas. This is achieved by reacting the biomass at high temperatures

(700–1000 °C) with a controlled amount of a gasifying agent, which can be air, oxygen, steam, carbon dioxide, or a mixture. Unlike combustion (excess oxygen) or pyrolysis (no oxygen), gasification involves partial oxidation [66].

10.4.1 Gasification Process and Operating Conditions

The gasification process involves several sequential or overlapping steps within the reactor:

a. **Drying:** Removal of moisture content present in biomass.
b. **Pyrolysis/Devolatilization (~300–700 °C):** Biomass is heated in the absence of oxygen to produce char, volatiles (tars), and gases.
c. **Gasification Reactions (>700 °C):** The char and the volatile tars react with the gasifying agent at high temperatures through endothermic and exothermic reactions:
   - Partial Oxidation: $C + 1/2\ O_2 \rightarrow CO$ (Exothermic)
   - Combustion: $C + O_2 \rightarrow CO_2$ (Exothermic – provides heat for other reactions)
   - Boudouard Reaction: $C + CO_2 \rightleftharpoons 2CO$ (Endothermic)
   - Steam Gasification: $C + H_2O \rightleftharpoons CO + H_2$ (Endothermic)
   - Water–Gas Shift Reaction: $CO + H_2O \rightleftharpoons CO_2 + H_2$ (Exothermic/Endothermic depending on T)
   - Methanation: $C + 2\ H_2 \rightleftharpoons CH_4$ (Exothermic)

     The gasifying agent and temperature determine the syngas composition (H_2/CO ratio, presence of N_2 if air is used) and heating value.
   - Ash Formation: Non-combustible inorganic material accumulates as ash, which may melt or agglomerate depending on temperature and composition (slagging).

In gasification technology, the type of reactor is an essential part, and based on feeding, there are three basic types of reactors, which are described in Figure 10.5.

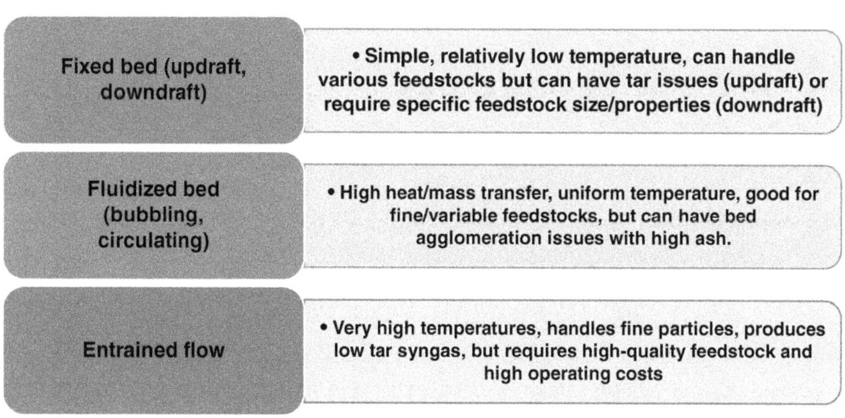

Figure 10.5 Reactor types utilized in gasification process. Adapted from [67].

10.4.2 Biochar Production from Gasification

Biochar is the solid carbonaceous residue remaining after the char has undergone gasification reactions. The yield is typically lower than in slow pyrolysis (5–15%) because more carbon is converted to gas at higher temperatures. Gasification biochar is produced at higher temperatures and in the presence of oxidizing agents, leading to a more complete removal of volatile matter and a higher degree of carbonization and aromaticity. However, it often has a significantly higher ash content (10–30% or more, depending on feedstock and process) and potentially lower specific surface area compared to pyrolysis biochar, as higher temperatures can sinter pores or mineral matter can block access. The surface chemistry may have fewer oxygen-containing functional groups. While gasification biochar can still be used as a soil amendment (particularly if ash is nutrient-rich and non-toxic), its higher ash content and potentially lower porosity may limit some high-value soil applications or contaminant adsorption capabilities compared to tailored pyrolysis biochars. It is often recycled back into the gasifier as feedstock or used for energy recovery due to its high carbon content. It can also serve as a precursor for activated carbon production [68]. Table 10.11 summarizes the biochar characteristics derived from processes of pyrolysis and gasification.

10.4.3 Advantages and Disadvantages of Gasification

- **Advantages:** Produces syngas, a valuable energy carrier or chemical building block. Can handle a wider range of feedstock moisture contents compared to dry pyrolysis. Higher potential for energy recovery from the process (syngas can be burned or used in engines/turbines). Can process feedstocks with high ash content that might cause issues in some pyrolysis reactors.
- **Disadvantages:** Lower biochar yield. Gasification biochar often has higher ash content and potentially lower surface area, limiting certain applications unless further processed (e.g., activated). Higher operating temperatures (requiring more energy and specialized materials). More complex reactor designs and process control compared to simple slow pyrolysis. Syngas requires cleaning (removal of particulates, tars, sulfur compounds) before use.

10.5 HTC: A Wet Biomass Solution

HTC, also known as wet pyrolysis or aqueous carbonization, is a thermochemical process that converts biomass into a carbonaceous solid (hydrochar) in hot, pressurized water (subcritical or supercritical conditions). The typical temperature range is lower than pyrolysis or gasification, usually 180–250 °C, under corresponding saturation pressures (2–6 MPa). Oliveira et al. [70] reported that HTC-biochar, a solid fuel comparable to brown coal, is produced by the HTC process applied on agricultural waste, and blending of various agricultural wastes resulted in successful carbonization, and residues such as corn silage and dough (rich in carbohydrates) produce biochar of better quality.

Table 10.11 Comparison of biochar properties derived from pyrolysis vs. gasification. Adapted from [69].

Property	Pyrolysis biochar	Gasification biochar	Explanation
Ash content	Lower (typically <10–15% for wood/straw, higher for husk/manure)	Higher (often 10–30% or more, depending on feedstock/process)	Higher temperatures and presence of oxidizing agents in gasification lead to more complete carbon consumption relative to ash accumulation
Carbon content	Higher (typically >60–70%)	Lower than pyrolysis biochar (can be <70%, depending on gasification extent/agent)	More carbon converted to gas phase (CO, CO_2, CH_4) in gasification
Surface area	Variable (can be high, e.g., 100–300 m^2/g)	Generally, lower (e.g., 10–100 m^2/g), higher only if specifically optimized/ activated	Higher gasification temperatures can lead to pore sintering or ash blocking within the char
pH	Variable (often alkaline, depending on ash)	Generally, more alkaline (due to higher concentration of alkali metals in ash)	Concentration of base cations (K, Ca, Mg) from the feedstock's inorganic fraction relative to carbon
O : C ratio	Variable (0.1–0.4)	Lower (more carbonized, 0.05–0.2)	Higher gasification temperatures remove more oxygen-containing functional groups via oxidation and decarboxylation
Volatility	Variable (5–30% V.M.)	Lower (often <10% V.M.)	Higher temperatures remove more volatile matter

10.5.1 HTC Process and Operating Conditions

The key distinction of HTC is the use of water as the reaction medium. Water at elevated temperatures and pressures behaves as a polar organic solvent and an acidic/basic catalyst due to changes in its dielectric constant and ionic product. The process involves several key reactions:

1. **Hydrolysis:** Water breaks down the glycosidic linkages in polysaccharides (cellulose, hemicellulose) into smaller sugars and sugar derivatives.
2. **Dehydration:** Removal of water molecules from the biomass structure, leading to the formation of furan derivatives and ultimately aromatic structures.
3. **Decarboxylation:** Removal of carboxyl groups (—COOH), primarily from organic acids produced during hydrolysis/dehydration, releasing CO_2.
4. **Polymerization/Condensation:** Soluble intermediates (sugars, phenolics) repolymerize and condense to form the solid hydrochar.
5. **Aromatization:** Formation of fused aromatic rings, particularly from lignin decomposition products and sugar derivatives.

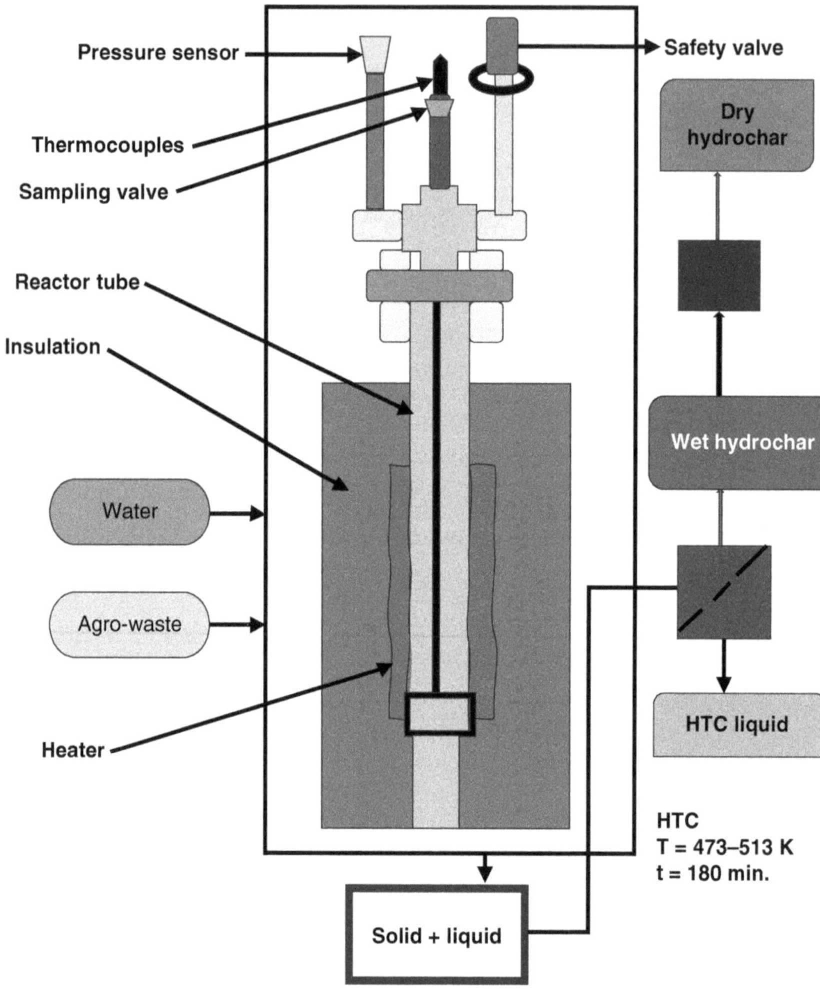

Figure 10.6 Hydrothermal carbonization process. Adapted from [71].

Thus, the process consists of mixing the wet feedstock with water (or processing it in its own water if moisture content is high, e.g., >70%), heating it in a sealed reactor (autoclave, stirred vessel) to the desired temperature, maintaining the temperature for a residence time (minutes to hours), and then cooling and separating the solid hydrochar from the aqueous phase and gases. Schematic presentation of HTC process is shown in Figure 10.6.

According to Phadtare et al. [71], hydrochar (coal-like) is produced as a product, while aqueous (high in nutrients) and gas phases (CO_2 mainly) are produced as a byproduct by HTC process, which takes place at temperature ~200–280 °C for approximately 5–230 minutes. HTC is generally carried out in batch autoclaves, which are common for lab-scale studies, while continuous reactors (e.g., plug flow, stirred tank) are needed for larger-scale operation. The process parameters involve temperature (primary driver of reaction rate and degree of carbonization), residence time (allows reactions to proceed to completion), water-to-biomass ratio (affects solubility of intermediates, reaction medium properties), pressure (determined by temperature; ensures water remains liquid/supercritical), and pH (can be self-catalyzed by organic acids or externally adjusted).

10.5.2 Hydrochar Production and Properties from HTC

The solid product of HTC is called hydrochar. While structurally similar to pyrolysis biochar, it has distinct properties due to the aqueous reaction environment and lower temperatures:

a. **Higher Oxygen Content:** Due to lower processing temperatures and reaction pathways, hydrochar retains more oxygen-containing functional groups (e.g., —OH, C=O, —COOH) than pyrolysis char. This gives it a lower O : C ratio than raw biomass but higher than typical pyrolysis biochar.
b. **Lower Carbon Content:** Typically, lower fixed carbon and total carbon content than high-temperature pyrolysis biochar.
c. **Lower Surface Area:** Generally, it has lower specific surface area and microporosity compared to activated biochar or high-temperature pyrolysis biochar, although it can have significant mesoporosity and macroporosity.
d. **Nutrient Retention:** Crucially, nutrients like N and P are better retained in hydrochar than in high-temperature dry pyrolysis/gasification, as they are less volatile in water. This enhances its value as a fertilizer or soil amendment.
e. **Surface Chemistry:** Higher density of surface functional groups (carboxyl, hydroxyl) compared to pyrolysis biochar, leading to higher polarity and potentially different adsorption mechanisms (e.g., metal complexation).
f. **Structure:** Often has a more spherical or agglomerated particle morphology compared to the more fractured/cellular structure of pyrolysis biochar, depending on feedstock.

Typically, 40–70% (mass basis on dry feedstock) of hydrochar, higher than pyrolysis yield from similar wet feedstock, is obtained without drying. HTC of sewage sludge at 200 °C for 6 hours yielded 50% hydrochar with high P content and good dewaterability, suitable for land application [72]. A study on food waste HTC explored the effect of temperature (180–230 °C) on hydrochar properties for energy storage applications, showing increased conductivity at higher temperatures [73].

10.5.3 Advantages and Disadvantages of HTC

- **Advantages:** Effectively processes wet biomass without energy-intensive pre-drying, making it highly suitable for feedstocks like manures, sewage sludge, and food waste (>50% moisture). Better retention of nutrients (N, P) in the solid product compared to high-temperature dry processes. Lower operating temperatures compared to pyrolysis/gasification require less energy input (though pressure adds cost). Solid product (hydrochar) is more hydrophobic than raw wet biomass, improving dewaterability and storage.
- **Disadvantages:** Lower surface area and potentially lower carbon content in hydrochar compared to high-temperature pyrolysis biochar, which may limit

certain high-value adsorption applications. Requires pressurized reactors, increasing capital costs and safety considerations. Produces a complex aqueous phase ("process water") containing dissolved organic compounds, which requires treatment and can be a significant environmental burden if not handled properly [74].

10.6 Comparative Analysis and Future Trends

Each thermochemical conversion technology offers a unique approach to valorizing agro-waste, with distinct advantages, disadvantages, product distributions, and resulting solid product properties. The choice of technology is a complex decision driven by feedstock characteristics (especially moisture and ash content), desired product emphasis (biochar, bio-oil, syngas, specific chemicals, hydrochar for nutrient recovery), desired biochar properties, scale of operation, energy integration potential, and economic feasibility.

Advancing agro-waste-to-biochar conversion requires continued research and development, particularly focused on overcoming technical and economic barriers:

1. **Process Optimization and Kinetics:** Developing detailed kinetic models for the thermal decomposition of various agro-waste components under different thermochemical conditions. This enables precise prediction of product yields and properties and facilitates optimal reactor design and control strategies.
2. **Advanced Reactor Design:** Innovation in reactor technology to improve energy efficiency, heat transfer, and mass transfer, especially for handling challenging feedstocks (e.g., high moisture, high ash, variable particle size) and achieving continuous, large-scale operation with minimal fouling and emissions. Examples include microwave-assisted pyrolysis, catalytic pyrolysis, or optimized fluid dynamics in gasifiers/HTC reactors.
3. **Catalytic Conversion:** Exploring the use of catalysts (both inherent in ash and added externally) to steer reaction pathways toward desired products (e.g., higher quality bio-oil fractions, specific syngas compositions) and potentially lower operating temperatures or increase reaction rates. This requires understanding catalyst–feedstock interactions [75].
4. **Biochar Property Tailoring and Modification:** Developing methods to customize biochar properties *during* or *after* production (e.g., activation, functionalization with specific chemical groups) for enhanced performance in targeted applications like high-capacity adsorbents, catalysts, or electrode materials. This involves detailed characterization techniques (BET, FTIR, XPS, SEM-EDX, TGA, NMR).
5. **Integrated Biorefinery Concepts:** Moving beyond single-product processes to integrated systems that maximize the value extracted from agro-waste by co-producing biochar alongside bio-oil, syngas, biochemicals, and renewable energy. This requires complex process integration, energy system analysis, and separation technologies.
6. **Techno-economic Analysis (TEA) and Life Cycle Assessment (LCA):** Conducting rigorous TEA and LCA studies for specific agro-waste-to-biochar

pathways and applications to evaluate their economic feasibility, environmental footprint, and overall sustainability under various geographical and market conditions. This provides crucial data for decision-making and policy development [76].

7. **Feedstock Pretreatment Optimization:** Further research into cost-effective and energy-efficient pretreatment methods (drying, size reduction, washing, torrefaction) tailored to specific agro-waste types and downstream conversion technologies.
8. **Standardization and Market Development:** Establishing standardized methods for biochar characterization and performance evaluation to build market confidence and facilitate wider adoption, particularly in agricultural and environmental sectors.

10.7 Conclusion

Thermochemical conversion technologies offer compelling and sustainable pathways for transforming vast quantities of agricultural waste into valuable biochar and other co-products. Pyrolysis (slow, fast, flash), gasification, and HTC each provide unique advantages and disadvantages, making the optimal technology choice dependent on feedstock characteristics, desired product spectrum, and site-specific economic and environmental factors.

A comprehensive understanding of the underlying chemical principles, process kinetics, mass and energy balances, reactor design, and the influence of operating parameters is essential for optimizing biochar yield and tailoring its physiochemical properties for specific applications, ranging from soil amendment and carbon sequestration to advanced materials for environmental remediation and energy storage.

While significant progress has been made in laboratory and pilot-scale studies, challenges remain in scaling up these technologies for large-scale, continuous operation, managing feedstock variability, optimizing process energy efficiency, and ensuring the consistent quality and safety of the resulting biochar. Future research integrating chemical engineering principles with advanced material science, catalysis, and system-level techno-economic and LCAs is crucial to unlock the full potential of agro-waste-to-biochar conversion as a key component of a circular bioeconomy, contributing significantly to sustainable waste management, climate change mitigation, and resource recovery.

References

1 Patel, M.R. and Panwar, N.L. Biochar from agricultural crop residues: environmental, production, and life cycle assessment overview. *Resour. Conserv. Recycl. Adv.* 2023.

2 Agrawal, P. and Rokhum, S.L. Supported metal/metal oxide catalysts in biodiesel production: an overview. In: *Biodiesel Production: Feedstocks, Catalysts, and Technologies*. Wiley, 2022, pp. 119–141.

3 Ao, S., Rashid, U., Shi, D., Rokhum, S.L., Thuy. L.T.G., Alahmadi, A.T., et al. Synthesis and utilization of biomass-derived sulfonated heterogeneous catalyst-BT-SO3H for microalgal biodiesel production. *Environ. Res.* 2024.

4 Ao, S., Gouda S.P., Selvaraj, M., Boddula, R., Al-Qahtani, N., Mohan, S., et al. Transesterification of Jatropha curcas oil to biodiesel using highly porous sulfonated biochar catalyst: optimization and characterization dataset. *Data Br.* 2024.

5 Park, J.H., Wang, J.J., Meng, Y., Wei, Z., DeLaune, R.D., and Seo, D.C. Adsorption/desorption behavior of cationic and anionic dyes by biochars prepared at normal and high pyrolysis temperatures. *Colloids Surfs. A: Physicochem. Eng. Asp.* 2019.

6 Leng, L., Xiong, Q., Yang, L., Li, H., Zhou, Y., Zhang, W., et al. An overview on engineering the surface area and porosity of biochar. *Sci. Total Environ.* 2021.

7 Wang, T.H., Lin, Y.C., Hung, C.J., Liao, Y., and Xu, R.X. Adopting abundant seawater as green chemical activators for preparing high surface area biochar. *Bioresour. Technol. Rep.* 2023.

8 Ashfaq, A., Nadeem, R., Gong, H., Rashid, U., Noreen, S., Rehman, S.U., et al. Fabrication of novel agrowaste (banana and potato peels)-based biochar/TiO2 nanocomposite for adsorption of Cr(VI), statistical optimization via RSM approach. *Polymers (Basel).* 2022.

9 Luskar, L., Polanšek, J., Hladnik, A., and Čeh, B. On-farm composting of hop plant green waste—chemical and biological value of compost. *Appl. Sci.* 2022.

10 Li, X., Shen, Q., Zhang, D., Mei, X., Ran, W., Xu, Y., et al. Functional groups determine biochar properties (pH and EC) as studied by two-dimensional 13C NMR correlation spectroscopy. *PLoS One.* 2013.

11 Romero Millán, L.M., Sierra Vargas, F.E., and Nzihou, A. Characterization of steam gasification biochars from lignocellulosic agrowaste towards soil applications. *Waste Biomass Valori.* 2021.

12 Pal, D. and Gupta, V. Biochar as an adsorbent for removal of pharmaceutical compounds from wastewater. In: *BioChar: Applications for Bioremediation of Contaminated Systems.* De Gruyterv, 2022, pp. 309–328.

13 Khalil, A.M., Michely, L., Pires, R., Bastide, S., Jlassi, K., Ammar, S., et al. Copper/nickel-decorated olive pit biochar: one pot solid state synthesis for environmental remediation. *Appl. Sci.* 2021.

14 Sudalaimuthu, P. and Sathyamurthy, R. Forecast sustainable and renewable hydrogen production via circular bio-economy of agro waste. *Int. J. Hydrog. Energy.* 2024.

15 Flammini, A., Brundin, E., Grill, R., and Zellweger, H. Supply chain uncertainties of small-scale coffee husk-biochar production for activated carbon in Vietnam. *Sustain.* 2020.

16 Van Nguyen, T.T., Phan A.N., Nguyen, T.A., Nguyen, T.K., Nguyen, S.T., Pugazhendhi, A., et al. Valorization of agriculture waste biomass as biochar: as first-rate biosorbent for remediation of contaminated soil. *Chemosphere.* 2022.

17 Yaashikaa, P.R., Kumar, P.S., Varjani, S., and Saravanan, A. A critical review on the biochar production techniques, characterization, stability and applications for circular bioeconomy. *Biotechnol. Rep.* 2020.

18 Agrawal, P.S., Belkhode, P.N., and Rokhum, S.L. Agro-waste-produced biofuels. In: *Sustainability in Biofuel Production Technology*. Wiley-Blackwell, 2022, pp. 164–186.

19 Filiberto, D.M. and Gaunt, J.L. Practicality of biochar additions to enhance soil and crop productivity. *Agric.* 2013.

20 Shaheen, S.M., Antoniadis, V., Shahid, M., Yang, Y., Abdelrahman, H., Zhang, T., et al. Sustainable applications of rice feedstock in agro-environmental and construction sectors: a global perspective. *Renew. Sustain. Energy Rev.* 2022.

21 O'Boyle, M., Mohamed, B.A., and Li, L.Y. Co-pyrolysis of sewage sludge and biomass waste into biofuels and biochar: a comprehensive feasibility study using a circular economy approach. *Chemosphere.* 2024.

22 Seo, J.Y., Tokmurzin, D., Lee, D., Lee, S.H., Seo, M.W. and Park, Y.K. Production of biochar from crop residues and its application for biofuel production processes – an overview. *Bioresour. Technol.* 2022.

23 Lima, I.M. and Beacorn, J.A. Targeting a sustainable sugar crops processing industry: a review (Part I)—by-product applications. *Sugar Tech.* 2022.

24 van der Weerden, T.J., Noble, A.N., Beltran, I., Hutchings, N.J., Thorman, R.E., de Klein C.A.M., et al. Influence of key factors on ammonia and nitrous oxide emission factors for excreta deposited by livestock and land-applied manure. *Sci. Total Environ.* 2023.

25 Nahm, K.H. Factors influencing nitrogen mineralization during poultry litter composting and calculations for available nitrogen. *World's Poult. Sci. J.* 2005.

26 Odell, J.D., Essington, M.E., and Howard, D.D. Surface application of liquid swine manure: chemical variability. *Commun. Soil Sci. Plant Anal.* 1995.

27 Penido, E.S., de Oliveira, M.A., Sales, A.L.R., Ferrazani, J.C., Magalhães, F., Bianchi, M.L., et al. Biochars produced from various agro-industrial by-products applied in Cr(VI) adsorption-reduction processes. *J. Environ. Sci. Health A Tox. Hazard Subst. Environ. Eng.* 2021.

28 Awogbemi, O. and Von Kallon, D.V. Pretreatment techniques for agricultural waste. *Case Stud. Chem. Environ. Eng.* 2022.

29 Ajien, A., Idris, J., Md Sofwan, N., Husen, R., and Seli, H. Coconut shell and husk biochar: a review of production and activation technology, economic, financial aspect and application. *Waste Manag. Res.* 2023.

30 Cao, X., Chen, F., Cen, K., Zhang, J., Chen, D., and Li, Y. Effect of torrefaction on the pyrolysis behavior, kinetics, and phenolic products of lignin. *Biomass Convers. Biorefin.* 2023.

31 Zhang, L., Larsson, A., Moldin, A., and Edlund, U. Comparison of lignin distribution, structure, and morphology in wheat straw and wood. *Ind. Crops Prod.* 2022.

32 Wang, S., Hu, Y., and Wang, Y. Pyrolysis of rice straw for bioenergy: kinetics, reaction mechanism, and by-products using TG and Py-GC/MS analysis. *J. Therm. Sci. Technol.* vol. 20, pp. 24–395, Apr. 2025.

33 McKendry, P. Energy production from biomass (Part 1): overview of biomass. *Bioresour. Technol.* vol. 83, pp. 37–46, 2002.

34 Kameyama, K., Shinogi, Y., Miyamoto, T., Ueda, T., and Agarie, K. Estimation of carbon sequestration potential with farmland application of bagasse-char:

life cycle inventory analysis through a pilot pyrolysis plant P-095. *Aust. J. Soil Res.* 2010.
35 Behera, S.S. and Ray, R.C. Solid state fermentation for production of microbial cellulases: recent advances and improvement strategies. *Int. J. Biol. Macromol.* 2016.
36 Sommers, L.E. Chemical composition of sewage sludges and analysis of their potential use as fertilizers. *J. Environ. Qual.* 1977.
37 Moore, P.A., Daniel, T.C., Edwards, D.R., and Miller, D.M. Effect of chemical amendments on ammonia volatilization from poultry litter. *J. Environ. Qual.* 1995.
38 Azzaz, A.A., Jcguirim, M., Jellali, S., and Ghimbeu C. Hydrothermal carbonization and slow pyrolysis as two thermal techniques for the production of carbon rich, added-value materials using olive milling byproduct: quid optimus? In: *2020 11th International Renewable Energy Congress (IREC)*. IREC Xplore, 2020.
39 Bednárek, J., Matejová, L., Jankovská, Z., Vaštyl, M., Sokolová, B., Peikertová, P., et al. The influence of structural properties on the adsorption capacities of microwave-assisted biochars for metazachlor removal from aqueous solutions. *J. Environ. Chem. Eng.* 2022.
40 Cueva, Z.L.L., Griffin, G.J., Ward, L.P., Madapusi, S., Shah, K.V., Parthasarathy, R. A study of chemical pre-treatment and pyrolysis operating conditions to enhance biochar production from rice straw. *J. Anal. Appl. Pyrolysis.* 2022.
41 Jangsawang, W. Utilization of biomass gasifier system for drying applications. In: *Energy Procedia*. 2017.
42 Saharudin, D.M., Jeswani, H.K., and Azapagic, A. Biochar from agricultural wastes: environmental sustainability, economic viability and the potential as a negative emissions technology in Malaysia. *Sci. Total Environ.* 2024.
43 Murugan, P., Dhanushkodi, S., Sudhakar, K., and Wilson, V.H. Industrial and small-scale biomass dryers: an overview. *Energy Eng.: J. Assoc. Energy Eng.* 2021.
44 Aljabri, H., Cherif, M., Siddiqui, S.A., Bounnit, T., and Saadaoui, I. Evidence of the drying technique's impact on the biomass quality of *Tetraselmis subcordiformis* (Chlorophyceae). *Biotechnol. Biofuels Bioprod.* 2023.
45 Pradhan, S., Abdelaal, A.H., Mroue, K., Al-Ansari, T., Mackey, H.R., and McKay, G. Biochar from vegetable wastes: agro-environmental characterization. *Biochar.* 2020.
46 Ganesapillai, M., Mehta, R., Tiwari, A., Sinha, A., Bakshi, H.S., Chellappa, V., et al. Waste to energy: a review of biochar production with emphasis on mathematical modelling and its applications. *Heliyon.* 2023.
47 Deng, L., Zhang, T., and Che, D. Effect of water washing on fuel properties, pyrolysis and combustion characteristics, and ash fusibility of biomass. *Fuel Process. Technol.* 2013.
48 Cen, K., Zhang, J., Ma, Z., Chen, D., Zhou, J., and Ma, H. Investigation of the relevance between biomass pyrolysis polygeneration and washing pretreatment under different severities: water, dilute acid solution and aqueous phase bio-oil. *Bioresour. Technol.* 2019.
49 Zhang, S. and Xiong, Y. Washing pretreatment with light bio-oil and its effect on pyrolysis products of bio-oil and biochar. *RSC Adv.* 2016.

50 Kumar, A., Bhattacharya, T., Shaikh, W.A, Roy, A., Chakraborty, S., Vithanage, M., et al. Multifaceted applications of biochar in environmental management: a bibliometric profile. *Biochar*. 2023.

51 Enaime, G., Baçaoui, A., Yaacoubi, A., and Lübken, M. Biochar for wastewater treatment-conversion technologies and applications. *Appl. Sci. (Switzerland)*. 2020.

52 Su, K., Qin, Q., Yang, J., Li, L., and Deng, S. Recent advance on torrefaction valorization and application of biochar from agricultural waste for soil remediation. *J. Renew. Mater*. 2022.

53 Nakason, K., Pathomrotsakun. J., Kraithong. W,, Khemthong, P., and Panyapinyopol, B. Torrefaction of agricultural wastes: influence of lignocellulosic types and treatment temperature on fuel properties of biochar. *Int. Energy J*. 2019.

54 Lee, K.T., Chen, W.H., Sarles, P., Park, Y.K., and Ok, Y.S. Recover energy and materials from agricultural waste via thermalchemical conversion. *One Earth*. 2022.

55 Venkatesa Prabhu, S., Jayakumar, M., Varadharajan, V., Gokulakrishnan, S.A., and Manivannan, S. Contemporary approaches for biochar production from agro-waste and its current and prognostic applications in environment health. In: *Agricultural Waste to Value-Added Products*. Springer Nature Singapore Pvt Ltd, 2023, pp. 239–254.

56 Suman, S. and Gautam, S. Biochar derived from agricultural waste biomass act as a clean and alternative energy source of fossil fuel inputs. In: *Energy Systems and Environment*. IntechOpen, 2018, pp. 225–236.

57 Wang, W., Huang, J., Wu, T., Ren, X., and Zhao, X. Research on the preparation of biochar from waste and its application in environmental remediation. *Water (Switzerland)*. 2023.

58 Kwaku Armah, E., Chetty, M., Adebisi Adedeji, J., Erwin Estrice, D., Mutsvene, B., Singh, N., et al. Biochar: production, application and the future. In: *Biochar – Productive Technologies, Properties and Applications*. IntechOpen, 2023, pp. 1–26.

59 Setyawan, H.Y., Sunyoto, N.M.S., Choirun, A., Musyaroh, M., and Arwani, M. The potential of biochar-slurry fuel from agricultural wastes in Indonesia. *Cogent Eng*. 2024.

60 Ghosh, S., Nandasana, M., Webster, T.J., and Thongmee, S. Agrowaste-generated biochar for the sustainable remediation of refractory pollutants. *Front. Chem*. 2023.

61 Gabhane, J.W., Bhange, V.P., Patil, P.D., Bankar, S.T., and Kumar, S. Recent trends in biochar production methods and its application as a soil health conditioner: a review. *SN Appl. Sci*. 2020.

62 Li, Y., Gupta, R., Zhang, Q., and You, S. Review of biochar production via crop residue pyrolysis: development and perspectives. *Bioresour. Technol*. 2023.

63 Ibitoye, S.E., Jen, T.C., Mahamood, R.M., and Akinlabi, E.T. Generation of sustainable energy from agro-residues through thermal pretreatment for developing nations: a review. *Adv. Energy Sustain. Res*. 2021.

64 Dun, Q., Chen, Z., Huang, F., Zhou, Y., Yu, J., Gao, S., et al. Influences of temperature and residence time on secondary reactions of volatiles from coal pyrolysis. *Guocheng Gongcheng Xuebao/Chin J. Process Eng*. 2018.

65 Yogalakshmi, K.N., Poornima, D.T., Sivashanmugam, P., Kavitha, S., Yukesh Kannah, R., Varjani, S., et al. Lignocellulosic biomass-based pyrolysis: a comprehensive review. *Chemosphere*. 2022.

66 Rex, P., Mohammed Ismail, K.R., Meenakshisundaram, N., Barmavatu, P., and Sai Bharadwaj, A.V.S.L. Agricultural biomass waste to biochar: a review on biochar applications using machine learning approach and circular conomy. *Chem. Eng*. 2023.

67 Gao, Y., Wang, M., Raheem, A., Wang, F., Wei, J., Xu, D., et al. Syngas production from biomass gasification: influences of feedstock properties, reactor type, and reaction parameters. *ACS Omega*. 2023.

68 Amalina, F., Razak, A.S.A., Krishnan, S., Sulaiman, H., Zularisam, A.W., and Nasrullah, M. Biochar production techniques utilizing biomass waste-derived materials and environmental applications – a review. *J. Hazard. Mater. Adv*. 2022.

69 Fryda, L. and Visser, R. Biochar for soil improvement: evaluation of biochar from gasification and slow pyrolysis. *Agric*. 2015.

70 Oliveira, I., Blöhse, D., and Ramke, H.G. Hydrothermal carbonization of agricultural residues. *Bioresour. Technol*. 2013.

71 Phadtare, P. and Kalbande, S. Biochar production technologies from agricultural waste, its utilization in agriculture and current global biochar market: a comprehensive review. *Int. J. Environ. Clim. Chang*. 2022.

72 Hejna, M., Świechowski, K., and Białowiec, A. Study on the effect of hydrothermal carbonization parameters on fuel properties of sewage sludge hydrochar. *Materials (Basel)*. 2023.

73 Le, H.S., Chen, W.H., Forruque Ahmed, S., Said, Z., Rafa, N., Le, A.T., et al. Hydrothermal carbonization of food waste as sustainable energy conversion path. *Bioresour. Technol*. 2022.

74 Román, S., Libra, J., Berge, N., Sabio, E., Ro, K., Li, L., et al. Hydrothermal carbonization: modeling, final properties design and applications: a review. *Energies*. 2018.

75 Cheng, D., Ngo, H.H., Guo, W., Pandey, A., Varjani, S., Zhang, Z., et al. Sustainability considerations of biochar production in biowaste management. In: *Current Developments in Biotechnology and Bioengineering: Biochar Towards Sustainable Environment*. Elsevier, 2023, pp. 41–62.

76 Carvalho, J., Nascimento, L., Soares, M., Valério, N., Ribeiro, A., Faria, L., et al. Life cycle assessment (LCA) of biochar production from a circular economy perspective. *Processes*. 2022.

11

Waste Valorization for Biogas Production: A Pathway to a Circular Economy

11.1 Introduction

The intensifying global push for a circular and sustainable economic paradigm has markedly amplified the importance of anaerobic digestion (AD) for agricultural waste (agro-waste). This process is increasingly viewed as a cornerstone in the shift toward more sustainable agricultural practices, offering a versatile and environmentally friendly energy alternative [1]. Notably, biogas generated from agro-waste offers an environmentally friendlier and renewable energy alternative, providing substantial advantages at both local and national levels. The dominant "take–make–dispose" linear economic approach has proven unsustainable, contributing markedly to resource depletion, environmental damage, and climate change impacts. The circular economy (CE) framework aims to improve resource efficiency, reduce waste production, extend the use phase of products, and keep materials in circulation longer, thereby conserving virgin resources.

Agro-waste, a pervasive byproduct of farming activities, presents a significant challenge regarding its disposal and associated environmental consequences. Conventional methods like open burning or landfilling contribute to air and water pollution and generate greenhouse gas (GHG) emissions. However, when considered within a CE perspective, agro-waste transforms from a liability into a valuable asset [2]. Sources of agro-waste include animal manure, food-processing residues (such as fruit/vegetable peels and slaughterhouse byproducts), and crops specifically cultivated on marginal lands for energy purposes.

AD stands out as a well-established biological process capable of transforming organic materials into biogas, a valuable source of renewable energy, and digestate, a nutrient-rich co-product [3–5]. This technology offers a viable pathway for the valorization of agro-waste, converting it into useful resources that mitigate environmental impacts while simultaneously generating economic benefits. The AD process operates by a complex community of microorganisms breaking down organic matter (OM) in the absence of oxygen. It captures methane (CH_4), a potent GHG with a global warming potential (GWP) 25 times higher than that of carbon dioxide (CO_2). By utilizing AD, CH_4 emissions from decomposing organic waste are

Agro-Waste Management and Valorization, First Edition. Pratibha S. Agrawal, et al.

prevented from entering the atmosphere. Furthermore, the digestate produced during AD can be beneficially utilized as a soil amendment and biofertilizer. It supplies vital plant nutrients and enhances soil health, potentially significantly reducing or eliminating the need for synthetic fertilizers [6]. Overall, AD technology represents a crucial element in transitioning toward a circular and sustainable economy by turning agro-waste into valuable energy and nutrient resources.

11.2 Biogas Production – A Versatile Energy Source

Biogas production from agro-waste is increasingly gaining prominence as a versatile energy source that addresses global challenges related to energy security, environmental sustainability, and resource management. This section specifically examines how biogas from agro-waste serves as a renewable, cleaner energy option when compared to conventional fossil fuels, simultaneously bolstering the CE and strengthening energy security at both local and national levels [7]. Additionally, this review explores the various factors essential for the successful implementation of agro-waste-to-biogas systems, such as digester design and setup, process optimization, and thorough feedstock management.

Biogas produced from agro-waste offers a persuasive and environmentally sound alternative to fossil fuel tanks [8]. Burning fossil fuels for energy releases GHGs into the atmosphere, contributing significantly to climate change and air pollution. In contrast, biogas production through AD captures CH_4, a potent GHG, before it escapes into the atmosphere, thereby reducing net GHG emissions. Biogas can power heat and electricity generation through combustion. Alternatively, it can be refined into biomethane, a high-purity product comparable to natural gas, suitable for use as a transportation fuel or injection into existing gas pipelines. Compared to fossil fuels, biogas derived from agro-waste offers several key advantages [9, 10]:

- **Renewable Resource:** Agro-waste is a continuously generated resource from agricultural activities, ensuring a sustainable and replenishable supply for biogas production.
- **Lower GHG Emissions:** While burning biogas does produce CO_2, this CO_2 is considered biogenic; it was originally absorbed from the atmosphere by plants during photosynthesis. Unlike burning fossil fuels, which release CO_2 that has been sequestered underground for millions of years, burning biogenic carbon is part of the natural carbon cycle. Furthermore, AD prevents CH_4 emissions from decomposing organic waste, and as CH_4 is a more potent GHG than CO_2, capturing and utilizing it for energy generation significantly lowers the overall climate impact compared to uncontrolled decomposition.
- **Enhanced Energy Security:** Utilizing domestically available agro-waste for biogas production contributes to energy independence and diversification of the energy mix, reducing reliance on imported fossil fuels [11]. The decentralized nature of many agro-waste sources allows for distributed energy generation, which can strengthen the resilience of energy grids and reduce transmission losses.

- **Waste Management and Environmental Benefits:** Converting agro-waste-to-biogas addresses waste disposal challenges and provides environmental benefits by reducing pollution, odors, and pathogen levels in the waste. Digestate, the solid residue from AD, can be used as a nutrient-rich biofertilizer, closing the nutrient loop and reducing the need for synthetic fertilizers, which have associated environmental costs.

The comparative environmental impact of biogas compared to fossil fuels using life cycle assessment (LCA) is a crucial consideration. LCA consistently shows that biogas derived from sustainable feedstocks generally has a lower GWP than traditional fossil fuels. Moreover, a well-managed AD system, combined with digestate application methods that store carbon in the soil, can result in overall carbon neutrality or even contribute to actively removing CO_2 from the atmosphere [12]. Stakeholders, including smallholder farmers and cooperatives, food-processing companies, and energy providers, play a crucial role in promoting the adoption of agro-waste-to-biogas technology. Their involvement creates opportunities for developing and deploying sustainable technologies and necessary infrastructure, including biogas generators, gas pipelines, storage tanks, and other small-scale equipment for biogas producers. Figure 11.1 shows a basic setup that allows smallholder producers to create both biogas and nutrient-rich fertilizer using AD.

For electricity generation from biogas, the basic requirements and equipment typically include a biogas digester, a gas purification system (to remove H_2S and CO_2), a biogas reservoir or storage bag, a gas engine or turbine connected to a generator, a battery bank for energy storage, and a main distribution board for power distribution. Addressing the need for sustainable energy sources and effective waste management necessitates a rapid transition toward low-carbon energy sources. Biogas produced from agro-waste holds significant potential in achieving this transition by providing a renewable and environmentally friendly energy alternative (Figure 11.1). By converting organic waste into biogas, it prevents the release of CH_4, a potent GHG, into the atmosphere, thus reducing its GWP to approximately 25 times higher than that of CO_2. While CO_2 is still released during biogas combustion, it is biogenic carbon, part of the natural carbon cycle [13]. In contrast, uncontrolled decomposition of organic waste in open lagoons or piles releases substantial quantities of CH_4 into the atmosphere through natural decomposition processes. AD effectively captures this CH_4 and converts it into a useful energy source, releasing biogenic CO_2 into the atmosphere while simultaneously offsetting the need for fossil fuel consumption.

The digestate resulting from AD can be beneficially utilized as a soil amendment and biofertilizer. It supplies vital plant nutrients and enhances soil health, potentially significantly reducing or eliminating the need for synthetic fertilizers. Digestate

Figure 11.1 Simple setup for biogas production.

provides essential plant nutrients and improves soil structure by promoting the aggregation of soil particles, increasing porosity, and enhancing aeration and water infiltration. It also enhances water retention capacity, making soils more resilient to drought and improving nutrient retention [14]. Furthermore, the OM acts as a food source for soil microorganisms, stimulating biological activity. This leads to improved soil health, increased nutrient availability, and enhanced plant growth and yield. The application of digestate contributes to a circular flow of nutrients, reducing reliance on mined minerals for fertilizer production. Additionally, storing biogenic carbon in the soil through digestate application is considered a carbon sequestration strategy, contributing to mitigating climate change. Overall, the utilization of both biogas and digestate from agro-waste AD systems contributes significantly to energy security, environmental sustainability, and economic development in rural areas. However, realizing its full potential necessitates ongoing innovation, strategic investment, and supportive policy structures to facilitate technology adoption and system optimization.

Agricultural residues are intrinsically energy-rich resources suitable for generating biofuels through various conversion processes. However, a significant portion of agricultural residues, such as crop residues and animal manure, are often managed through practices like open burning or improper disposal [15]. Such practices are harmful, causing pollution, wasting resources, and exacerbating climate change effects. Transforming these residues into valuable products, like biogas, is a key aspect of waste valorization and CE principles. Moreover, the biomass residue left after biofuel production still contains valuable components that can be further processed or used for other applications, such as producing biochemicals, animal feed, or organic fertilizers. This cascading utilization of agricultural residues enhances resource efficiency and economic value while reducing environmental impacts. This chapter will detail the pretreatment methods and process technologies used to convert agro-waste into biogas.

11.3 Agro-waste as a Resource

Agro-waste, a broad term encompassing various byproducts from agricultural activities, is increasingly recognized as a valuable resource for generating energy and producing value-added products. It includes crop residues, animal manure, and processing byproducts. The specific composition and quantity of agro-waste vary depending on factors such as the type of crop or animal production, farming practices, and local conditions. Examples of agro-waste from agricultural output, including resources such as paddy straw, sugarcane tops, maize stalks, coconut shells, and palm oil bunches [16, 17].

Agro-waste is naturally biodegradable and, over time, decomposes and is reintegrated into the soil. However, the uncontrolled decomposition of organic waste can lead to the release of GHGs, including CH_4 and nitrous oxide (N_2O), into the atmosphere. These emissions contribute significantly to climate change [18, 19]. Analyzing agro-waste streams yields critical data needed for developing effective valorization strategies. This analysis usually includes recording observations related to different aspects of how agro-waste is generated and utilized:

- Volume of agro-waste generated per year from different crops and livestock.
- Volume of agro-waste currently utilized, for example, as animal feed or compost.
- Volume of agro-waste left unused or disposed of without beneficial treatment per year.
- Volume of animal manure generated per year (e.g., manure) and its nutrient content (nitrogen [N], phosphorus [P], potassium [K]).
- Volume of plant residues generated per year (e.g., corn stover, wheat straw) and its composition (lignin, cellulose, hemicellulose).
- Volume of animal waste left unused or disposed of without beneficial treatment per year.
- Total combined volume of agro-waste (plant + animals) left unused or disposed of without beneficial treatment per year, estimated total volume per year.
- Potential market price of each main agricultural product and its secondary products (residues).

Figure 11.2 depicts potential valorization pathways for agricultural waste and residues.

Agro-waste is a biodegradable material that naturally decomposes over time, returning nutrients to the soil. However, uncontrolled decomposition can release harmful GHGs into the atmosphere. Proper management and treatment of agro-waste can mitigate negative environmental impacts [20]. Effective strategies include utilizing it as a valuable resource for energy generation or producing other beneficial products. Analyzing agro-waste streams is essential for developing

Figure 11.2 Some of the agro-waste and its byproduct.

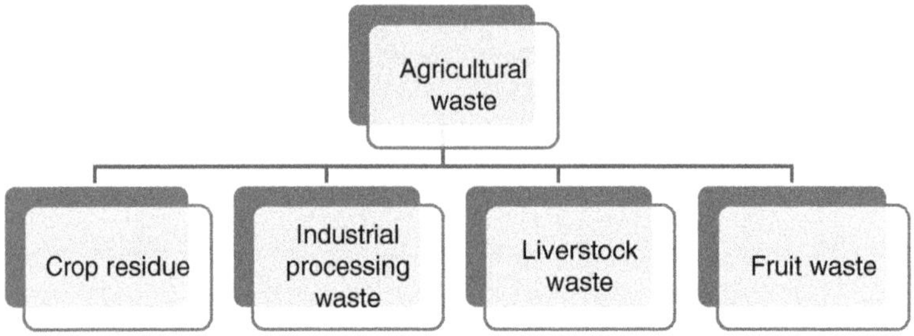

Figure 11.3 Major agricultural waste and residues. Adapted from [24, 25].

effective valorization strategies, involving recording observations related to waste generation and utilization. Implementing proper management practices and utilizing agro-waste for biogas production can help reduce GHG emissions and prevent environmental pollution, thereby contributing to environmental health and sustainability [21, 22].

Agricultural residues are broadly categorized into two types: primary and secondary [23, 24]. Primary residues are obtained directly from the harvesting process, such as paddy straw, sugarcane tops, maize stalks, and coconut shells [9]. Secondary residues result from processing agricultural products, including rice husk obtained during rice milling, corn cob from maize processing, parts obtained from coconuts like shell and husk, and sawdust from wood processing. Figure 11.3 illustrates the major agro-waste and residues [24, 25] (Figure 11.3).

During harvesting, agricultural residues (also termed agro residues) are acquired as byproducts. These include stalks, canes, seeds, leaves, and roots, all rich in cellulose, often found alongside hemicellulose and lignin [26]. In contrast, secondary biofuels are produced from processed biomass feedstocks via various conversion methods like fermentation or transesterification. Primary biofuels originate directly from natural sources and undergo minimal processing, exemplified by firewood used for heating or cooking.

11.4 Pretreatment of Agro-waste

Increased fossil fuel consumption has resulted in elevated GHG emissions and subsequent environmental pollution. The production of biodiesel from food crops is not broadly adopted, primarily because of social concerns and discussions surrounding the use of food crops for fuel [27, 28]. Treating and utilizing agro-waste for producing biofuels and chemicals is becoming increasingly important, given its potential to help meet energy demands and address environmental concerns. However, utilizing agro-waste requires pretreatment to effectively break down its complex lignocellulosic structure and enhance the accessibility of its components for subsequent conversion processes.

Lignocellulosic biomass, composed primarily of cellulose, hemicellulose, and lignin, is a common feedstock for producing biofuels [29, 30]. These components are arranged in a complex matrix that resists degradation by microorganisms and

Table 11.1 Pretreatment for agricultural residues. Adapted from [35–37].

Method	Pretreatments	Methods
Residues pretreatment method	Physical	Mechanical milling
		Microwave
		Ultrasonic
	Chemical	Acid
		Alkali
		Organo solvent
		Ozonolysis
	Physicochemical	Explosion of steam
		Subcritical water
		Supercritical CO_2
		Explosion of ammonia fiber
	Biological	Microbial delignification

enzymes [31]. Effectively separating these components is essential for the efficient production of biofuels or biogas. The process is designed to remove lignin, break down hemicellulose, and reduce the crystallinity of cellulose, thus making the polysaccharides more accessible to enzymes for hydrolysis. Various pretreatment methods have been developed and applied to agricultural residues to boost their digestibility, fermentability, and overall efficiency in valorization [32]. Pretreating agricultural residues involves dismantling the lignocellulosic structure and releasing fermentable sugars for subsequent conversion processes [33]. Selecting the most suitable pretreatment method depends on factors like the desired processing speed, associated costs, and potential environmental impact. Achieving optimal performance often relies on controlling parameters like temperature, pressure, reaction time, and pH, and selecting the most appropriate process for the specific feedstock [34]. Pretreatment technologies for agro-waste valorization are broadly categorized into physical, chemical, and physiochemical methods, as outlined in Table 11.1.

11.4.1 Physical Pretreatment

Physical pretreatment techniques concentrate on reducing particle size and disrupting the biomass's physical structure, thereby increasing the surface area available for subsequent biological or chemical reactions [38]. This reduction in size is critical for enhancing enzymatic hydrolysis and digestion efficiency by improving particle size and uniformity [39, 40] (Figure 11.4). Physical pretreatment methods include milling, grinding, chipping, and shredding. Milling and grinding reduce biomass particle size to improve accessibility, while chipping and shredding are used for larger biomass materials. Physical pretreatment can also include techniques such as microwave irradiation, electro-mechanical radiation, and ultrasound, which use energy to break down biomass structure and enhance component accessibility [41].

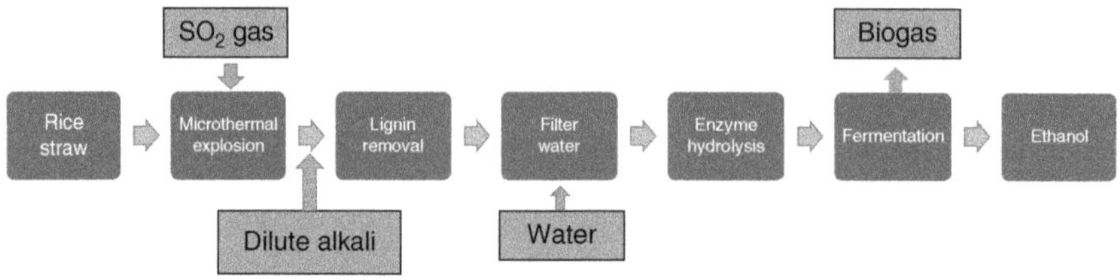

Figure 11.4 Pretreatment process for rice straw. Adapted from [42].

While generally energy-intensive, physical pretreatment can effectively reduce particle size and increase the surface area for subsequent processing steps.

11.4.2 Chemical Pretreatment

Chemical pretreatment methods involve combined chemical and physical actions to break down the lignocellulosic structure [43]. These methods often employ high temperature and pressure, sometimes with catalysts or oxidizing agents. The process is designed to remove lignin, break down hemicellulose, and reduce the crystallinity of cellulose, thus making the polysaccharides more accessible to enzymes for hydrolysis. Common chemical pretreatment methods include dilute acid hydrolysis, alkaline hydrolysis, organosolv, supercritical water treatment, supercritical CO_2 treatment, and ammonia fiber explosion (AFEX) [44].

Dilute acid hydrolysis, typically using sulfuric acid (H_2SO_4) or hydrochloric acid (HCl), is carried out at high temperatures and pressures to break down hemicellulose [45] into fermentable sugars. This method is effective at hydrolyzing hemicellulose but can also lead to the formation of inhibitory compounds (e.g., furfural and hydroxymethylfurfural, HMF) that can inhibit microbial activity during subsequent fermentation or digestion steps.

Alkaline pretreatment, commonly using sodium hydroxide (NaOH) or potassium hydroxide (KOH), effectively breaks the ester bonds between lignin and hemicellulose and cleaves linkages within the lignin–carbohydrate complex. This increases biomass porosity and accessibility, making hemicellulose more available. Alkaline pretreatment is typically considered effective for lignocellulosic biomass, especially feedstocks with high lignin content. However, disposing of alkaline waste streams from alkali pretreatment methods can present environmental concerns.

Organosolv pretreatment uses organic solvents (e.g., ethanol, methanol, acetone) or mixtures of organic solvents and water to break down lignocellulosic biomass. The solvents dissolve lignin and hemicellulose, separating them from cellulose. Organosolv pretreatment can produce a lignin-rich stream and a cellulose-rich residue, which can be further processed for biofuel or biochemical production [46].

Chemical pretreatment methods are being developed to mitigate the disadvantages of conventional methods [47]. Ionic liquids are liquid salts with unique properties, such as high thermal stability, low vapor pressure, and the ability to dissolve lignocellulosic biomass. Ionic liquids can effectively break down the complex structure and extract lignin and hemicellulose. Organic acids, ionic liquids, and deep eutectic solvents are emerging as promising green solvents for extracting lignin and fractionating biomass, providing benefits such as recyclability and lower toxicity compared to traditional organic solvents [48].

11.4.3 Physiochemical Pretreatment

Physiochemical pretreatment methods combine physical and chemical actions to break down the lignocellulosic structure. These methods often employ high temperature and pressure, sometimes in the presence of catalysts or oxidizing agents. Common physiochemical pretreatment methods include steam explosion, hydrothermal treatment, and AFEX.

In steam explosion, biomass is treated with high-pressure saturated steam for a short duration, followed by a rapid pressure drop. The sudden drop in pressure causes the steam trapped inside the biomass to expand explosively, disrupting the fiber structure and releasing lignocellulosic components. Steam explosion is effective at breaking down hemicellulose and disrupting lignin, making cellulose more accessible for subsequent hydrolysis.

Hydrothermal treatment involves treating biomass with hot water or steam at high temperatures and pressures [49, 50]. This process breaks down hemicellulose and lignin through hydrolysis and decomposition reactions. Hydrothermal treatment can produce a liquid stream rich in hemicellulose-derived sugars and a solid residue rich in cellulose and lignin.

Supercritical CO_2 pretreatment uses CO_2 beyond its critical point (31.1 °C and 7.38 MPa), where it displays characteristics of both liquid and gas. Supercritical CO_2 can permeate the biomass structure and dissolve lignocellulosic components, increasing their accessibility for enzymatic hydrolysis. Studies have shown that adding agents like ammonia (NH_3) during supercritical CO_2 treatment enhances the removal of hemicelluloses and lignin, leading to higher yields [51].

AFEX treats biomass with liquid NH_3 under moderate temperatures and pressures. A rapid pressure drop causes the NH_3 to expand and disrupt the lignocellulosic structure, also partially hydrolyzing hemicellulose. AFEX is especially effective for grasses and herbaceous feedstocks.

Subcritical water treatment utilizes water at high temperatures (typically 175–374 °C) and pressures (above boiling point but below critical point). Under these conditions, water acts as a solvent and reaction medium, facilitating the breakdown of lignocellulosic biomass. Hydrothermal hydrolysis, conducted on wheat straw and corn stover samples at temperatures around 175–225 °C and pressures between 5 and 7 bar, has demonstrated good performance for treating agro-waste-like lignocellulosic materials.

11.4.4 Biological Pretreatment

Biological pretreatment methods employ microorganisms, usually fungi or bacteria, or their enzymes, to break down the lignocellulosic matrix. This method is considered environmentally benign, as it generally does not necessitate harsh chemicals [52]. The main goals are to remove lignin and decompose hemicellulose.

Certain fungi, especially white-rot fungi, produce enzymes (like lignin peroxidases, manganese peroxidases, and laccases) capable of effectively degrading lignin. Brown-rot fungi primarily degrade cellulose and hemicellulose [53]. Certain bacteria, often cellulolytic and hemicellulolytic types, can also be used for biological pretreatment to enhance polysaccharide accessibility.

Biological pretreatment offers several benefits, such as lower energy use and reduced chemical consumption, compared to chemical and physiochemical methods. However, it usually requires longer treatment durations, and the rate of biomass decomposition can be slower. The effectiveness of biological pretreatment is influenced by factors like the selection of microbial species, the initial composition of the biomass, and optimizing environmental conditions (temperature, pH, moisture content, nutrient availability) for peak activity [54]. Using biological methods to pretreat agricultural residues has been shown to enhance the yield of subsequent hydrolysis steps for biofuel or biogas production.

11.5 Process Technology – Agro-waste to Bioenergy

Agricultural residues serve as inherently energy-rich resources suitable for generating bioenergy through various conversion processes. Transforming these residues into valuable products is a key aspect of waste valorization and CE principles. High-yield bioethanol production from lignocellulosic biomass is another significant pathway. This is typically achieved by effectively removing lignin and significantly reducing cellulose crystallinity, thereby making the material more accessible to hydrolytic enzymes. After hydrolysis, the released fermentable sugars from cellulose and hemicellulose are then fermented by yeasts or bacteria to produce ethanol.

11.5.1 Hydrolysis

Hydrolysis is a crucial stage that breaks down complex polysaccharides (cellulose and hemicellulose) in pretreated biomass into fermentable sugars (such as glucose, xylose, and other simple sugars). This step is necessary before fermentation (e.g., for bioethanol production) or further microbial conversion via AD or gasification. Hydrolysis can be performed using either chemical methods (acid hydrolysis) or biological methods (enzymatic hydrolysis).

In acid hydrolysis, acids serve as catalysts to break the glycosidic bonds found in cellulose and hemicellulose. Both dilute and concentrated acids are applicable. Dilute acid hydrolysis is generally conducted at high temperatures and pressures and is effective for breaking down hemicellulose. Concentrated acid hydrolysis uses higher acid concentrations (e.g., >30% H_2SO_4) at lower temperatures and pressures, enabling more complete breakdown of both cellulose and hemicellulose into sugars. However, concentrated acid hydrolysis processes can be more energy-intensive and necessitate careful handling due to the corrosive nature of concentrated acids [55]. Additionally, disposal of acidic waste streams can present environmental challenges.

Enzymatic hydrolysis employs enzymes (such as cellulases and hemicellulases) to break down polysaccharides into fermentable sugars. This method typically operates under mild conditions (usually 40–50 °C and neutral pH), lowering the risk of equipment corrosion and minimizing inhibitory compound formation [56]. However, enzymatic hydrolysis can be slower and costlier than chemical methods, and its efficiency can be reduced by factors like enzyme inactivation caused by inhibitory compounds in the hydrolysate (such as lignin or furfural) or product inhibition. Research efforts are concentrated on developing more efficient and affordable enzyme formulations, optimizing process parameters, and overcoming inhibition challenges. Combining enzymatic hydrolysis with simultaneous saccharification and fermentation (SSF) or simultaneous saccharification and co-fermentation (SSCF) can enhance overall process efficiency and decrease enzyme inhibition. SSF/SSCF allows the conversion of sugars to ethanol to occur concurrently with hydrolysis, reducing product inhibition on enzymes and improving overall process kinetics.

11.5.2 Anaerobic Digestion

AD is a complex, multi-stage biological process happening without oxygen, performed by a diverse consortium of anaerobic microorganisms. It is a highly effective method for treating organic waste streams and converting them into biogas and digestate. The process is typically divided into four main stages:

1. **Hydrolysis:** Complex organic polymers (carbohydrates, proteins, lipids) in the feedstock are broken down into simpler molecules by hydrolytic enzymes (like cellulases, hemicellulases, proteases, lipases) that cleave the chemical bonds in these polymers. This stage prepares the material for subsequent microbial conversion. Hydrolysis yields fermentable sugars, amino acids and peptides, and lipids broken down into fatty acids and glycerol [57]. Hydrolysis is often considered the rate-limiting step for solid feedstocks, particularly those rich in lignocellulosic biomass, as the complex structure resists enzymatic and microbial degradation.
2. **Acidogenesis:** During this stage, the products from hydrolysis (monosaccharides, amino acids, fatty acids, and glycerol) are transformed into simpler organic acids, alcohols, hydrogen (H_2), and CO_2 by fermentative bacteria. Typical products encompass volatile fatty acids (VFAs) like acetate, propionate, butyrate, and lactate, as well as alcohols (ethanol, methanol), ketones, H_2, and CO_2. This stage is marked by a pH decrease if not properly controlled, resulting in the buildup of organic acids. Acidogenesis typically proceeds quickly compared to the preceding hydrolysis and subsequent methanogenesis stages. The specific products produced during acidogenesis depend on the feedstock type and the composition of the microbial community.
3. **Acetogenesis:** Acetogenic bacteria convert intermediate products from acidogenesis (like VFAs, longer-chain fatty acids, and alcohols) into acetate, H_2, and CO_2 [32]. This stage is vital for methanogens, as it supplies the necessary substrates for CH_4 production, and typically occurs in close metabolic partnership

with methanogenic archaea. Acetogenic bacteria produce H_2, which inhibits their own metabolic processes. However, methanogenic archaea consume H_2 for CH_4 production through hydrogenotrophic methanogenesis, thus enabling acetogenesis to continue. This delicate syntrophic relationship is crucial for efficient AD.

4. **Methanogenesis:** This is the final stage where methanogenic archaea convert acetate, H_2, and CO_2 into CH_4 and CO_2 [58]. Two primary pathways exist: acetoclastic methanogenesis, where acetate is split into CH_4 and CO_2, and hydrogenotrophic methanogenesis, where H_2 and CO_2 are converted into CH_4 and water (the latter pathway accounting for most CH_4 production). Optimal operation of the methanogenesis stage relies on factors such as temperature, pH, and the presence of inhibitory substances (e.g., NH_3, hydrogen sulfide [H_2S]) [59]. Understanding the diversity and activity of methanogenic communities is vital for optimizing biogas yield and ensuring process stability.

The overall efficiency and stability of the AD process are determined by the effective functioning and balance of the microbial community involved in these four stages. Monitoring and control are essential to maintain optimal conditions for the diverse microbial consortium.

11.5.3 Dark Fermentation

Dark fermentation is an anaerobic biological process where organic substrates are converted into H_2 gas, VFAs, and alcohols, predominantly in the absence of light. This technology is particularly well-suited and efficient for producing H_2 from various agricultural residues, especially those high in carbohydrates like sugars, starch, and hemicellulose. H_2 production typically follows a fermentative pathway catalyzed by bacteria that produce H_2. The main metabolic products are typically acetate and butyrate, along with the production of H_2 and CO_2. Other VFAs and alcohols may also be produced depending on the specific microbial community and feedstock composition. While dark fermentation produces H_2, its theoretical yield is limited (typically 4 moles of H_2 per mole of glucose) compared to other processes like photo fermentation (up to 12 moles of H_2 per mole of glucose). Furthermore, the process often produces a mixed stream of H_2 and CO_2, requiring subsequent purification steps to obtain high-purity H_2, which can still contain significant chemical energy. This VFA-rich effluent can serve as a suitable feedstock for subsequent AD (producing CH_4) or photo fermentation (producing additional H_2), contributing to an integrated biorefinery concept. This approach combines dark fermentation with other biological processes, such as AD or photo fermentation, to maximize energy recovery from organic waste.

11.5.4 Transesterification

Transesterification is a chemical reaction process primarily used to create biodiesel, a renewable fuel derived from triglycerides (fats and oils) [60]. This process involves triglycerides reacting with an alcohol, usually methanol or ethanol, in the presence of a catalyst to yield fatty acid alkyl esters (biodiesel) and glycerol as a byproduct.

The overall process involves exchanging the alkoxy group of an ester compound with another alcohol. This reaction is reversible, so using excess alcohol or removing one of the products (glycerol) can shift the equilibrium toward biodiesel production [10, 61, 62]. The reaction rate is influenced by factors such as the type of alcohol used, the alcohol-to-oil molar ratio, reaction temperature, pressure (if applicable), and reaction time.

Transesterification is generally favored for biodiesel production because it is a relatively simple process and can handle diverse lipid feedstocks. Catalysts can be either homogeneous (dissolved in the reaction mixture, such as alkali catalysts like NaOH, KOH, or acid catalysts like H_2SO_4, HCl) or heterogeneous (solid catalysts like metal oxides, enzymes) [63]. Homogeneous catalysts are commonly used but can be corrosive and require removal from the product. Heterogeneous catalysts offer benefits such as simpler separation from the reaction mixture, recyclability, and reduced corrosion, although they can be less reactive than homogeneous catalysts. Enzyme catalysts are becoming more popular due to their ability to function under very mild conditions and high substrate specificity, but they are often costlier and less active than chemical catalyst.

Pretreatment is often necessary for lipid feedstocks, especially those high in free fatty acid (FFA) or moisture content, to avoid saponification or other undesirable reactions during transesterification. This pretreatment typically involves esterifying FFAs with methanol (using an acid catalyst), followed by alkali-catalyzed transesterification of triglyceride.

11.5.5 Creating Wealth from the Agro-waste

Transforming agro-waste, often seen as a disposal issue, into valuable products and resources is a way of creating wealth [64]. This approach aligns with CE principles, where waste is viewed as a feedstock for new value streams. Agro-waste-to-biogas technology is a prime example of how waste valorization can generate economic value while simultaneously addressing environmental concerns. The primary products of this technology, biogas and digestate, offer multiple opportunities for creating wealth through various pathways. Examples of wealth creation pathways include:

- **Energy Production:** Producing biogas (for heat, electricity, or transport fuel) or bioethanol from biomass.
- **Biofertilizers:** Converting animal manure and AD digestate into nutrient-rich organic fertilizers, reducing the need for synthetic fertilizers.
- **Biochemicals:** Extracting valuable compounds like cellulose, hemicellulose, lignin, antioxidants, or other bioactive molecules for use in various industries (e.g., pulp and paper, chemicals, pharmaceuticals).
- **Biomaterials:** Converting residues into materials like biochar (for soil amendment or carbon sequestration), bio-based plastics, composites, or building materials.
- **Animal Feed Supplements:** Producing protein-rich animal feed or bioproducts into animal feed components, provided they meet safety and nutritional standards [65, 66].

- **Cultivation Substrates:** Using certain residues (e.g., straw, sawdust) as substrates for cultivating edible or medicinal mushrooms [67].
- **Platform Chemicals:** Converting biomass sugars into platform chemicals that serve as building blocks for a range of bio-based products.

Each valorization pathway requires specific technologies (e.g., AD, fermentation, extraction, pyrolysis) and careful economic assessment to determine viability. Successful implementation often involves integrating multiple processing streams and optimizing the recovery of various co-products to maximize value from the incoming biomass. This creates economic activity, generates revenue, and contributes to rural development by providing new income sources and employment opportunities.

11.5.6 Economic Valuation of Agro-waste

The economic valuation of agro-waste involves conducting a cost–benefit analysis to assess the financial feasibility and profitability of utilizing agro-waste for biogas production. This analysis is crucial within the framework of the bioeconomy, which is intrinsically linked to the sustainable use of biological resources for economic development. The bioeconomy aligns with the goal of optimizing resource utilization and implementing CE principles to achieve integrated system benefits aimed at enhancing sustainable growth and development through the synergistic interaction between environmental stewardship and societal well-being.

Effective implementation of policies is necessary to foster the bioeconomy by promoting the sustainable use of biological resources and facilitating the development of integrated systems. The bioeconomy holds potential for creating green jobs, stimulating rural development, and enhancing environmental quality. However, its full potential faces challenges, including resource sustainability, technological advancements, economic viability, and social acceptance. Addressing these challenges requires policy support, strategic investments, and stakeholder collaboration. The escalating global population and increasing demand for food and energy pose significant challenges, leading to potential failures in achieving a sustainable bioeconomy. Therefore, effective management of agro-waste is vital for promoting a CE. Optimizing agro-waste utilization for biogas production contributes to a sustainable bioeconomy by recovering valuable resources, generating renewable energy, and reducing environmental impacts. This contributes to resource efficiency, reduces reliance on fossil fuels, and supports sustainable agriculture and rural development. Furthermore, valorizing agro-waste through biogas production aligns with CE principles by closing nutrient and energy loops, reducing waste, and creating new value streams within the agricultural sector. Quantifying the environmental footprint of agro-waste-to-biogas systems using LCA helps identify hotspots for improvement. Mitigation strategies focusing on energy efficiency, resource recycling, and sustainable sourcing can reduce the environmental impact, contributing to achieving a sustainable bioeconomy and fostering green growth. Implementing supportive policies and regulatory measures is essential for promoting environmental sustainability and incentivizing the adoption of sustainable practices and technologies in the agro-waste-to-biogas sector. Policy implications include establishing environmental standards, providing financial incentives, and engaging stakeholders to build support for sustainable initiatives. Overall, the economic valuation of agro-waste within the bioeconomy framework underscores its potential as a valuable resource for creating wealth, promoting sustainable development, and addressing global challenges related to energy, environment, and resource management.

Analyzing agro-waste streams provides crucial data needed for developing effective valorization strategies. This analysis usually includes recording observations related to different aspects of how agro-waste is generated and utilized:

- Volume of agro-waste generated per year from different crops and livestock.
- Volume of agro-waste currently utilized, for example, as animal feed or compost.
- Volume of agro-waste left unused or disposed of without beneficial treatment per year.
- Volume of animal manure generated per year (e.g., manure) and its nutrient content (N, P, K).
- Volume of plant residues generated per year (e.g., corn stover, wheat straw) and its composition (lignin, cellulose, hemicellulose).
- Volume of animal waste left unused or disposed of without beneficial treatment per year.
- Total combined volume of agro-waste (plant + animals) left unused or disposed of without beneficial treatment per year, estimated total volume per year.
- Potential market price of each main agricultural product and its secondary products (residues).

This data is essential for evaluating the economic feasibility of different valorization pathways and selecting the most suitable option for a specific region or feedstock. Conducting a comprehensive analysis of agro-waste streams, including their quantity, composition, and current utilization patterns, is the first step in developing effective valorization strategies. This information can be used to identify potential feedstocks for biogas production, estimate potential biogas yields, and assess the economic viability of establishing biogas plants. Analyzing agro-waste streams provides crucial data for effective valorization strategies. This analysis typically involves recording observations related to various aspects of agro-waste generation and utilization: It helps in identifying opportunities for setting up manufacturing units that convert these residues into valuable bioproducts at a lower cost, thereby becoming a source of revenue generation.

11.6 Factors Influencing the Efficiency of Agro-waste-to-biogas Systems

The efficient and economically feasible operation of agro-waste-to-biogas systems depends on a holistic approach that carefully considers digester design, process optimization, and feedstock management. Major factors influencing the efficiency of agro-waste-to-biogas systems include the following, with their relative importance and specific considerations depending on the feedstock characteristics, desired scale of operation, and end-use of the biogas and digestate.

11.6.1 Digester Design and Configuration

The selection of digester type is a crucial decision, influenced by numerous factors as outlined follows:

- **Batch Digesters:** These are simple and relatively inexpensive, suitable for small-scale or decentralized applications and situations with intermittent feedstock availability. A batch digester is loaded, sealed for digestion, and then emptied for reloading. This design is simple but less efficient for continuous operations and large-scale facilities compared to continuous digesters [68].
- **Continuous Stirred Tank Reactors (CSTRs):** This is the most common type for large-scale applications, especially for liquid or semi-solid feedstocks such as manure. CSTRs operate continuously, featuring constant feeding and product removal. They offer excellent mixing (using mechanical stirrers or gas injection) to ensure uniform distribution of feedstock, microorganisms, and nutrients, maximizing contact and reaction rates. Mixing prevents stratification and the formation of dead zones.
- **Plug Flow Reactors (PFRs):** These are suitable for semi-solid feedstocks where material moves through the reactor in a plug-like manner, with minimal back mixing. PFRs are suitable for solid or semi-solid feedstocks, particularly those with high total solids (TS) content, such as agricultural residues or food waste. They are less common than CSTRs but can be more efficient for certain types of feedstocks, as they provide a longer retention time for solid particles compared to CSTRs.
- **Anaerobic Sequencing Batch Reactors (ASBRs):** A type of batch reactor where multiple steps (feeding, digestion, settling, decanting) happen sequentially in the same tank [9]. ASBRs provide operational flexibility, suit various feedstocks (including those with high solids content or inhibitory substances), and allow better control over different microbial stages [69].
- **Upflow Anaerobic Sludge Blanket (UASB) Reactors:** Highly efficient digesters frequently used for treating wastewater or liquid manure [70]. UASB reactors encourage the formation of granular sludge, which consists of dense microbial aggregates providing a large surface area for microbial activity and settling well, thereby minimizing washout. UASB reactors are energy-efficient, requiring less mixing, but are primarily suitable for liquid feedstocks.
- **Membrane Bioreactors (MBRs):** These combine AD with membrane filtration technology. MBRs integrate AD with a membrane separation step (e.g., ultrafiltration or microfiltration) to separate solids from the liquid effluent. This enables higher biomass concentration in the reactor, improved process kinetics, and the production of a high-quality effluent. MBRs are suitable for treating various feedstocks (e.g., wastewater, agricultural residues) and offer higher overall efficiency and stability compared to conventional digesters (with less volume for potentially higher efficiency and stability), and their modular or containerized designs facilitate easier deployment and scalability.

- **Dry AD Systems:** Designed for feedstocks with high TS content (e.g., solid manure, crop residues). Dry AD systems operate with less water than wet AD systems, making them suitable for feedstocks that are difficult to dewater or transport.
- **Two-stage AD Systems:** Involve separating the acidogenesis and methanogenesis stages into different reactors. This can improve overall process efficiency and stability, particularly for complex feedstocks.

The selection of the most suitable digester type is influenced by many factors, such as the type and characteristics of the feedstock (including solids content, biodegradability, and the presence of inhibitors), the target processing rate, environmental conditions, and specific project objectives and site conditions. Optimizing the process is vital for maximizing biogas production efficiency and yield, and ensuring stable, reliable operation while minimizing costs. Key process parameters that need to be optimized include:

- **Organic Loading Rate (OLR):** This is the amount of OM (measured as volatile solids, VS) introduced into the digester per unit volume per day (kg VS/m^3/day). Higher OLRs can potentially boost biogas production capacity per unit reactor volume, improving economic viability. However, excessively high OLRs can cause digester instability, leading to the accumulation of organic acids (VFAs), a drop in pH, and potential inhibition of methanogenesis.
- **Hydraulic Retention Time (HRT):** This refers to the average duration the liquid or influent feedstock stays inside the digester. HRT affects the extent of digestion and biogas yield. Longer HRTs typically result in higher biogas yields but require larger digester volumes, increasing capital costs. Optimal HRT depends on the feedstock characteristics and digester type.
- **Temperature:** Proper temperature control is crucial for optimizing enzyme activity and microbial metabolism during AD. AD can operate under mesophilic (typically 30–40 °C) or thermophilic (typically 50–60 °C) conditions. Mesophilic digestion is typically more stable and less prone to inhibition but operates at a slower rate. Thermophilic digestion operates at higher temperatures (typically 50–60 °C), leading to faster reaction rates and higher biogas yields. It also offers better pathogen inactivation but can be less stable and more sensitive to changes in operating conditions or feedstock characteristics.
- **Mixing:** Adequate mixing is essential in most digester types (especially CSTRs) to ensure uniform distribution of feedstock, microorganisms, and nutrients, maximizing contact and reaction rates [71]. Mixing prevents the formation of settled or floating layers (scum), which can reduce active digester volume and hinder gas release. However, excessive mixing can consume significant energy and may disrupt microbial communities.
- **pH Control:** Controlling pH to stay within the optimal range is critical for the different microbial groups involved in AD. Acidogenic bacteria flourish at slightly acidic pH, while methanogenic archaea prefer neutral to slightly alkaline conditions. Deviations from the optimal pH range can inhibit microbial activity and reduce biogas production. pH can be controlled by adjusting the feedstock composition, adding buffering agents (e.g., lime, sodium bicarbonate), or recirculating digestate.

- **NH_3 Inhibition:** Elevated concentrations of NH_3 or ammonium (NH_4^+) can inhibit or be toxic to methanogenic archaea, potentially reducing biogas production or causing digester failure [72]. This is particularly relevant for feedstocks high in N, such as animal manure. Strategies to mitigate NH_3 inhibition include optimizing feedstock C/N ratio, co-digestion with low-N feedstocks, or adjusting pH.
- **Inhibitor Management:** It is vital to identify and mitigate the effects of various inhibitory substances that may be present in agro-waste feedstocks (e.g., NH_3, H_2S formation, long-chain fatty acids [LCFAs], phenolic compounds from lignin breakdown, and heavy metals) to maintain digester stability and optimize biogas production [73]. Pretreatment methods can help remove or reduce inhibitory substances.
- **Monitoring and Control:** Implementing strategies (e.g., using predictive models or artificial intelligence) can help detect process disruptions early and adjust operating parameters to maintain optimal performance [74]. This includes monitoring parameters like temperature, pH, OLR, VFA concentrations, and biogas composition.

Effective optimization of these parameters necessitates a comprehensive understanding of the specific characteristics of the feedstock and the dynamic nature of the microbial community within the digester. Effective feedstock management is crucial for ensuring the long-term sustainability, economic feasibility, and consistent performance of agro-waste-to-biogas systems. Its major aspects include:

- **Feedstock Collection and Storage:** Efficient collection methods are required to gather agro-waste from various sources while minimizing contamination. Appropriate storage conditions are also vital to prevent biomass decomposition and environmental problems such as leachate production and uncontrolled CH_4 emissions. Storage methods should consider factors like moisture content, temperature, and the intended storage duration [75]. Examples include covered lagoons for liquid manure, bunkers or piles (ideally covered) for solid residues, or silos for crops.
- **Feedstock Characterization**: Thorough characterization of the incoming feedstock is essential before it is fed into the digester. This provides vital information for assessing its suitability for AD, estimating potential biogas yield, and optimizing operating parameters (e.g., OLR, HRT) [76]. Primary parameters for analysis include TS, VS – the biologically degradable fraction of TS, moisture content, nutrients (e.g., carbon, N, P, K), C/N ratio, pH, alkalinity, and the presence of potential inhibitory substances (e.g., heavy metals, sulfur, VFAs).
- **Feedstock Pre-processing:** Depending on the type of feedstock and the selected digester configuration, pre-processing steps may be necessary to enhance its handling characteristics and optimize its conversion to biogas. Pre-processing steps might include particle size reduction (shredding, grinding) to improve flow and surface area, screening (to remove inert materials like stones or plastics), mixing (to homogenize the feedstock or blend different feedstocks for co-digestion), and pasteurization or thermal treatment (to eliminate pathogens, especially in animal byproducts or food waste, and sometimes enhance biodegradability).

11.7 Digestate Utilization and Impact

Besides generating biogas, AD also produces digestate, a nutrient-rich material that serves as a valuable co-product. Digestate can be effectively used in agriculture as a biofertilizer and soil amendment, supplying vital plant nutrients, improving soil health, and boosting crop production. The beneficial use of digestate offers a significant opportunity to create a circular nutrient flow, reducing reliance on synthetic fertilizers and promoting sustainable farming practices.

11.7.1 Digestate Composition and Variability

The composition of digestate can vary considerably based on the feedstock characteristics and the AD process operating parameters. These factors include the type of feedstock used in AD, operating parameters (like temperature, pH, and hydraulic retention time), and the presence of inhibitory substances (e.g., in batch or continuous systems). However, a consistent feature of digestate is its abundance of essential plant nutrients and OM.

- **N:** Digestate contains a significant amount of N, mostly in the form of NH_4^+. This readily available N form is easily absorbed by plant roots, being especially beneficial for early plant growth and development [77]. However, high concentrations of NH_4^+ can also inhibit plant growth.
- **P:** Digestate contains P in both organic and inorganic forms. The AD process can solubilize some P initially bound in the feedstock, making it more accessible to plants.
- **K:** K in digestate is generally in a readily available ionic form (K^+), as it is not significantly changed during the AD process. K is vital for plant growth, enzyme activation, and water uptake; it is readily mobile within the plant and supports enzyme activation. Its availability from digestate is typically high and less affected by soil conditions compared to N or P.
- **Micronutrients:** Besides macronutrients, digestate also contains essential micronutrients needed for healthy plant growth. These include elements like iron (Fe), manganese (Mn), zinc (Zn), copper (Cu), boron (B), and molybdenum (Mo). Micronutrients are essential for various plant physiological processes, such as photosynthesis, respiration, and nutrient assimilation. Deficiencies in micronutrients can severely restrict plant growth and yield, especially in areas with nutrient-poor or intensively farmed soils. The presence of these micronutrients in digestate further boosts its value as a biofertilizer, contributing to balanced nutrition for crops. However, it is important to monitor the heavy metal content in digestate from certain feedstocks (e.g., industrial organic waste) to prevent the accumulation of toxic levels in the soil.
- **OM:** Digestate is rich in OM, offering numerous benefits for soil health and fertility. Adding digestate to soil improves soil structure by promoting particle aggregation, increasing porosity, and enhancing aeration and water infiltration. OM also improves water retention capacity, making soils more resilient to drought conditions. Furthermore, the OM acts as a food source for soil microorganisms, stimulating biological activity. OM contributes to soil carbon

Table 11.2 Typical nutrient content of digestate from different feedstocks.

Feedstock	Total nitrogen (N) (kg/ton)	Total phosphorus (P_2O_5) (kg/ton)	Total potassium (K_2O) (kg/ton)	Dry matter (%)
Cattle manure	3–7	2–5	2–6	4–10
Pig manure	4–8	2–6	3–7	3–8
Poultry manure	8–15	6–12	4–8	20–30
Crop residues	5–10	1–3	5–12	15–25
Food waste	8–15	4–8	5–10	10–20
Mixed agro-waste	6–12	3–7	4–9	8–15

sequestration, which helps mitigate climate change and improves long-term soil productivity. Digestate from agricultural residues (via composting or drying) may have more stabilized OM than raw digestate. Table 11.2 depicts the variation in nutrient contents with feedstocks.

11.7.2 Impact of Digestate on Soil Health and Crop Productivity

Applying digestate to agricultural soils offers numerous benefits due to its unique composition of nutrients, OM, and microbial communities. These combined effects contribute significantly to improving soil health and boosting crop productivity. The readily available N and K in digestate offer crops quick access to nutrients. P in digestate is present in both organic and inorganic forms, and the AD process can solubilize some P initially bound in the feedstock, making it more accessible to plants.

OM in digestate improves soil structure by promoting aggregation, increasing porosity, and enhancing water infiltration and aeration (facilitating plant root respiration) [78]. OM also enhances water retention capacity, making soils more resilient to drought and improving nutrient retention. Adding digestate stimulates soil microbial activity, which is essential for nutrient cycling, OM decomposition, and soil health. A healthy and active soil microbial community helps suppress soil-borne diseases and promotes plant growth through various mechanisms. The composition of the microbial community introduced or stimulated by applying digestate can create a more resilient and disease-suppressive soil ecosystem [79]. Applying digestate has shown beneficial effects on crop yield and quality across various cropping systems.

Compared to mineral fertilizers, digestate offers a more holistic approach to soil health and crop nutrition. Digestate contains essential plant nutrients (N, P, K, and micronutrients) and OM, offering a more balanced and complete nutrient source for crops compared to mineral fertilizers, which usually supply only a few specific nutrients [80]. The OM in digestate provides numerous soil health benefits, such as improved structure, water retention capacity, and microbial activity, none of which are provided by mineral fertilizers [81]. Using digestate as a biofertilizer can decrease reliance on synthetic fertilizers, leading to cost savings for farmers and reducing environmental impacts linked to synthetic fertilizer production and use [82].

Careful management of digestate application is important to optimize nutrient use efficiency and minimize potential environmental risks like nutrient runoff, NH_3 emissions, or the introduction of heavy metals or organic pollutants if the feedstock is contaminated. Soil testing and nutrient planning are crucial to determine appropriate digestate application rates based on crop needs and soil nutrient status.

11.8 Conclusion

AD of agro-waste represents a vital and promising pathway toward achieving a truly circular and sustainable economy. By effectively converting organic agro-waste into valuable resources – biogas and digestate – AD technology provides multiple environmental, economic, and social benefits. AD technology includes:

- **Reduced GHG Emissions:** AD captures CH_4 from decomposing organic waste, preventing its release into the atmosphere, and the biogas produced can displace fossil fuels, leading to a net reduction in GHG emissions.
- **Enhanced Energy Security:** Utilizing domestically available agro-waste for biogas production contributes to energy independence and diversification of the energy mix.
- **Improved Waste Management:** AD provides an environmentally sound solution for managing large volumes of agricultural organic waste, reducing reliance on landfilling or incineration.
- **Sustainable Rural Development:** Establishing agro-waste-to-biogas systems can create green jobs, stimulate rural economic activity, and empower local communities.
- **Nutrient Recycling and Soil Health:** The digestate co-product is a valuable biofertilizer that recycles essential nutrients back to the soil, improves soil fertility and structure, and enhances crop productivity, promoting sustainable agricultural practices.

The broad adoption of agro-waste-to-biogas technology faces several challenges that must be addressed for its full potential to be achieved. Overcoming these obstacles necessitates continued investment in research and development to enhance technological efficiency and cost-effectiveness. Supportive policy frameworks, financial incentives, favorable regulations, and standards for biogas and digestate quality arc also crucial for fostering a supportive environment for sector growth. Promoting research and development: ongoing investment in research and development is essential for enhancing the efficiency, reliability, and cost-effectiveness of agro-waste-to-biogas technology.

With the correct approach, involving collaboration between researchers, policymakers, industry stakeholders, and rural communities, agro-waste-to-biogas technology can play a crucial role in building a more sustainable and resilient future. This shift toward a circular and sustainable economy, powered by the valorization of agro-waste, offers multiple benefits for both the environment and society. By carefully considering the diverse factors influencing the efficiency of agro-waste-to-biogas systems and actively addressing the existing technical, economic, and policy challenges, we can collectively work toward realizing the full potential of this technology. Continued efforts in research, development, policy support, and stakeholder engagement are essential for scaling up agro-waste-to-biogas technology and maximizing its contribution to a sustainable and circular bioeconomy.

References

1 Sher, F., Smječanin Omerbegović, N., Hrnjić, H., Karadža, A., Omanović, R., Šehović, E., et al. Emerging technologies for biogas production: a critical review on recent progress, challenges and future perspectives. *Process Saf. Environ. Prot.* vol. 188, pp. 834–859, 2024. https://doi.org/10.1016/j.psep.2024.05.138.

2 Rocha-Meneses, L., Bergamo, T.F., and Kikas, T. Potential of cereal-based agricultural residues available for bioenergy production. *Data Br.* 2019. https://doi.org/10.1016/j.dib.2019.103829.

3 Agrawal, P.S., Belkhode, P.N., and Rokhum, S.L. Advanced biofuels – alternatives to biofuels. In: *Sustainability in Biofuel Production Technology*. John Wiley & Sons, Ltd, 2022. https://doi.org/10.1002/9781119888864.ch3.

4 Agrawal, P.S., Belkhode, P.N., and Rokhum, S.L. Biofuel production technologies –an overview. In: *Sustainability in Biofuel Production Technology*. John Wiley & Sons, Ltd, 2022. https://doi.org/10.1002/9781119888864.ch4.

5 Agrawal, P.S., Belkhode, P.N., and Rokhum, S.L. Greener catalytic processes in biofuel production. In: *Sustainability in Biofuel Production Technology*. John Wiley & Sons, Ltd, 2022. https://doi.org/10.1002/9781119888864.ch12.

6 Kirchherr, J., Yang, N.H.N., Schulze-Spüntrup, F., Heerink, M.J., and Hartley, K. Conceptualizing the circular economy (revisited): an analysis of 221 definitions. *Resour. Conserv. Recycl.* 2023. https://doi.org/10.1016/j.resconrec.2023.107001.

7 Piadeh, F., Offie, I., Behzadian, K., Rizzuto, J.P., Bywater, A., Córdoba-Pachón, J.R., et al. A critical review for the impact of anaerobic digestion on the sustainable development goals. *J Environ. Manage.* 2024. https://doi.org/10.1016/j.jenvman.2023.119458.

8 Hahn, H., Hartmann, K., Bühle, L., and Wachendorf, M. Comparative life cycle assessment of biogas plant configurations for a demand oriented biogas supply for flexible power generation. *Bioresour. Technol.* 2015. https://doi.org/10.1016/j.biortech.2014.12.007.

9 Chhandama, M.V.L., Chetia, A.C., Satyan, K.B., Ao, S., Ruatpuia, J.V., and Rokhum, S.L. Valorisation of food waste to sustainable energy and other value-added products: a review. *Bioresour. Technol. Rep.* 2022. https://doi.org/10.1016/j.biteb.2022.100945.

10 Agrawal, P.S., Belkhode, P.N., and Rokhum, S.L. Chemically produced biofuels. In: *Sustainability in Biofuel Production Technology*. John Wiley & Sons, Ltd, 2022. https://doi.org/10.1002/9781119888864.ch5.

11 Ferreira, J.A. and Taherzadeh, M.J. Improving the economy of lignocellulose-based biorefineries with organosolv pretreatment. *Bioresour. Technol.* 2020. https://doi.org/10.1016/j.biortech.2019.122695.

12 Ao, S., Gouda, S.P., Selvaraj, M., Boddula, R., Al-Qahtani, N., Mohan, S., et al. Transesterification of Jatropha curcas oil to biodiesel using highly porous sulfonated biochar catalyst: optimization and characterization dataset. *Data Br.* 2024. https://doi.org/10.1016/j.dib.2024.110096.

13 Ao, S., Gouda, S.P., Selvaraj, M., Boddula, R., Al-Qahtani, N., Mohan, S., et al. Active sites engineered biomass-carbon as a catalyst for biodiesel production: process optimization using RSM and life cycle assessment. *Energy Convers. Manag.* 2024. https://doi.org/10.1016/j.enconman.2023.117956.

14 Azadbakht, M., Safieddin Ardebili, S.M., and Rahmani, M. A study on biodiesel production using agricultural wastes and animal fats. *Biomass Convers. Biorefin.* 2023. https://doi.org/10.1007/s13399-021-01393-1.

15 Onu, P. and Mbohwa, C. (Editors) Sustainable agricultural waste diversity: advances in green energy and materials production. In: *Agricultural Waste Diversity and Sustainability Issues.* Academic Press, 2021. https://doi.org/10.1016/b978-0-323-85402-3.00008-5.

16 Wan, J.Q., Wang, Y., and Xiao, Q. Effects of hemicellulose removal on cellulose fiber structure and recycling characteristics of eucalyptus pulp. *Bioresour. Technol.* 2010. https://doi.org/10.1016/j.biortech.2010.01.026.

17 Tian, D., Shen, F., Hu, J., Huang, M., Zhao, L., He, J., et al. Complete conversion of lignocellulosic biomass into three high-value nanomaterials through a versatile integrated technical platform. *Chem. Eng. J.* 2022. https://doi.org/10.1016/j.cej.2021.131373.

18 Viccaro, M., Cozzi, M., Rocchi, B., and Romano, S. Conservation agriculture to promote inland biofuel production in Italy: an economic assessment of rapeseed straight vegetable oil as a self-supply agricultural biofuel. *J. Clean. Prod.* 2019. https://doi.org/10.1016/j.jclepro.2019.01.251.

19 Wasiak, A.L. Effect of biofuel production on sustainability of agriculture. *Procedia Eng.* 2017. https://doi.org/10.1016/j.proeng.2017.03.192.

20 Cadoche, L. and López, G.D. Assessment of size reduction as a preliminary step in the production of ethanol from lignocellulosic wastes. *Biol. Wastes.* 1989. https://doi.org/10.1016/0269-7483(89)90069-4.

21 Pattanaik, L., Pattnaik, F., Saxena, D.K., and Naik, S.N. Biofuels from agricultural wastes. In: A. Basile and F. Dalena (Eds.) *Second and Third Generation of Feedstocks: The Evolution of Biofuels.* Elsevier Science Publishing Co Inc, 2019. https://doi.org/10.1016/B978-0-12-815162-4.00005-7.

22 Toldrá-Reig, F., Mora, L., and Toldrá, F. Trends in biodiesel production from animal fatwaste. *Appl. Sci.* 2020. https://doi.org/10.3390/app10103644.

23 Lubwama, M., Yiga, V.A., and Lubwama, H.N. Effects and interactions of the agricultural waste residues and binder type on physical properties and calorific values of carbonized briquettes. *Biomass Convers. Biorefin.* 2022. https://doi.org/10.1007/s13399-020-01001-8.

24 Loulidi, I., Boukhlifi, F., Ouchabi, M., Amar, A., Jabri, M., Kali, A., et al. Adsorption of crystal violet onto an agricultural waste residue: kinetics, isotherm, thermodynamics, and mechanism of adsorption. *Sci. World J.* 2020. https://doi.org/10.1155/2020/5873521.

25 Lubwama, M., Yiga, V.A., and Lubwama, H.N. Effects and interactions of the agricultural waste residues and binder type on physical properties and calorific values of carbonized briquettes. *Biomass Convers. Biorefin.* 2020. https://doi.org/10.1007/s13399-020-01001-8.

26 Hassan, S.S., Williams, G.A., and Jaiswal, A.K. Emerging technologies for the pretreatment of lignocellulosic biomass. *Bioresour. Technol.* 2018. https://doi.org/10.1016/j.biortech.2018.04.099.

27 Singh, S., Singh, B., Mishra, B.K., Pandey, A.K., and Nain, L. Microbes in agrowaste management for sustainable agriculture. In: S. Tulasi and B. Johri (Eds.) *Microorganisms in Sustainable Agriculture and Biotechnology*. Springer, 2013. https://doi.org/10.1007/978-94-007-2214-9_8.

28 Srivastava, N., Mishra, P.K., and Upadhyay, S.N. (Editors) Pectinases: significance in the digestion of pectin-rich agrowaste. In: *Industrial Enzymes for Biofuels Production*. Elsevier, 2020. https://doi.org/10.1016/b978-0-12-821010-9.00009-7.

29 Khan, H.M., Iqbal, T., Yasin, S., Ali, C.H., Abbas, M.M., Jamil, M.A., et al. Application of agricultural waste as heterogeneous catalysts for biodiesel production. *Catalysts*. 2021. https://doi.org/10.3390/catal11101215.

30 Awogbemi, O., Von Kallon, D.V., and Aigbodion, V.S. Trends in the development and utilization of agricultural wastes as heterogeneous catalyst for biodiesel production. *J. Energy Inst.* 2021. https://doi.org/10.1016/j.joei.2021.06.017.

31 Sánchez, Ó.J. and Cardona, C.A. Trends in biotechnological production of fuel ethanol from different feedstocks. *Bioresour. Technol.* 2008. https://doi.org/10.1016/j.biortech.2007.11.013.

32 Chen, S.T., Kuo, H.I., and Chen, C.C. The relationship between GDP and electricity consumption in 10 Asian countries. *Energy Policy*. 2007. https://doi.org/10.1016/j.enpol.2006.10.001.

33 Joale, P.C., Ricardo, P.F., and Medina, T.G.I. Sizing of electric power generation systems from biogas through the generator staggering methodology. *SN Appl. Sci.* 2020. https://doi.org/10.1007/s42452-020-1968-0.

34 Gitarskiy, M.L. The refinement to the 2006 IPCC guidelines for national greenhouse gas inventories. *Fundam Appl Climatol.* 2019. https://doi.org/10.21513/0207-2564-2019-2-05-13.

35 Kushwaha, S., Soni, H., Ageetha, V., and Padmaja. P. An insight into the production, characterization, and mechanisms of action of low-cost adsorbents for removal of organics from aqueous solution. *Crit. Rev. Environ. Sci. Technol.* 2013. https://doi.org/10.1080/10643389.2011.604263.

36 Jamaldheen, S.B., Sharma, K., Rani, A., Moholkar, V.S., and Goyal, A. Comparative analysis of pretreatment methods on sorghum (*Sorghum durra*) stalk agrowaste for holocellulose content. *Prep. Biochem. Biotechnol.* 2018. https://doi.org/10.1080/10826068.2018.1466148.

37 Thite, V.S. and Nerurkar, A.S. Valorization of sugarcane bagasse by chemical pretreatment and enzyme mediated deconstruction. *Sci. Rep.* 2019. https://doi.org/10.1038/s41598-019-52347-7.

38 Kannisto, M.S., Mangayil, R.K., Shrivastava-Bhattacharya, A., Pletschke, B.I., Karp, M.T., and Santala, V.P. Metabolic engineering of *Acinetobacter baylyi* ADP1 for removal of *Clostridium butyricum* growth inhibitors produced from lignocellulosic hydrolysates. *Biotechnol. Biofuels*. 2015. https://doi.org/10.1186/s13068-015-0389-6.

39 Piasecka, A., Krzemińska, I., and Tys, J. Physical methods of microalgal biomass pretreatment. *Int. Agrophys*. 2014. https://doi.org/10.2478/intag-2014-0024.

40 Nunes, L.J.R., De Oliveira Matias, J.C., and Da Silva Catalão, J.P. (Editors) Physical pretreatment of biomass. In: *Torrefaction of Biomass for Energy Applications*. Academic Press, 2018. https://doi.org/10.1016/b978-0-12-809462-4.00002-x.

41 Behera, S., Arora, R., Nandhagopal, N., and Kumar, S. Importance of chemical pretreatment for bioconversion of lignocellulosic biomass. *Renew. Sustain. Energy Rev*. 2014. https://doi.org/10.1016/j.rser.2014.04.047.

42 Sharma, H.K., Xu, C., and Qin, W. Biological pretreatment of lignocellulosic biomass for biofuels and bioproducts: an overview. *Waste Biomass Valori*. 2019. https://doi.org/10.1007/s12649-017-0059-y.

43 Keller, F.A., Hamilton, J.E., and Nguyen, Q.A. Microbial pretreatment of biomass. *Biotechnol. Fuels Chem*. 2003. https://doi.org/10.1007/978-1-4612-0057-4_3.

44 Halder, P., Kundu, S., Patel, S., Setiawan, A., Atkin, R., Parthasarthy, R., et al. Progress on the pre-treatment of lignocellulosic biomass employing ionic liquids. *Renew. Sustain. Energy Rev*. 2019. https://doi.org/10.1016/j.rser.2019.01.052.

45 Soares, J.F., Confortin, T.C, Todero, I., Mayer, F.D., and Mazutti, M.A. Dark fermentative biohydrogen production from lignocellulosic biomass: technological challenges and future prospects. *Renew. Sustain. Energy Rev*. 2020. https://doi.org/10.1016/j.rser.2019.109484.

46 Stams, A.J.M. Metabolic interactions between anaerobic bacteria in methanogenic environments. *Antonie Van Leeuwenhoek*. 1994. https://doi.org/10.1007/BF00871644.

47 Zhao, R., Liu, Y., Zhang, Z., Guo, S., Tseng, M.L., and Wu, K.J. Enhancing eco-efficiency of agro-products' closed-loop supply chain under the belt and road initiatives: a system dynamics approach. *Sustain*. 2018. https://doi.org/10.3390/su10030668.

48 Da Silva, A.S.A., Espinheira, R.P., Teixeira, R.S.S., De Souza, M.F., Ferreira-Leitão, V., and Bon, E.P.S. Constraints and advances in high-solids enzymatic hydrolysis of lignocellulosic biomass: a critical review. *Biotechnol. Biofuels*. 2020. https://doi.org/10.1186/s13068-020-01697-w.

49 Saikia, K., Rajkumari, K., Moyon, N.S., Basumatary, S., Halder, G., Rashid, U., et al. Sulphonated biomass-based catalyst for solketal synthesis by acetalization of glycerol – a byproduct of biodiesel production. *Fuel Process. Technol*. vol. 238, pp. 107482, 2022. https://doi.org/10.1016/J.FUPROC.2022.107482.

50 Changmai, B., Sudarsanam, P., and Rokhum, L. Biodiesel production using a renewable mesoporous solid catalyst. *Ind. Crops. Prod*. vol. 145, p. 111911, 2020. https://doi.org/10.1016/j.indcrop.2019.111911.

51 Tobita, H., Komatsu, M., Harayama, H., Yazaki, K., Kitaoka, S., and Kitao, M. Effects of combined CO_2 and O_3 exposures on net CO_2 assimilation and biomass allocation in seedlings of the late-successional Fagus crenata. *Climate*. 2019. https://doi.org/10.3390/cli7100117.

52 Bala, S., Garg, D., Sridhar, K., Inbaraj, B.S, Singh, R., Kamma, S., et al. Transformation of agro-waste into value-added bioproducts and bioactive

compounds: micro/nano formulations and application in the agri-food-pharma sector. *Bioengineering*. 2023. https://doi.org/10.3390/bioengineering10020152.
53 Soni, S.K., Sharma, A., and Soni, R. Microbial enzyme systems in the production of second generation bioethanol. *Sustain*. 2023. https://doi.org/10.3390/su15043590.
54 Agrawal, P.S, Belkhode, P.N., and Rokhum, S.L. Microalgae –biofuel production trends. In: *Sustainability in Biofuel Production Technology*. John Wiley & Sons, Ltd, 2022. https://doi.org/10.1002/9781119888864.ch6.
55 Tiwari, R., Agrawal, P.S., Belkhode, P.N., Brijpuriya, D., Devase, A., and Samarth, P. Utilization of waste transformer oil for the synthesis of biodiesel and glyptal. In: *Lecture Notes in Civil Engineering*. 2024. https://doi.org/10.1007/978-981-99-4186-5_7.
56 Islam, S., Basumatary, B., Rokhum, S.L., Mochahari, P.K., and Basumatary, S. Advancement in utilization of nanomaterials as efficient and recyclable solid catalyst for biodiesel synthesis. *Clean. Chem. Eng*. 2022. https://doi.org/10.1016/j.clce.2022.100043.
57 Dutta, N., Usman, M., Luo, G., and Zhang, S. An insight into valorization of lignocellulosic biomass by optimization with the combination of hydrothermal (HT) and biological techniques: a review. *Sustain. Chem*. 2022. https://doi.org/10.3390/suschem3010003.
58 Ochando-Pulido, J.M., Vuppala, S., García-López, A.I., and Martínez-Férez, A. A focus on anaerobic digestion and co-digestion strategies for energy recovery and digestate valorization from olive-oil mill solid and liquid by-products. *Sep. Purif. Technol*. 2024. https://doi.org/10.1016/j.seppur.2023.125827.
59 Cisneros de la Cueva, S., Veana Hernández, F., Arjona López, M.A., Alvarez Guzmán, C.L., and Pérez Vega, S.B. Optimization of variables from the anaerobic digestion process of cheese whey in biogas production. *Rev. Int. Contam. Ambient*. 2021. https://doi.org/10.20937/RICA.53879.
60 Suzuki, S., Shibata, Y., Hirose, D., Endo, T., Ninomiya, K., Kakuchi, R., et al. Cellulose triacetate synthesis via one-pot organocatalytic transesterification and delignification of pretreated bagasse. *RSC Adv*. 2018. https://doi.org/10.1039/c8ra03859g.
61 Agrawal, P. and Rokhum, S.L. Supported metal/metal oxide catalysts in biodiesel production: an overview. In: *Biodiesel Production: Feedstocks, Catalysts, and Technologies*. John Wiley & Sons, Ltd, 2022. https://doi.org/10.1002/9781119771364.ch7.
62 Das, A., Li, H., Kataki, R., Agrawal, P.S., Moyon, N.S., Gurunathan, B., et al. Terminalia arjuna bark – a highly efficient renewable heterogeneous base catalyst for biodiesel production. *Renew. Energy*. 2023. https://doi.org/10.1016/j.renene.2023.05.066.
63 Leung, D.Y.C., Wu, X., and Leung, M.K.H. A review on biodiesel production using catalyzed transesterification. *Appl. Energy*. 2010. https://doi.org/10.1016/j.apenergy.2009.10.006.
64 Atinkut, H.B., Yan, T., Arega, Y., and Raza, M.H. Farmers' willingness-to-pay for eco-friendly agricultural waste management in Ethiopia: a contingent valuation. *J. Clean. Prod*. 2020. https://doi.org/10.1016/j.jclepro.2020.121211.

65 Rajkumari, K., Das, D., Pathak, G., and Rokhum, L. Waste-to-useful: a biowaste-derived heterogeneous catalyst for a green and sustainable Henry reaction. *New J. Chem.* 2019. https://doi.org/10.1039/c8nj05029e.

66 Rao, P. and Rathod, V. Valorization of food and agricultural waste: a step towards greener future. *Chem. Rec.* 2019. https://doi.org/10.1002/tcr.201800094.

67 Yamakawa, C.K., Qin, F., and Mussatto, S.I. Advances and opportunities in biomass conversion technologies and biorefineries for the development of a bio-based economy. *Biomass Bioenerg.* 2018. https://doi.org/10.1016/j.biombioe.2018.09.007.

68 Mata-Alvarez, J., Dosta, J., Romero-Güiza, M.S., Fonoll, X., Peces, M., and Astals, S. A critical review on anaerobic co-digestion achievements between 2010 and 2013. *Renew. Sustain. Energy Rev.* 2014. https://doi.org/10.1016/j.rser.2014.04.039.

69 De la Rubia, M.A., Villamil, J.A., Rodriguez, J.J., Borja, R., and Mohedano, A.F. Mesophilic anaerobic co-digestion of the organic fraction of municipal solid waste with the liquid fraction from hydrothermal carbonization of sewage sludge. *Waste Manag.* 2018. https://doi.org/10.1016/j.wasman.2018.02.046.

70 Mao, C., Feng, Y., Wang, X., and Ren, G. Review on research achievements of biogas from anaerobic digestion. *Renew. Sustain. Energy Rev.* 2015. https://doi.org/10.1016/j.rser.2015.02.032.

71 Ngan, N.V.C., Chan, F.M.S., Nam, T.S., Van Thao, H., Maguyon-Detras, M.C., Hung, D.V., et al. Anaerobic digestion of rice straw for biogas production. In: M. Gummert, N. Hung, P. Chivenge, B. Douthwaite (Eds.) *Sustainable Rice Straw Management*. Springer, 2019. https://doi.org/10.1007/978-3-030-32373-8_5.

72 Sutaryo, S., Ward, A.J., and Møller, H.B. Ammonia inhibition in thermophilic anaerobic digestion of dairy cattle manure. *J. Indones. Trop. Anim. Agric.* 2014. https://doi.org/10.14710/jitaa.39.2.83-90.

73 Zahedi, S., Dahunsi, S.O., Perez, M., and Solera, R. Assessment of chemical inhibitor addition to improve the gas production from biowaste. *Waste Biomass Valori.* 2019. https://doi.org/10.1007/s12649-017-0189-2.

74 Yan, P., Gai, M., Wang, Y., and Gao, X. Review of soft sensors in anaerobic digestion process. *Processes.* 2021. https://doi.org/10.3390/pr9081434.

75 Holm-Nielsen, J.B., Al Seadi, T., and Oleskowicz-Popiel, P. The future of anaerobic digestion and biogas utilization. *Bioresour. Technol.* 2009. https://doi.org/10.1016/j.biortech.2008.12.046.

76 Zhang, F., Wang, J., Liu, S., Zhang, S., and Sutherland, J.W. Integrating GIS with optimization method for a biofuel feedstock supply chain. *Biomass Bioenergy.* 2017. https://doi.org/10.1016/j.biombioe.2017.01.004.

77 Goedhart, P.W., Mosquera, J., and Huijsmans, J.F.M. Estimating ammonia emission after field application of manure by the integrated horizontal flux method: a comparison of concentration and wind speed profiles. *Soil Use Manag.* 2020. https://doi.org/10.1111/sum.12564.

78 Rizwan, M.A., Bhagat, M., Singh, S., Arisutha, S., Suresh, S., Verma, S., et al. Agricultural waste: an exploration of the innovative possibilities in the pursuit of long-term sustainability. In: S. Verma, R. Khan, M. Mili, S.A.R. Hashmi, and A.K. Srivastava (Eds.) *Advanced Materials from Recycled Waste*. Elsevier, 2022. https://doi.org/10.1016/B978-0-323-85604-1.00001-9.

79 Machado, L.T.S., Leroy, K.A., Oliveira, K.S.M., and Sánchez Román, R.M. Characterization of residential organic waste digested and viability of use in irrigated agriculture. *Irriga*. 2023. https://doi.org/10.15809/irriga.2023v28n1p48-59.

80 Doyeni, M.O., Barcauskaite, K., Buneviciene, K., Venslauskas, K., Navickas, K., Rubezius M., et al. Nitrogen flow in livestock waste system towards an efficient circular economy in agriculture. *Waste Manag. Res.* 2023. https://doi.org/10.1177/0734242X221123484.

81 Matjuda, D.S., Tekere, M., and Thaela-Chimuka, M.J. Characterization of the physicochemical composition of anaerobically digested (digestate) high throughput red meat abattoir waste in South Africa and the determination of its quality as a potential biofertilizer. *Heliyon*. 2023. https://doi.org/10.1016/j.heliyon.2023.e21647.

82 Möller, K. and Müller, T. Effects of anaerobic digestion on digestate nutrient availability and crop growth: a review. *Eng. Life Sci.* 2012. https://doi.org/10.1002/elsc.201100085.

12

Digitalization for Agro-waste Management

12.1 Introduction

The concept of digitalization has significantly advanced agricultural waste management processes. Utilizing digital tools facilitates improved monitoring, communication, and decision-making across various stages. Agricultural waste, including crop residues, livestock manure, and processing byproducts, poses significant environmental sustainability and economic viability challenges. Traditional waste management approaches have often proven inadequate in addressing the complexities associated with dispersed waste sources, fluctuating volumes, and the inherent variability of organic materials. In this regard, the digital infrastructure presents a compelling opportunity to transform agricultural waste management from a reactive approach into a proactive, data-driven, and streamlined system. Recent advancements in connectivity and computational power pave the way for innovative solutions that optimize resource utilization, minimize environmental impacts, and potentially unlock economic value from agricultural waste. A key transformation lies in the control offered by digital infrastructure, enabling comprehensive monitoring across all facets of the waste stream.

Digital technologies serve as tools to control and enhance various waste management techniques and practices. Digitalization, as defined by [1, 2], refers to the application of digital technologies to improve organizational processes. By enabling the collection of valuable resources from waste streams, these technologies help reduce dependency on virgin materials and mitigate associated environmental and climate impacts. Key digital technologies involved include data recognition, blockchain, Internet of Things (IoT) applications, and artificial intelligence (AI) [3, 4]. Given the considerable volume of waste, the implementation of an intelligent waste management system, utilizing these technologies, offers an optimal solution [3]. Smart waste management addresses these issues through the use of sensors, intelligent monitoring systems, and mobile applications [3, 4]. The effective utilization of digital technologies is crucial for transitioning waste management toward a more sustainable approach to material management [5]. Digital technologies enhance the recycling process, facilitate the utilization of recycled materials

Agro-Waste Management and Valorization, First Edition. Pratibha S. Agrawal, et al.

by manufacturers, increase consumer awareness regarding sustainable purchasing and sorting, and broaden the options for recyclers sourcing waste materials [5].

Emerging techniques like waste e-commerce platforms, waste-specific software, and business analytics are becoming increasingly important across the waste collection sector, with the potential to transform how waste is managed. Digital technologies have transformed waste collection, encompassing tasks like organizing, planning, dispatching personnel, and managing vehicles [6]. Utilizing digital tools can enhance these processes by facilitating the storage, processing, analysis, and optimization of necessary information [6]. Information generated during collection can be monitored to identify irregularities, surpassing the limitations of previous manual documentation, communication, and billing methods. Similar to other sectors, the shift from paper-based to digital management systems improves process efficiency and information flow. Digital identification tags on smart bins and containers enable digital order processing, billing, and payment, and support the development of user interfaces for consumer communication. Data gathered from these processes provides valuable insights, facilitating advancements in the circular economy (CE) by revealing temporal patterns of waste production.

Waste encompasses various types, including construction and demolition materials, hazardous waste, industrial sludge, and municipal solid waste like food waste, paper, cardboard, plastic, and textiles [7–9]. Managing this diverse waste presents growing challenges globally [10]. Digitalization is fundamentally important for controlling waste management processes and aligning waste management plans with the development of a CE [11]. The CE is an economic model focused on eliminating waste and maximizing resource utilization [12]. It advocates for regenerating products and materials at the end of their lifecycle, creating a closed-loop system that minimizes waste and conserves resources. Key principles include designing for durability and recyclability, promoting renewable energy sources, and fostering systems thinking in resource management [12].

The agricultural-based economy transformed during the Industrial Revolution into one dominated by large-scale industry, mechanized manufacturing, and the factory system. Elements of the third industrial revolution included resource scarcity, new reserves, IoT and system integration, renewable energy, oil scarcity, robotics, and the CE. The CE poses a significant challenge across all industrial sectors. It is considered a primary approach to address resource scarcity and ensure efficient use of critical raw materials [13]. This concept centers on the current and potential circularities of materials before they become waste. The CE model promotes the production and utilization of materials and products, aiming to allocate, reuse, repair, and recycle them for as long as possible. This extends product life and minimizes waste. It protects the environment by slowing the consumption of natural resources, reducing landscape and habitat disruption, and helping to limit biodiversity loss. It decreases reliance on virgin materials, creates jobs, and saves consumers' money. The circular flow model illustrates how money facilitates the transformation of production factors into goods and services, which are then exchanged with consumers, enabling firms to continue production, increase output, and generate profit.

12.2 Infrastructure for Agro-waste Management

The synergy of emerging technologies, increased connectivity, and computational power has enabled innovative solutions to optimize resource utilization, minimize environmental impacts, and unlock potential economic value from agricultural waste [14]. This transformation is founded on the capability of digital infrastructure to facilitate comprehensive monitoring of the waste stream. This review critically examines the role of digital infrastructure in enhancing agricultural waste management processes, focusing on four key areas: (i) improved resource allocation using geospatial tools, (ii) enhanced monitoring of waste generation and collection via IoT sensors, (iii) increased efficiency in waste transportation through smart logistics, and (iv) better decision-making enabled by real-time data and predictive analytics [15]. Each area is explored in detail, considering relevant technologies, potential benefits, existing challenges, and future research opportunities. Investing in robust waste management infrastructure yields significant environmental, health, economic, and social benefits. These benefits are as follows:

12.2.1 Environmental Benefits

The implementation of robust and effective waste management infrastructure is a critical prerequisite for minimizing the environmental footprint of anthropogenic waste generation and transitioning toward more sustainable resource management practices. At a fundamental level, well-designed infrastructure facilitates the systematic collection, transportation, and managed disposal of waste streams, fundamentally altering practices from environmentally detrimental methods like open burning or illegal dumping, which directly release pollutants into the atmosphere, soil, and water bodies [16]. Managed disposal, particularly in properly engineered landfills, incorporates critical environmental controls such as leachate collection systems to prevent groundwater contamination and landfill gas (primarily methane [CH_4] and carbon dioxide [CO_2]) capture systems, which significantly reduce the emission of potent greenhouse gases compared to unmanaged sites. Captured CH_4 can often be utilized for energy recovery, further enhancing the environmental performance [2]. Beyond basic disposal, effective infrastructure is essential for enabling higher levels of the waste hierarchy. It supports the operation of sorting and transfer stations that prepare waste for recycling, composting facilities that transform organic waste into valuable soil amendments, and processing plants (e.g., anaerobic digesters, material recovery facilities) that recover energy and materials [3]. By diverting biodegradable waste from landfills to composting or anaerobic digestion, infrastructure helps mitigate CH_4 generation and promotes nutrient cycling [4]. Recycling infrastructure reduces the demand for virgin raw materials, conserving natural resources and reducing the energy and environmental impacts associated with their extraction and processing [4]. Ultimately, comprehensive waste management infrastructure provides the necessary physical and logistical framework to support a shift from a linear "take-make-dispose" model to a CE, protecting ecosystems from pollution, conserving biodiversity, and contributing to climate change mitigation by reducing greenhouse gas emissions from waste sources and displacing resource-intensive production processes [6, 10]. The presence and quality of this infrastructure are thus direct determinants of a region's environmental sustainability performance in waste management.

12.2.2 Health Benefits

Inadequate or non-existent waste management infrastructure poses severe and pervasive risks to public health, particularly in densely populated or underserved areas [11]. Accumulation of uncollected waste in residential areas serves as breeding grounds for vectors such as rodents, insects (flies, mosquitoes), and other pests that transmit diseases [12]. Exposure to decomposing waste can lead to respiratory illnesses, infections, and skin problems due to the presence of pathogens, hazardous chemicals, and bioaerosols [13]. Uncontrolled burning of waste, often practiced where formal collection is lacking, releases a complex mixture of toxic air pollutants, including particulate matter (PM), dioxins, furans, carbon monoxide (CO), and volatile organic compounds (VOCs), which are linked to a range of health issues including respiratory diseases, cardiovascular problems, and certain cancers [14]. Improper disposal of hazardous or medical waste alongside general solid waste in unmanaged dumpsites creates risks of exposure to dangerous substances, contaminating soil and water sources used for drinking or agriculture [15]. Effective waste management infrastructure directly mitigates these health hazards by ensuring the regular and timely collection of waste, removing it from living environments before it attracts vectors or decomposes extensively [17]. Containment within engineered landfills or controlled processing facilities (composting, digestion, incineration with proper emission controls) prevents direct exposure to waste, reduces the spread of pathogens, and minimizes the release of harmful pollutants into the environment [18]. Providing access to safe disposal options reduces the incidence of dangerous informal practices like open burning or scavenging [19]. The presence of functional waste management infrastructure is therefore fundamental to public health, serving as a critical component of sanitation systems that prevent disease transmission, reduce exposure to harmful substances, and improve the overall health and well-being of communities [20]. The lack of equitable access to such infrastructure is often a significant contributor to health disparities in low-income areas and developing regions [21].

12.2.3 Economic Benefits

Investing in and maintaining effective waste management infrastructure yields significant economic benefits, extending beyond simple cost avoidance in disposal to encompass value creation and job generation within a more CE framework [22]. A direct economic advantage is the reduction in the volume of waste sent to landfills. As landfill space becomes increasingly scarce and subject to stricter environmental regulations, the costs associated with landfilling (tipping fees, transportation to potentially distant sites, closure, and post-closure care) continue to rise [23]. Infrastructure that supports waste reduction, reuse, recycling, and energy recovery directly diverts waste from landfills, resulting in substantial savings on these disposal costs [24]. Furthermore, reducing reliance on landfilling lessens the need to site and develop new landfill facilities, which is often a costly and socially contentious process involving land acquisition, engineering, and permitting expenses [25]. Beyond cost avoidance, effective infrastructure facilitates the recovery and valorization of materials and energy embedded in waste. Recycling infrastructure enables the collection and processing of valuable secondary raw materials (e.g., paper, plastics, metals, glass) that can be sold back into manufacturing supply chains, creating revenue streams and reducing industry's dependence on more expensive or volatile virgin materials markets [26]. Organic waste infrastructure (composting, anaerobic digestion) produces compost, digestate, and biogas, which can be sold as fertilizers, soil amendments, or renewable energy, generating additional revenue [26]. Energy-from-waste facilities, supported by controlled waste streams, can produce electricity or heat, providing an alternative energy source with associated economic benefits [27]. Moreover, the development and operation of comprehensive waste management infrastructure stimulate economic activity, creating jobs in collection, sorting, processing, maintenance, and administration [28]. The transition to a CE model, underpinned by this infrastructure, fosters innovation and can lead to the development of new green businesses and industries focused on repair, remanufacturing, and value-added processing of secondary resources, contributing to overall economic growth and resilience [29].

12.2.4 Social Benefits

The presence of reliable and effective waste management infrastructure contributes significantly to the social well-being and quality of life within communities, impacting aspects from daily living conditions to civic pride and equity [30]. A fundamental social benefit is the improvement of living environments. Regular waste collection and proper disposal eliminate unsightly waste accumulations, reduce unpleasant odors, and create cleaner, more aesthetically pleasing surroundings [10, 31]. These improvements directly enhance the quality of daily life, fostering a sense of dignity and comfort in residential areas. Clean environments contribute to community pride and can encourage civic engagement and collective action for further local improvements [32]. Waste management infrastructure is also intrinsically linked to social equity. Ensuring that all residents, regardless of socio-economic status or geographic location, have equitable access to reliable waste collection and safe disposal options is crucial for promoting social justice [33]. Lack of infrastructure disproportionately affects marginalized communities, often leading to the proliferation of informal dumpsites and associated environmental and health hazards in these areas [21]. Providing equitable service is a key component of inclusive urban and rural development [33]. Furthermore, well-managed waste systems can contribute to social inclusion by formalizing the roles of informal waste pickers or recyclers, providing them with safer working conditions, fair compensation, and access to social services [34]. Engaging the community in waste management initiatives, such as source separation programs or local composting efforts, can build social capital, enhance environmental awareness, and foster a sense of shared responsibility [35]. By creating cleaner, healthier,

Table 12.1 Identification of wastes.

Wastes/problems	Category of the problem
Coal seam height thin and thick difficult to excavate; coal cannot be left unmined	C
Coal spillage and rehandling	A
Overall equipment effectiveness (OEE)	B
Pumping system in underground mine	B
Bottleneck (underground bunker) of coal transportation system	C
Proper tool bit profile during re-sharpening process	A
Human energy loss due to heavy working conditions	C

and more equitable living environments, supported by local employment opportunities and community engagement, effective waste management infrastructure underpins the social dimension of sustainable development, contributing to safer, more livable, and cohesive communities [36].

Collectively, the benefits of proper waste management infrastructure are evident. Investment in these systems allows communities to protect the environment, enhance public health, support economic growth, and build resilience. This research highlights both the potential benefits and associated challenges [13]. A critical review of potential problems leads to a categorization based on team capability, management, and external involvement, as discussed in Table 12.1:

- "A" category problem – Requires minimal involvement from other departments.
- "B" category problem – Requires involvement from other departments.
- "C" category problem – Requires management sanction for implementation.

This study approaches the topic from multiple perspectives, including business management, environmental science, and industrial production. This interdisciplinary approach fills a gap in existing knowledge. The study examines the future of the CE within the context of business management, focusing specifically on recycling and waste management. It emphasizes the four key areas shown in Figure 12.1.

12.3 Improved Resource Allocation with Geospatial Tools

Geospatial technologies, such as Geographic Information Systems (GIS) and remote sensing, provide powerful capabilities for visualizing, analyzing, and managing spatial data related to agricultural waste [18]. These tools enable a comprehensive understanding of waste generation patterns, distribution, and proximity to potential processing or disposal sites. Integrating geospatial data with other relevant information (e.g., soil types, water resources, land use) allows for optimizing

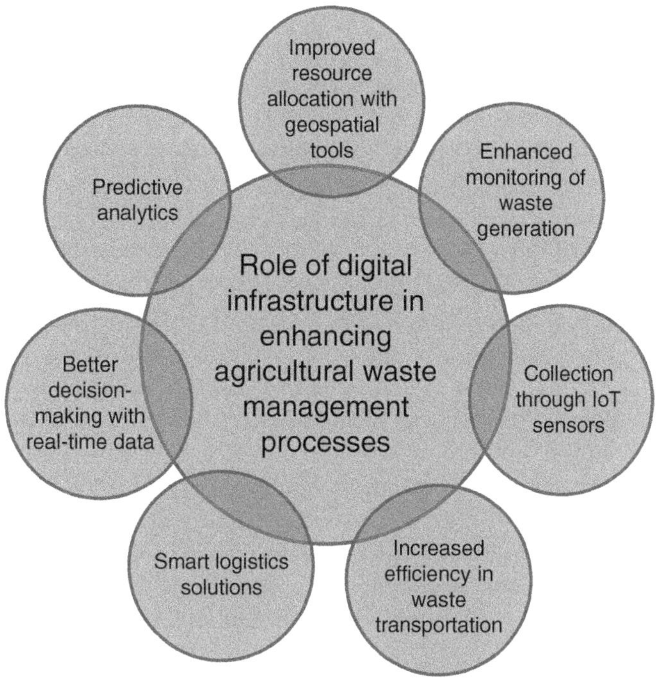

Figure 12.1 Role of digital infrastructure.

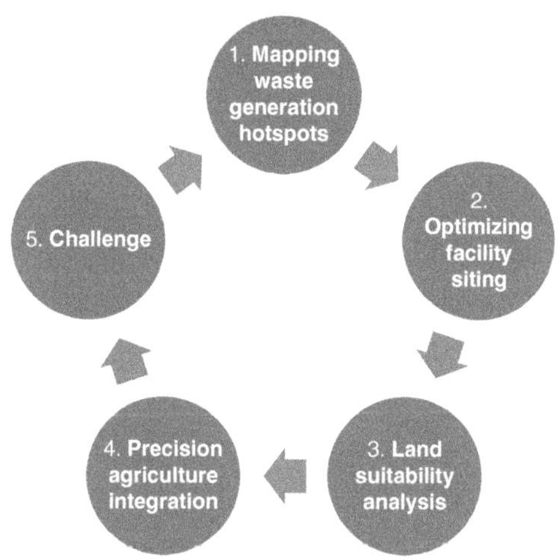

Figure 12.2 Sustainable waste management strategies.

resource allocation and developing more sustainable waste management strategies, as depicted in Figure 12.2.

- **Mapping Waste Generation Hotspots:** GIS can create detailed maps of agricultural regions by overlaying data on crop yields, livestock densities, and processing plant locations. These maps help identify areas with high waste generation potential ("hotspots"), enabling targeted interventions and resource allocation. For instance, areas with high concentrations of livestock manure can be prioritized for biogas production or composting.

- **Optimizing Facility Siting:** The optimal placement of composting facilities, anaerobic digesters, or other waste-processing plants is crucial for minimizing transportation costs and environmental impacts. GIS can evaluate potential sites based on factors like proximity to waste sources, transportation network accessibility, environmental sensitivity, and regulatory constraints. Multi-criteria decision analysis (MCDA) techniques can be integrated within GIS to rank sites based on weighted criteria.
- **Land Suitability Analysis:** For land application of agricultural waste (e.g., manure or compost), GIS can assess land suitability based on soil properties, slope, drainage, and proximity to water bodies. This analysis ensures environmentally sound land application practices and minimizes risks of nutrient runoff or groundwater contamination. Remote sensing data, such as satellite imagery, can monitor vegetation health and evaluate land application effectiveness.
- **Precision Agriculture Integration:** The principles of precision agriculture, which involve using data and technology to optimize agricultural inputs and outputs, can extend to waste management. Geospatial data on soil nutrient levels, crop yields, and environmental conditions can tailor manure application rates to meet crop needs while minimizing nutrient losses.

A significant challenge is the availability and quality of data (Figure 12.2). Reliable data on agricultural waste generation and distribution is often scarce or incomplete in many regions. Furthermore, accessing high-resolution remote sensing data can be costly. Developing robust data collection and sharing mechanisms is essential for realizing the full potential of geospatial technologies in this domain.

12.4 Enhanced Monitoring of Waste Generation and Collection through IoT Sensors

The IoT refers to a network of interconnected devices capable of collecting and exchanging data [19]. In the context of agricultural waste management, IoT sensors can be deployed to monitor waste generation rates, storage conditions, and collection logistics. Real-time data from these sensors enables optimization of waste management operations, detection of anomalies, and improved overall efficiency, as illustrated in Figure 12.3.

- **Waste Bin Monitoring:** Sensors installed in waste bins or containers can monitor fill levels, temperature, and moisture content. This information helps optimize collection schedules, reduce unnecessary trips, and prevent overflowing.
- **Manure Storage Monitoring:** IoT sensors can monitor manure storage facilities for temperature, gas emissions (e.g., ammonia, CH_4), and structural integrity. This data helps prevent leaks, reduce air pollution, and improve worker safety.
- **Livestock Monitoring for Manure Prediction:** Wearable or embedded sensors on livestock can track animal health, activity patterns, and feed intake.

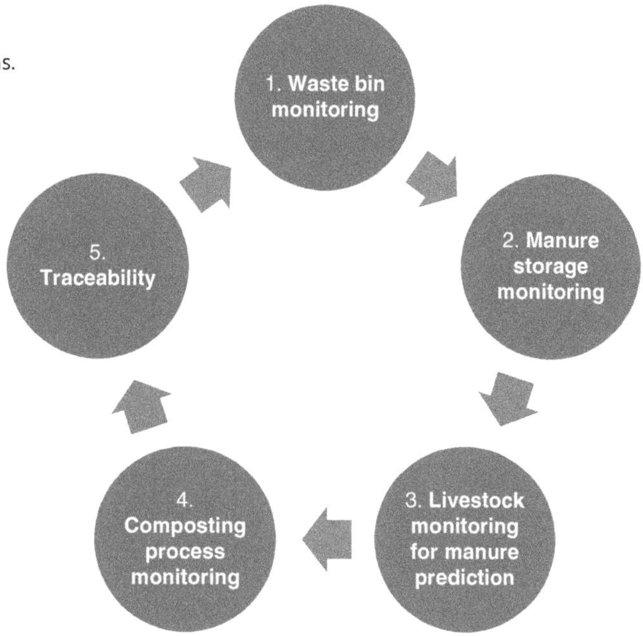

Figure 12.3 Network to optimize waste management operations.

This data can be correlated with manure production rates, enabling more accurate forecasting.

- **Composting Process Monitoring:** IoT sensors strategically placed within compost piles can monitor temperature, moisture, and oxygen levels. This data can optimize composting conditions, accelerating decomposition and improving final compost quality.
- **Traceability:** Attaching sensors or RFID tags to waste containers/pallets from the source allows waste to be tracked throughout the process, helping determine its origin, transport duration, and handling.

The widespread deployment of IoT sensors involves significant investment in hardware, software, and communication infrastructure, making it potentially expensive. Managing the high volume of generated data is a key issue. Data security and privacy are also major concerns, as IoT sensors can collect sensitive information about agricultural operations. Addressing these challenges and ensuring data interoperability are critical for successful IoT adoption in agricultural waste management. The development of low-cost sensors is also needed [20].

Effective agricultural waste management relies on robust data collection systems that can characterize waste generation, composition, and location with high spatio-temporal resolution [28]. Recent advances in sensor technologies, remote sensing platforms, and data analytics offer unprecedented opportunities to optimize strategies, shifting from reactive disposal practices toward proactive resource recovery. This review critically examines the current landscape of data collection systems in agricultural waste management, focusing on capabilities, limitations, and integration potential. IoT sensors, strategically deployed across agricultural fields, storage facilities, and processing units, provide real-time monitoring of critical parameters affecting waste degradation and potential hazards [28]. These sensors utilize diverse sensing modalities, including temperature, humidity, gas composition (e.g., CH_4, ammonia), and microbial activity, enabling early detection of spoilage, contamination, or anaerobic decomposition. The granularity of data acquired by IoT sensors facilitates precise management interventions, such as adjusting composting parameters, optimizing aeration, or triggering timely collection schedules to prevent environmental risks. Unmanned Aerial Vehicles (UAVs), equipped with multispectral, hyperspectral, or thermal cameras, offer a complementary approach. They enable rapid assessment of waste quantity, spatial distribution, and composition across expansive agricultural landscapes. High-resolution imagery from UAVs allows for accurate mapping of crop residue accumulation, identification of concentrated waste generation areas, and monitoring of composting site performance.

Agricultural waste, a pervasive byproduct of modern farming, presents both a significant environmental challenge and an opportunity for the CE. Efficient and cost-effective transportation is critical for minimizing negative impacts (e.g., greenhouse gas emissions, soil contamination, pathogen spread) while maximizing resource recovery potential (e.g., composting, anaerobic digestion). Traditional methods often suffer from inefficiencies like suboptimal routing, underutilized fleets, and reactive scheduling, leading to increased costs and environmental burdens. Smart logistics solutions, integrating real-time data, advanced sensors, sophisticated algorithms, and networked communication, offer a promising path to revolutionize agricultural waste transportation [21]. This section reviews the state-of-the-art, discussing underlying principles, practical challenges, and future potential.

12.5 The Imperative of Efficient Agricultural Waste Transportation

Agricultural waste encompasses a diverse range of materials, including crop residues (stalks, leaves, husks), animal manure, processing byproducts (fruit peels, vegetable trimmings, dairy whey), and discarded packaging. Its composition and quantity vary significantly based on agricultural activity, location, and season. Inefficient management can have profound environmental consequences. Uncontrolled decomposition of organic waste releases CH_4, a potent greenhouse gas with a global warming potential far exceeding that of CO_2 [18]. Runoff from improperly stored manure can contaminate water resources with nitrates, phosphates, and pathogens, posing health and ecological risks. Accumulated waste can also create breeding grounds for pests and vectors, spreading diseases and economic losses.

Efficient transportation is crucial for mitigating these negative impacts by facilitating timely collection, processing, and disposal. It minimizes waste residence time in the field, reducing emissions and preventing pollutant leaching. It also enables waste concentration at processing facilities for conversion into valuable resources like biogas, compost, or biofertilizers. Furthermore, efficient transportation reduces overall management costs, making sustainable practices more economically viable for farmers and waste management companies.

12.6 Smart Logistics: A Paradigm Shift in Waste Transportation

Smart logistics solutions leverage information and communication technologies (ICT) to optimize the flow of goods and materials [21]. In agricultural waste transportation, this involves integrating various technologies, as shown in Figure 12.4:

- **Real-time Data Acquisition:** The foundation of a successful smart logistics system is the ability to collect real-time data on waste location, quantity, composition, vehicle status, and infrastructure. This data can be acquired through various sensors, including Global Positioning System (GPS) trackers, Radio Frequency Identification (RFID) tags, load cells, and chemical sensors.
- **Communication Networks:** Collected data must be transmitted reliably and securely to a central processing unit for analysis and informed decision-making. Networks can include cellular, satellite, and wireless local area networks (WLANs).
- **Data Analytics and Optimization Algorithms:** Sophisticated algorithms analyze collected data to optimize aspects of waste transportation, such as route planning, scheduling, and fleet management. These algorithms can consider factors like traffic conditions, weather forecasts, waste generation rates, and processing capacity.
- **Decision Support Systems:** The results of data analysis and optimization algorithms are presented to decision-makers via user-friendly interfaces. These systems enable informed decisions about transportation operations and can include web-based dashboards, mobile apps, and automated alerts.

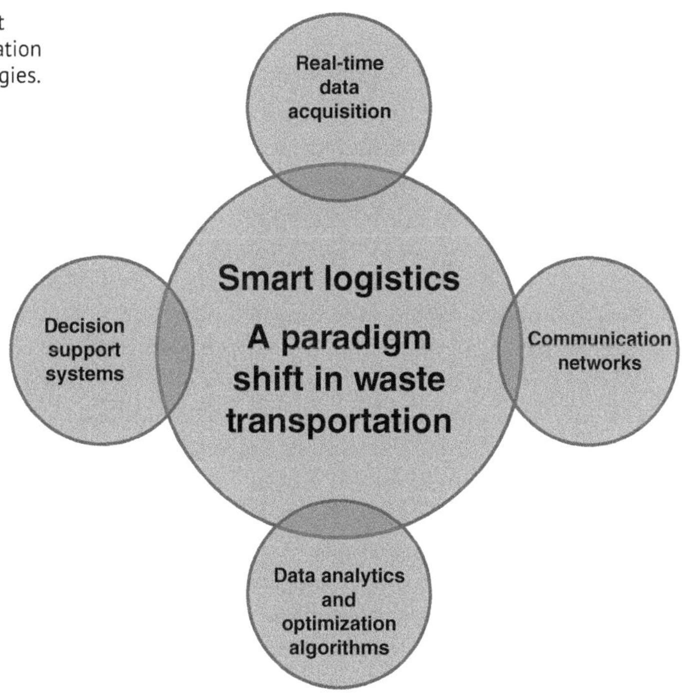

Figure 12.4 Smart logistics the integration of various technologies.

Applying smart logistics solutions to agricultural waste transportation can significantly enhance efficiency, cost-effectiveness, and environmental sustainability [28]. Optimizing routes, scheduling, and fleet management can reduce travel time, fuel consumption, and greenhouse gas emissions. It can also improve vehicle utilization, potentially reducing the need for additional vehicles and infrastructure. Furthermore, smart logistics can enable dynamic adjustments of collection schedules based on real-time data, ensuring waste is collected promptly and efficiently.

12.6.1 Components of Smart Logistics Solutions for Waste Transportation

1. **Route Optimization:** Route optimization is a critical component of smart logistics for waste transportation [6]. Its goal is to identify the most efficient routes for vehicles collecting and transporting waste from multiple sources to processing or disposal facilities. Traditional route planning often relies on static data (road maps, historical traffic) that may not accurately reflect real-time conditions. Smart logistics, however, uses real-time data on traffic congestion, weather, road closures, and vehicle locations to dynamically optimize routes.
   - **Scientific Principles:** Route optimization algorithms are based on graph theory, a branch of mathematics studying networks [22]. The road network is modeled as a graph where nodes represent intersections and edges represent road segments. Algorithms seek the shortest path between nodes, considering edge length and constraints like traffic or closures. Common algorithms include Dijkstra's, A* search, and genetic algorithms.
   - **Technological Implementation:** Real-time traffic data is typically obtained from GPS devices in vehicles or third-party providers. Weather data comes from weather stations or online services. Road closure information is sourced from traffic management agencies or online sources. Optimization algorithms continuously monitor these data sources to dynamically adjust routes, minimizing travel time and fuel consumption.
   - **Benefits:** Route optimization can significantly reduce travel time, fuel consumption, and greenhouse gas emissions [15]. Studies have shown potential reductions of up to 20% in fuel consumption and 30% in travel time. It can also enhance driver safety by helping avoid congested and hazardous conditions.
2. **Fleet Management:** Fleet management involves monitoring and controlling vehicles to ensure efficient and safe operation. Traditional methods rely on manual tracking, which can be time-consuming and inaccurate. Smart logistics uses GPS tracking and telematics to monitor vehicle location, speed, fuel consumption, and other parameters in real-time.
   - **Scientific Principles:** GPS tracking is based on triangulation, determining vehicle location by measuring distance to at least three GPS satellites. Telematics systems use sensors to measure parameters like engine performance, brake usage, and tire pressure. This data is wirelessly transmitted to a central monitoring system for analysis and problem identification.
   - **Technological Implementation:** GPS trackers are small electronic devices installed in vehicles, transmitting location data via cellular or satellite networks to a central server. Telematics systems typically consist of sensors, a

data logger, and a communication module that collects sensor data and transmits it wirelessly.

- **Benefits:** Fleet management can improve driver performance, prevent unauthorized vehicle use, and optimize maintenance schedules [3]. Monitoring driver behavior can identify training needs. Tracking location helps prevent unauthorized use and theft. Monitoring vehicle performance can identify potential maintenance issues proactively, reducing downtime and repair costs.

3. **Dynamic Scheduling:** Dynamic scheduling involves adjusting waste collection schedules in real-time based on data from IoT sensors and other sources [21]. Traditional schedules are often fixed. Smart logistics, however, uses data from waste bin sensors to monitor fill levels and dynamically adjust collection schedules.
   - **Scientific Principles:** IoT sensors utilize various technologies (ultrasonic, infrared, load cells) to measure waste bin fill levels [3]. This data is transmitted wirelessly to a central monitoring system for analysis to determine if a bin needs emptying. Scheduling algorithms incorporate bin fill levels, vehicle locations, and other factors to optimize collection schedules.
   - **Technological Implementation:** IoT sensors are small, low-power devices easily installed in bins. They transmit data via cellular or Long Range Wide Area Network (LoRaWAN) networks [2]. Scheduling algorithms are typically implemented in software and can be integrated with existing management systems.
   - **Benefits:** Dynamic scheduling can reduce unnecessary trips, saving fuel and reducing greenhouse gas emissions. It can also improve vehicle utilization, potentially reducing the need for additional vehicles. Furthermore, it ensures waste is collected promptly and efficiently, preventing overflow and reducing environmental contamination risk.
4. **Transportation Resource Sharing:** Transportation resource sharing involves pooling transport resources among multiple farms or waste generators to reduce costs and emissions. Traditional waste transportation often involves individual generators contracting separately. Smart logistics solutions can facilitate sharing by providing a platform for connecting farms/generators and coordinating collection.
   - **Scientific Principles:** Transportation resource sharing is based on economies of scale, where the cost per unit of waste transported decreases as volume increases. Pooling resources allows farms and generators to achieve economies of scale and reduce costs.
   - **Technological Implementation:** A sharing platform can be implemented as a web or mobile application. It allows farms/generators to register waste generation schedules and needs and waste management companies to register available capacity. The platform then matches generators with companies based on needs and availability.
   - **Benefits:** Transportation resource sharing can reduce transportation costs for participants. It can also improve vehicle utilization, reducing the need for additional vehicles. Furthermore, it can reduce greenhouse gas emissions by consolidating collection activities and decreasing the number of required trips.

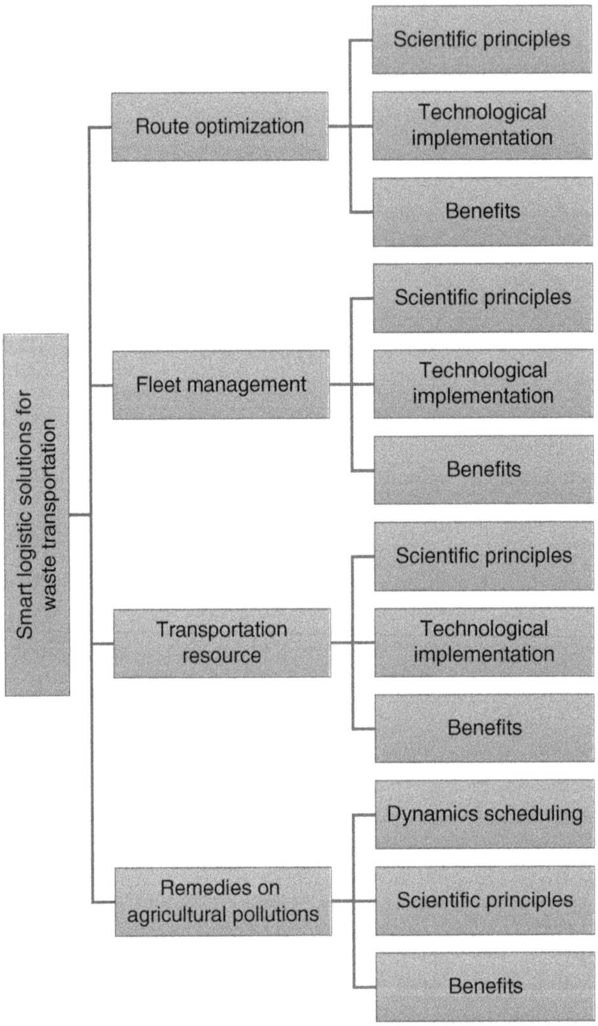

Figure 12.5 Smart logistics the integration of various technologies.

Figure 12.5 represents the smart logistics integration of various technologies, which will be the best solution for the coming future.

12.6.2 Challenges and Future Directions

While smart logistics solutions hold significant potential for improving agricultural waste transportation efficiency, several challenges must be addressed for successful implementation. These challenges are outlined in Figure 12.6.

While smart logistics promises significant advancements in agro-waste management efficiency, sustainability, and resource valorization by enabling data-driven decision-making and optimization across the value chain, its effective widespread implementation is currently hampered by a confluence of technical, economic, and systemic challenges that demand critical attention and innovative solutions for successful translation from conceptual models and pilot projects to

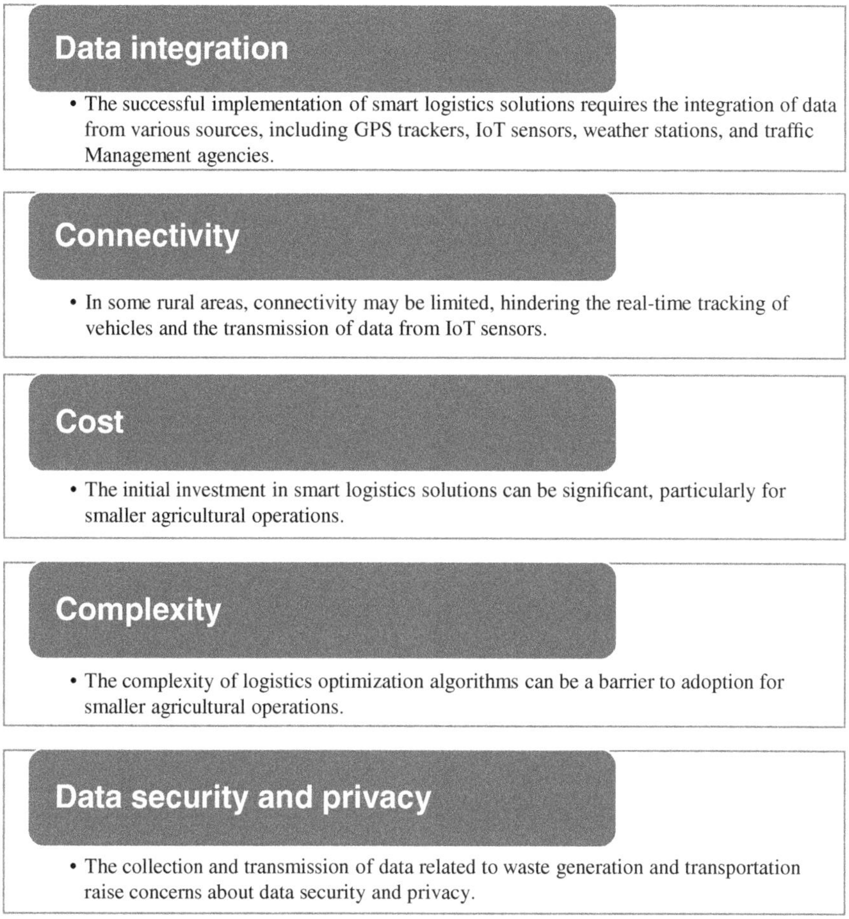

Figure 12.6 Smart logistics solutions.

routine operational practice. A foundational technical hurdle is the integration of data from diverse sources, a prerequisite for obtaining the comprehensive, real-time visibility necessary for intelligent logistics planning and execution. Smart logistics systems rely upon the aggregation and synthesis of information originating from numerous heterogeneous and often disparate sources, including real-time location and telematics data streamed from GPS units installed in collection vehicles, dynamic data generated by myriad IoT sensors deployed at various points – on waste bins to monitor fill levels or weight, within composting piles to track temperature and moisture, on processing equipment to gauge operational status, or even directly on farms to estimate waste generation volumes based on activity – as well as historical records of waste generation, meteorological data from weather stations crucial for predicting waste characteristics and accessibility, and potentially real-time traffic and road condition information from external agencies or platforms [1, 2]. The inherent challenge lies in the significant variability and lack of standardization across these data streams; they emanate from different device manufacturers using proprietary protocols, are transmitted via various communication technologies (cellular, Wi-Fi, LoRaWAN, satellite), are collected at different frequencies and granularities, are often stored in isolated data silos, and may have varying levels of quality, reliability, accuracy, and missing information [3, 4]. Achieving seamless, automated, and continuous ingestion, cleaning, transformation, and harmonization of this heterogeneous data into a unified, structured format suitable for analysis and algorithmic processing requires sophisticated middleware, robust Application Programming Interfaces (APIs), standardized data models (which are often lacking in the agricultural and waste sectors), and advanced data engineering capabilities [5, 6]. Without robust data integration infrastructure and processes, smart logistics platforms face fragmented information, leading to delayed or inaccurate insights, suboptimal optimization outcomes based on incomplete data, and hindering the ability to implement truly dynamic, real-time logistics strategies that respond to changing conditions and demands across the complex agro-waste landscape [10, 11].

Compounding the data integration challenge, particularly within the predominantly rural and often remote geographical context of agricultural production, is the significant and persistent issue of limited and unreliable connectivity. Effective smart logistics for agro-waste hinges upon the continuous, low-latency, and reliable transmission of data from distributed edge devices, such as IoT sensors in distant fields or GPS trackers on collection trucks operating outside of urban centers, back to centralized or cloud-based platforms for processing, analysis, and decision-making, and conversely, the reliable transmission of control signals or updated routes back to vehicles or automated equipment [12]. However, many agricultural areas globally, especially in developing regions, suffer from inadequate cellular network coverage, lack access to stable high-speed internet infrastructure, or face high costs for robust network access, creating significant connectivity gaps and zones of limited or intermittent service (often referred to as the "digital divide") [13, 14]. This poor or unreliable connectivity directly impedes the real-time flow of critical data, resulting in delayed data updates, significant data packet loss, communication failures that prevent the execution of dynamic route adjustments or remote monitoring, and potentially leading to operational disruptions or system failures if offline capabilities and data synchronization are not robustly implemented [15, 17]. While alternative or complementary communication technologies like LoRaWAN or satellite communication offer potential solutions for transmitting data from remote IoT sensors, these technologies may have limitations in terms of bandwidth, latency, or cost compared to traditional cellular or wired networks and ensuring their seamless integration and reliability across diverse and potentially challenging rural landscapes (e.g., hilly terrain, dense vegetation) remains a significant technical and practical hurdle for widespread and consistent deployment [18, 19]. The variability and unpredictability of connectivity quality in these areas necessitate the development of more resilient, edge-computing-enabled smart logistics architectures that can process data and make local decisions autonomously when network connectivity is lost, synchronizing later when a connection is re-established, adding further complexity to the system design and implementation [20].

Beyond these technical complexities, the initial investment required for implementing smart logistics solutions constitutes a substantial **cost** barrier, particularly for individual farmers or smaller agricultural operations that may operate on tighter margins compared to large-scale waste management companies [21, 22]. The total cost encompasses significant capital expenditure (CapEx) for acquiring and deploying the foundational digital hardware, including procurement and installation costs for IoT sensors, communication gateways, GPS tracking devices for vehicles, and potentially investing in basic IT infrastructure or network enhancements if existing connectivity is insufficient [23, 24]. Furthermore, substantial expenditure is required for obtaining software licenses for specialized logistics optimization platforms, data management systems, and potentially AI/analytics software, often involving recurring subscription fees [25]. Compounding the initial outlay are the ongoing operational expenses (OpEx), which include recurring costs for data transmission (cellular data plans, network fees), cloud computing service subscriptions for data storage, processing, and running optimization algorithms, software maintenance, technical support, and cybersecurity measures [26, 27]. Crucially, the cost of training personnel to operate and maintain these complex digital systems, or the necessity of hiring specialized technical expertise to manage the infrastructure and analyze the data, adds significantly to the operational budget, particularly in areas where such skills are scarce and command higher wages [28, 29]. While large waste management entities or agricultural cooperatives may be able to leverage economies of scale and consolidate operations to achieve a clear and potentially significant return on investment (ROI) through reduced fuel consumption, labor optimization, and improved efficiency, the fragmented nature of agro-waste generation across numerous small- to medium-sized farms means that the CapEx and OpEx per unit of waste managed can be disproportionately high for smaller entities [10, 30]. Demonstrating a compelling ROI based solely on logistical cost savings can be challenging when factoring in the total cost of ownership of the digital infrastructure and the need for continuous technological updates, making the initial financial outlay a significant deterrent for adoption by individual farms or small rural communities without targeted financial incentives or shared infrastructure models [32].

Moreover, the inherent complexity of smart logistics systems themselves, particularly the sophistication of the underlying optimization algorithms and the operational demands of integrated software platforms, presents a significant technical and cognitive barrier to widespread adoption, especially within agricultural sectors where technological expertise may vary widely [33, 34]. Effective smart logistics requires dynamic, real-time optimization algorithms capable of solving complex vehicle routing and scheduling problems that consider numerous interconnected variables and constraints simultaneously – this includes fluctuating waste generation volumes (derived from sensor data or predictive models), variable vehicle capacities, driver availability, real-time traffic conditions, time windows for pickups or deliveries to processing facilities, the varying types and characteristics of waste (e.g., wet vs. dry, hazardous vs. non-hazardous), and potentially different destination requirements for selective waste streams (e.g., manure to a digester, straw to a pellet plant) [35, 36]. These dynamic, multi-constraint optimization problems are computationally intensive and rely on advanced techniques from operations research, mathematical programming, and AI, including algorithms for route optimization, dynamic dispatching, and resource allocation [3, 37]. Developing, implementing, and fine-tuning these complex algorithms requires specialized expertise in data science, operations research, and computational optimization, skills that are often not readily available within standard agricultural or traditional waste management skill sets [38, 39]. Furthermore, while graphical user interfaces (GUIs) are designed to simplify the interaction with smart logistics platforms, the overall management, monitoring, and troubleshooting of integrated systems that ingest data from diverse sources, process it, run analytical models, execute optimization algorithms, and control operational workflows can be technically demanding for end-users [40, 41]. This perceived complexity and the need for specialized knowledge contrast sharply with traditional, simpler manual or paper-based logistics planning methods, even though traditional methods are significantly less efficient and lack the ability to respond dynamically to real-time changes [10]. The technical sophistication required to effectively understand, operate, and maintain these complex digital systems, along with the necessary interpretation of data analytics outputs for strategic decision-making, represents a significant hurdle to seamless adoption and effective utilization by agricultural operations lacking dedicated IT or data science support [28].

Finally, the inherent collection, transmission, and storage of granular operational data about farming activities and waste generation within smart logistics systems raise substantial concerns regarding data security and privacy, which can erode trust and act as a significant impediment to adoption among agricultural stakeholders [42, 43]. Data collected by IoT sensors on farms and GPS trackers on vehicles can include highly sensitive information about farming operations, such as precise locations of fields or livestock facilities, waste generation volumes linked to specific activities or production levels, timings of operations (planting, harvesting, feeding cycles), and potentially information about inputs used [44, 45]. This data, when transmitted wirelessly and stored on centralized or cloud-based platforms, is vulnerable to various cyber threats, including unauthorized access, data breaches, theft of proprietary information, malicious data manipulation, or denial-of-service attacks targeting the system [46, 47]. Risks associated with these security vulnerabilities include the potential loss of competitive intelligence for farming businesses (e.g., providing competitors with insights into operational scale, efficiency, or practices), operational disruption of waste management activities through cyberattacks on the logistics platform, or unauthorized surveillance [48, 49]. For farmers and agribusinesses, ensuring that their operational data is secure from external threats and that their privacy is protected regarding how their data is used and shared is paramount [50]. Addressing these concerns requires the implementation of robust cybersecurity measures across the entire system architecture, including strong data encryption protocols during transmission and storage, multi-factor authentication and secure access controls for platforms, regular security audits and vulnerability assessments, and clear, transparent policies regarding data ownership, consent for data usage, and data sharing agreements [51, 52]. Decentralized technologies like blockchain could potentially play a role in creating tamper-proof and transparent data trails for waste streams, enhancing security and trust [53, 54]. Without establishing a high level of trust and providing strong, verifiable guarantees regarding

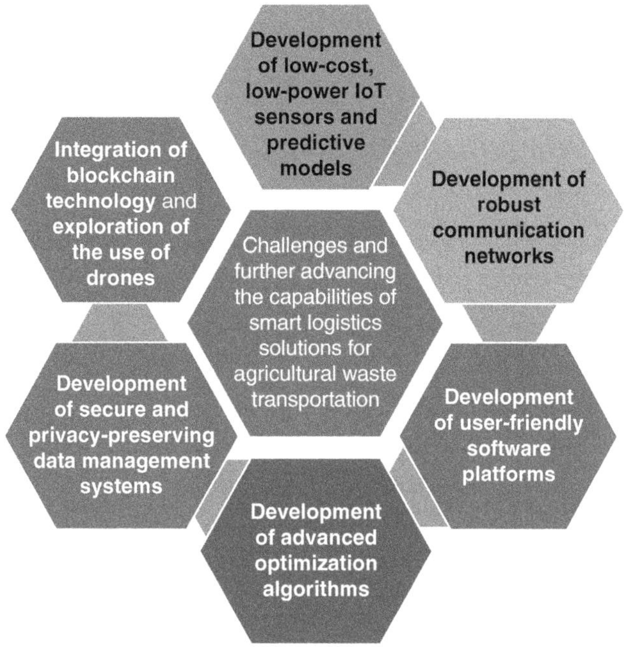

Figure 12.7 Advancing the capabilities of smart logistics solutions.

the security and privacy of their sensitive operational data, agricultural stakeholders are likely to remain hesitant to adopt smart logistics solutions, thereby limiting the overall potential for digital transformation in the agro-waste management sector [50].

Overcoming these multifaceted challenges requires a concerted effort involving technological innovation, targeted economic support, infrastructure development, capacity building, and the establishment of clear policy and regulatory frameworks that prioritize data security, privacy, and trust across the entire digital ecosystem for agro-waste management (Figure 12.7). Future research and development efforts should focus on addressing these challenges and further advancing the capabilities of smart logistics solutions for agricultural waste transportation [13]. Specific areas for research are as follows:

- Development of low-cost, low-power IoT sensors for monitoring waste bin fill levels and composition.
- Development of robust communication networks for transmitting data from remote agricultural areas.
- Development of user-friendly software platforms for managing and analyzing waste transportation data.
- Development of advanced optimization algorithms capable of handling complex waste transportation scenarios.
- Development of secure and privacy-preserving data management systems for protecting transportation data.
- Integration of blockchain technology to enhance transparency and traceability in the waste transportation supply chain.

- Exploration of using drones for waste surveillance and collection in remote areas.
- Development of predictive models to forecast waste generation rates and optimize transportation schedules.

12.6.3 Better Decision-making with Real-time Data and Predictive Analytics

The integration of digital infrastructure generates vast amounts of data related to agricultural waste management, as shown in Figures 12.8 and 12.9.

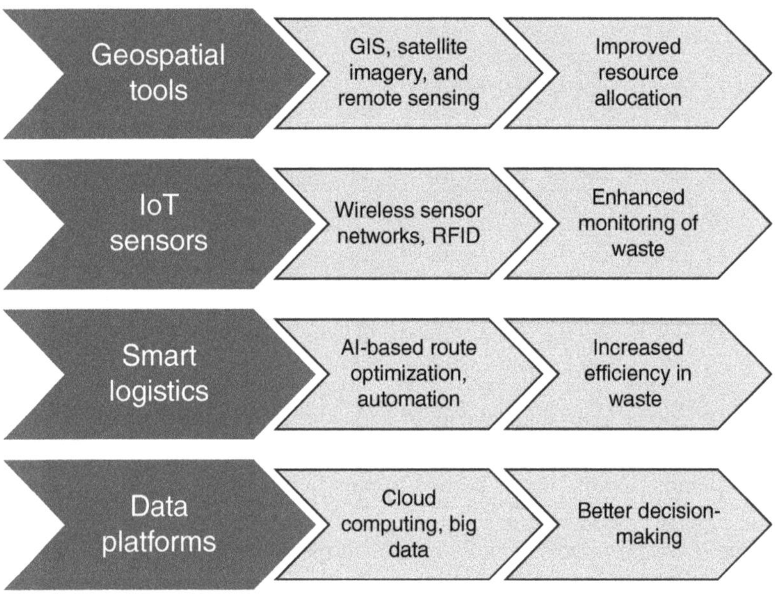

Figure 12.8 Digital infrastructure components in agro-waste management.

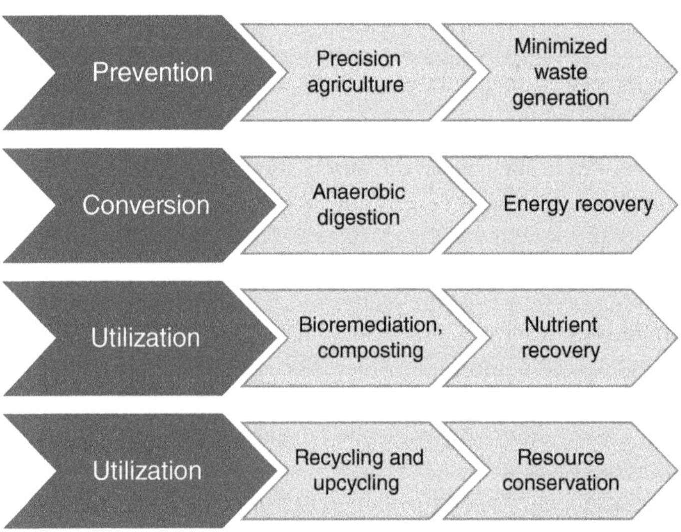

Figure 12.9 Technological applications in agro-waste management.

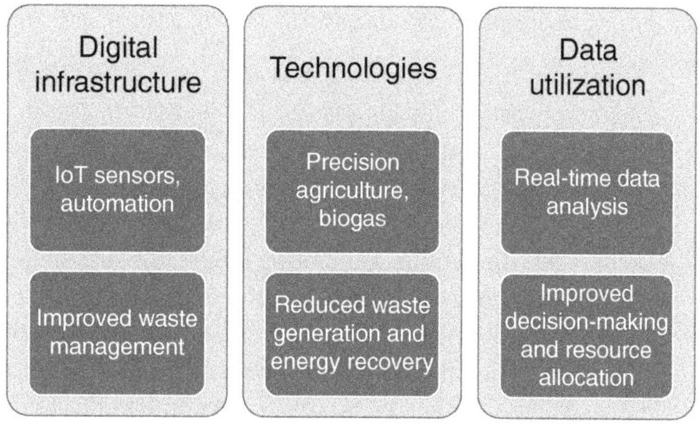

Figure 12.10 Data utilization and integration in agro-waste management.

This data can be analyzed using advanced analytics techniques, such as machine learning (ML) and AI, to derive insights, identify patterns, and facilitate improved decision-making, as illustrated in Figure 12.10.

- **Waste Generation Forecasting:** Predictive models can forecast waste generation rates based on historical data, weather patterns, crop cycles, and livestock populations. This information enables proactive planning of waste management activities and resource allocation.
- **Process Optimization:** Anomaly detection algorithms can identify deviations from normal operating conditions in waste-processing facilities. For example, a sudden temperature increase in a compost pile could indicate a fire risk, prompting immediate intervention.
- **Supply Chain Optimization:** Predictive analytics can optimize the agricultural waste supply chain, matching generation with demand for compost, biogas, or other value-added products.
- **Policy Making:** Policy decisions regarding regulations and subsidies for waste management can be continuously adjusted based on data analysis, ensuring policies remain effective and relevant.
- **Challenge:** Successful application of predictive analytics requires large, high-quality datasets and expertise in data science. Many agricultural operations lack the resources or expertise to develop and implement these models. Additionally, transparency and explainability in AI-powered decision-making are needed to ensure trust and understanding among stakeholders. For waste management, public acceptance and adoption are crucial, requiring clear communication of decisions to maximize approval.

12.7 Communication Networks

Reliable communication networks are essential for transmitting data collected from remote agricultural sites to centralized processing centers or cloud platforms. Key considerations include coverage, bandwidth, and latency, especially in rural areas where connectivity may be limited. Such applications also ensure that waste materials gain economic value [27, 28].

12.7.1 Coverage and Reliability

- **Fourth-generation (4G)/Fifth-generation (5G) Networks:** 4G and 5G cellular networks offer high-speed connectivity for transmitting large data volumes in real-time. 5G's lower latency and higher bandwidth capabilities are particularly promising for communication between IoT devices and central servers. This technology supports seamless monitoring and management of agro-waste operations, facilitating timely decision-making and intervention.
- **Satellite Communication:** In remote or geographically dispersed agricultural areas, satellite communication provides a reliable alternative to terrestrial networks. Satellite-enabled IoT devices ensure continuous data transmission even in areas with limited or no cellular coverage [3]. This capability is crucial for monitoring agro-waste in remote locations, improving operational efficiency, and reducing response times.

12.7.2 Benefits of Communication Networks

Communication networks form the indispensable backbone of digital transformation in agro-waste management, providing the critical infrastructure necessary for interconnectivity, data flow, and coordinated operations across the entire value chain, from farm generation to valorization facilities. A primary and significant benefit conferred by these networks is the enablement of remote monitoring, allowing for unprecedented real-time visibility into waste generation, accumulation, and characteristics regardless of geographical location or challenging terrain [43]. By facilitating the transmission of data from strategically deployed IoT sensors – monitoring parameters such as fill levels in waste bins, temperature and moisture in composting piles, or GPS locations of collection vehicles – networks empower waste managers and farmers to oversee operations from a central point without the necessity of frequent physical inspections [2, 15]. This capability is particularly valuable in the diffuse and often remote settings of agricultural production, overcoming spatial barriers and enabling timely detection of potential issues such as impending bin overflow, suboptimal decomposition conditions, or deviations from planned collection routes [26]. Building upon remote monitoring, the communication network ensures data accessibility, providing instant or near-real-time access to critical waste management data for relevant stakeholders and automated systems [17]. This immediate access to current information is crucial for facilitating data-driven decision-making, allowing for rapid and informed responses to dynamic changes in waste volumes, operational status, or environmental conditions [53]. Decision-support systems, predictive analytics models, and logistics optimization algorithms rely heavily on the timely availability of this data to generate actionable insights, optimize collection schedules dynamically, allocate resources efficiently, and trigger proactive interventions, thereby moving away from rigid, static planning toward a more responsive and adaptive management paradigm [1]. Furthermore, a key long-term advantage of communication networks in this context is their inherent scalability, which is essential for accommodating the growth and evolution of smart agro-waste management systems over time [51].

As operational needs expand, additional IoT sensors, vehicles, or processing units can be integrated into the system, generating increasing volumes of data and requiring communication infrastructure capable of handling this growing load without performance degradation [52]. A scalable network design ensures that the digital infrastructure can support phased deployments, integrate new types of sensors or devices as technologies advance, and ultimately accommodate the transition to more complex, large-scale operations or the aggregation of data from multiple farms or regions, future-proofing the investment in digital solutions and enabling the long-term expansion of smart waste management practices [55, 56]. In essence, communication networks are fundamental enablers that transform fragmented, static agro-waste handling into an integrated, dynamic, and data-informed system, leveraging remote visibility, enhancing data utilization for decision support, and providing a scalable foundation for future growth and technological integration.

12.7.3 Challenges

The realization of smart logistics for agro-waste management is fundamentally dependent on the availability of robust and pervasive communication network infrastructure, yet establishing and maintaining such networks presents significant technical and economic challenges, particularly in the geographically dispersed and often underserved rural areas where agricultural production is concentrated. A primary hurdle is the substantial infrastructure investment required to build communication networks capable of supporting the data transmission demands of smart agro-waste systems [1, 43]. Deploying dense networks of IoT sensors, enabling real-time GPS tracking of vehicles, and facilitating the transfer of large volumes of data for predictive analytics necessitate reliable, low-latency connectivity that is frequently absent or inadequate in rural agricultural landscapes [12]. Constructing or upgrading the necessary infrastructure – including cellular towers, fiber optic cables, LoRaWAN gateways, or satellite ground stations, along with ensuring consistent power supply to these assets and distributed edge devices – demands considerable initial capital outlay [14, 18]. Furthermore, the ongoing maintenance and operational costs of ensuring the reliability and uptime of this network infrastructure, especially in remote locations subject to environmental challenges, add a significant long-term financial burden [51]. This high investment cost can be prohibitive for individual farms, small agricultural cooperatives, or even regional waste management entities without dedicated funding mechanisms or public–private partnerships, creating a significant barrier to the foundational connectivity required for digital transformation in these areas [21].

Beyond the physical infrastructure, navigating the complexities of regulatory compliance poses additional administrative and financial challenges for deploying communication networks and devices within the smart agro-waste ecosystem [2]. The use of wireless communication technologies, including cellular networks, unlicensed spectrum for IoT devices (like LoRaWAN), and satellite communication systems, is governed by a complex web of regulations related to spectrum licensing, device certification, data transmission standards, and potentially cross-border data flow if operations span multiple regions [15, 17]. Compliance requires obtaining the necessary permits and licenses for transmitting data, ensuring that deployed IoT devices and communication equipment meet national and international technical standards for performance and electromagnetic compatibility, and adhering to data privacy and security regulations related to the collection and transmission of sensitive operational data [50]. The process of navigating these regulatory landscapes, which can vary significantly between jurisdictions, can be time-consuming, involve substantial administrative effort, and incur additional costs for testing, certification, and legal consultation [47]. Furthermore, changes in regulatory requirements or spectrum allocation policies can introduce uncertainty and necessitate costly updates to deployed infrastructure or devices [57]. These administrative complexities and compliance costs can hinder the rapid and widespread rollout of communication network solutions, particularly for smaller technology providers or agricultural businesses attempting to implement smart logistics systems across diverse regulatory environments. Together, the high cost of building and maintaining robust infrastructure and the administrative burden of regulatory compliance represent significant technical and economic limitations that impede the foundational connectivity essential for effective smart logistics in agro-waste management.

12.8 Cloud Computing and Storage

Cloud computing and storage platforms form the backbone of data-intensive applications in agro-waste management. These platforms provide scalable resources for storing, processing, and analyzing large datasets generated from various sensors and remote sensing technologies.

12.8.1 Availability and Utilization

- **Cloud Platforms:** Public, private, and hybrid cloud solutions offer flexible options for storing and managing agro-waste data. These platforms provide high availability, data redundancy, and scalability to accommodate fluctuating storage and computational demands. Agricultural cooperatives or waste management agencies, for instance, can utilize cloud-based applications for real-time analytics on waste generation trends, optimizing logistics and resource allocation [25].
- **Data-processing Capabilities:** Cloud computing enables advanced data-processing capabilities, including ML algorithms for predictive analytics and decision support [6]. By analyzing historical data on waste generation and environmental conditions, these algorithms can forecast future trends, enhancing operational efficiency and sustainability.

12.8.2 Benefits and Challenges of Cloud Computing and Storage

The effective implementation and scaling of smart logistics and data-driven decision-making in agro-waste management necessitate robust data storage and processing capabilities, with cloud computing emerging as a pivotal technological solution offering distinct advantages over traditional on-premises infrastructure. A primary benefit lies in the inherent **cost efficiency** of cloud services [1]. By adopting a cloud-based model, organizations can largely eliminate the need for significant upfront capital investments (CapEx) in physical servers, data centers, and associated networking hardware, alongside reducing ongoing operational costs related to power consumption, cooling, physical security, and the maintenance of complex IT infrastructure [25]. Instead, users leverage a flexible pay-as-you-go or subscription model, converting CapEx to OpEx and potentially leading to overall cost reductions, particularly beneficial for entities with fluctuating data-processing needs or limited capital resources. Furthermore, cloud computing offers exceptional scalability, a crucial advantage in dynamic systems like smart agro-waste management where data volumes and processing requirements can vary significantly [2]. Resources, including computational power, data storage capacity, and processing capabilities, can be easily scaled up rapidly to handle peak loads (e.g., during harvest seasons or data-intensive analytics runs) or scaled down during periods of lower demand, ensuring optimal performance while avoiding costly over-provisioning of resources [52]. This elasticity supports the seamless integration of expanding IoT deployments, the analysis of increasingly large datasets generated by diverse sensors and platforms, and adaptation to evolving operational needs without requiring manual infrastructure upgrades. Finally, cloud providers offer robust **data security** measures, often exceeding the capabilities affordable or manageable by individual agricultural operations or small waste management entities [50]. These measures typically include advanced data encryption protocols, both for data stored at rest on servers and data transmitted across networks, multi-factor authentication and fine-grained access controls to restrict unauthorized access, intrusion detection systems, and physically secure data center facilities [50]. These security features are designed to protect the confidentiality and integrity of sensitive agro-waste data, such as generation volumes linked to specific farms or operational schedules.

However, despite these compelling benefits, the adoption of cloud computing in agro-waste management is not without its challenges. A significant concern is **data sovereignty** and compliance with diverse data protection regulations, particularly when utilizing public cloud services hosted across multiple jurisdictions [4, 58]. Data stored in the cloud becomes subject to the laws and regulations of the country where the cloud provider's data centers are physically located, which may differ from the jurisdiction where the agro-waste is generated or managed [58]. Navigating these complexities to ensure compliance with national or regional data protection laws, such as requirements for data residency or limitations on cross-border data transfers, can pose significant administrative and legal challenges, potentially restricting the choice of cloud providers or requiring complex legal agreements [47]. Furthermore, **integration complexity** arises when connecting existing agricultural legacy systems and operational technologies with modern cloud-based platforms [34]. Many farms and waste management facilities utilize older software, hardware, and data formats that may not be designed for seamless interoperability with cloud services [33]. Integrating these disparate systems, ensuring consistent data flow, harmonizing heterogeneous data streams, and establishing reliable connections requires significant technical effort, custom middleware development, and ongoing maintenance, posing a technical barrier that can impede the smooth transition to a fully cloud-integrated smart system [5]. Addressing these challenges requires careful planning, potential updates or replacements of legacy systems, and the development of standardized APIs and data models specifically tailored for the agro-food and waste sectors.

12.9 Technological Applications in Agro-waste Management

Advancements in technology are revolutionizing agro-waste management, offering innovative solutions to optimize processes, enhance efficiency, and promote sustainability [8]. This section explores key technological applications, including monitoring and tracking systems, predictive analytics, and automation and robotics, within the context of agro-waste management [26].

12.9.1 Monitoring and Tracking Systems

Monitoring and tracking systems play a crucial role in ensuring effective management of agro-waste from generation through to disposal or valorization points [14, 15]. These technologies enable real-time visibility and data-driven decision-making, enhancing operational efficiency and resource utilization.

12.9.2 Adoption of Technologies

- **GPS:** GPS technology provides accurate location tracking of vehicles, containers, or equipment in agro-waste collection and transportation [29]. By tracking waste material movement, GPS systems optimize route planning, reduce transport times, and minimize fuel consumption, lowering carbon emissions and operational costs.
- **RFID:** RFID tags and readers facilitate automatic identification and tracking of individual waste bins, containers, or batches [22]. RFID technology improves inventory management, enhances traceability [5], and ensures compliance with regulations throughout the waste management chain. It enables stakeholders to monitor waste movement and status in real-time, from generation to processing or disposal.

12.9.3 Benefits and Challenges of Monitoring and Tracking Systems

Monitoring and tracking systems, notably GPSs for vehicle and asset location and RFID or sensor technologies for identifying waste containers and potentially waste streams, serve as foundational pillars for the development of smart waste management paradigms, enabling granular visibility and real-time control across the waste lifecycle from generation to disposal or valorization [2, 43].

A significant array of benefits stems from the strategic implementation of these systems. Primarily, they contribute to enhanced transparency, providing unparalleled real-time insight into waste management operations, allowing managers, regulators, and even the public to monitor the movement and status of waste collection vehicles, track when bins are serviced, and potentially follow the path of waste materials from the point of generation to final processing or disposal [15, 17]. This level of visibility is crucial for building trust among stakeholders, reducing the risk of unauthorized waste disposal (e.g., illegal dumping), preventing fraudulent reporting of collected waste volumes or service delivery, and ensuring accountability throughout the waste management chain, thereby contributing to more ethical and responsible practices [42, 43]. Furthermore, the integration of real-time tracking data is instrumental in achieving optimized logistics, a core objective of smart waste management systems [10, 36]. GPS data provides continuous updates on vehicle locations, speed, and operational status, which is fed into sophisticated route planning and scheduling algorithms [32]. By combining this with data on waste container fill levels (from sensors or RFID scans indicating servicing), systems can dynamically plan and adjust collection routes and schedules in response to actual demand, traffic conditions, or unforeseen delays [35]. This dynamic optimization minimizes overall travel distances, reduces fuel consumption and associated operational costs, decreases vehicle wear and tear, and enhances the overall efficiency of collection fleets and resource allocation, ensuring timely service delivery and preventing waste accumulation [11, 30]. Beyond operational efficiency, monitoring and tracking systems play a vital role in facilitating compliance and accountability with increasingly stringent waste disposal regulations and environmental standards. These systems create a verifiable, digital audit trail for each collection event or waste transfer, recording the time, location, and often the volume or weight of waste collected [27]. This detailed record-keeping is essential for demonstrating adherence to regulatory requirements, such as mandated collection frequencies or proper disposal at licensed facilities [59]. It provides the necessary documentation for reporting to regulatory bodies, supports environmental enforcement efforts by allowing the investigation of complaints or incidents, and ultimately reinforces accountability for ensuring waste streams are handled and processed in accordance with established protocols and environmental best practices [26, 42].

However, despite these considerable operational, administrative, and environmental advantages, the widespread implementation of robust monitoring and tracking systems in waste management, particularly in agricultural contexts or less developed regions, faces significant challenges, primarily related to the initial investment required and the complexities of data management. A key economic hurdle is the substantial initial investment needed to deploy the necessary hardware and software infrastructure [21, 22]. This includes the CapEx for purchasing and installing GPS tracking units in an entire fleet of collection vehicles, acquiring and tagging waste containers (bins, skips, etc.) with RFID tags or installing integrated sensors, procuring the necessary RFID reader hardware (either fixed readers at collection points or mobile readers on vehicles), and investing in the software platforms required for data collection, storage, visualization, analysis, and integration with other management systems [23, 24]. Furthermore, integrating this new tracking infrastructure with existing legacy IT systems or fleet management platforms can require significant technical effort and cost [34]. This upfront financial commitment can be a prohibitive barrier for smaller waste management entities, municipalities with limited budgets, or agricultural cooperatives that may lack access to sufficient capital or prioritize other infrastructure needs [25]. Compounding the cost is the technical challenge of data integration; ensuring seamless interoperability between data generated by various monitoring and tracking technologies from potentially different vendors and harmonizing this data into a unified, usable format is complex [60, 61]. Tracking data from GPS units may need to be combined with identification data from RFID readers, timestamps from collection events, and potentially weight or volume data from sensors [61]. These heterogeneous data streams, often using different data formats, protocols, and communication methods, must be integrated into a central database or cloud platform for analysis and utilization within smart logistics algorithms or reporting tools. Overcoming these data integration challenges requires developing custom middleware, utilizing standardized APIs if available, or investing in comprehensive data integration platforms, which adds complexity, cost, and requires specialized technical expertise. The absence of universally adopted data standards across the waste management and tracking technology sectors exacerbates these interoperability challenges, hindering the seamless flow of information essential for leveraging the full potential of monitoring and tracking systems. Addressing these investment and integration hurdles through targeted financial support, standardization efforts, and the development of more interoperable technological solutions is critical for unlocking the benefits of enhanced transparency, optimized logistics, and improved compliance in waste management operations.

12.10 Predictive Analytics

Predictive analytics leverages AI and ML algorithms to analyze historical data and forecast future waste generation patterns [19]. These models enable proactive decision-making, optimizing waste management strategies and resource allocation [27].

12.10.1 Implementation of Predictive Models

- **AI/ML Algorithms:** AI and ML algorithms analyze large datasets [26], including weather patterns, crop yields, and historical waste generation rates. By identifying trends and patterns, predictive models forecast peak waste generation periods, allowing waste management agencies to anticipate demand for collection services and adjust operational schedules accordingly [28].
- **Decision Support Systems:** Integrated with IoT sensors and monitoring systems, predictive analytics systems provide actionable insights to stakeholders. For example, municipalities can predict the volume of agricultural residues during harvest seasons and efficiently deploy resources for timely collection and processing.

12.10.2 Benefits and Challenges of Predictive Analytics

Predictive analytics, leveraging statistical methods and ML algorithms to forecast future outcomes based on historical and real-time data, offers transformative potential for enhancing efficiency and sustainability in agro-waste management by shifting from reactive responses to proactive strategies. A core benefit lies in resource optimization, as predictive models can anticipate fluctuations in waste generation volumes, compositional changes, and spatial distribution patterns, enabling more proactive and informed planning of collection schedules, vehicle deployment, sorting operations, and downstream processing capacities [43]. By forecasting waste availability, managers can optimize feedstock supply to valorization facilities, ensuring consistent operation and maximizing resource recovery yields [62]. This foresight directly contributes to cost reduction, minimizing OpEx through optimized scheduling and more efficient resource allocation [30]. Predicting waste generation allows for right-sizing collection efforts, avoiding unnecessary trips or overtime, while forecasting equipment maintenance needs enables preventative interventions, reducing costly unexpected breakdowns and associated downtime [63]. Furthermore, predictive analytics contributes to reducing the environmental impact by optimizing logistical operations, particularly transport routes and frequency [32, 64]. More accurate predictions of waste availability enable more efficient collection planning, reducing overall vehicle miles traveled, fuel consumption, and consequently, the emission of greenhouse gases and other air pollutants, directly lowering the carbon footprint of waste management activities [11]. Anticipating changes in waste characteristics, such as increased moisture content or potential for rapid decomposition based on weather forecasts, allows for proactive adjustments in handling and processing, potentially mitigating uncontrolled emissions of CH_4 or odors [65].

However, realizing the full potential of predictive analytics in this domain is constrained by significant challenges. Foremost among these is the critical issue of **data quality and availability** [6]. Accurate predictive modeling is fundamentally dependent on access to high-quality, consistent, and sufficiently granular historical data covering various relevant factors, including crop yields, livestock numbers, farming practices, weather patterns, and detailed records of past waste generation volumes, composition, and disposal or valorization outcomes [44]. In many agricultural settings, such comprehensive and standardized data may be scarce, inconsistently recorded, fragmented across different sources, or of questionable quality, introducing noise and uncertainty that limit the accuracy and reliability of predictive models [43]. Furthermore, even with sufficient historical data, the inherent variability and complexity of biological systems and external environmental factors can make precise long-term forecasting challenging [66]. Secondly, the development and implementation of effective predictive models are limited by model complexity and the requirement for specialized expertise [67]. Designing, training, validating, and continuously fine-tuning AI/ML algorithms capable of accurately predicting nuanced patterns in agro-waste generation and system behavior demands advanced knowledge in data science, statistics, and agricultural systems [38]. These skills are often scarce, particularly within traditional agricultural or waste management sectors [28]. Moreover, predictive models require continuous validation against real-world operational data to ensure their accuracy and reliability over time as conditions change, necessitating ongoing technical support and expertise for monitoring performance and updating models [29]. This complexity and the need for specialized knowledge can pose a significant barrier to entry and widespread adoption of predictive analytics solutions, particularly for smaller operations lacking dedicated technical resources [68].

12.11 Automation and Robotics

Automation and robotics are increasingly recognized as pivotal technologies capable of fundamentally transforming agro-waste management practices, driving a paradigm shift away from labor-intensive and often inefficient manual processes toward streamlined, highly controlled, and technologically advanced operations that enhance overall efficiency, resource recovery, and environmental sustainability [69]. The integration of automated systems and robotics within the agro-waste value chain extends across various critical stages, from initial sorting and characterization to subsequent processing and handling. A key area of application is in **automated sorting technologies**, where sophisticated systems employ a combination of sensors (e.g., optical, near-infrared, hyperspectral, X-ray), mechanical separators (e.g., conveyors, air jets), and intelligent algorithms, often powered by AI and ML for real-time image processing and pattern recognition, to accurately identify and segregate different types of materials within complex, mixed agro-waste streams based on properties such as size, shape, color, density, and material composition [3, 43, 69]. This precise sorting capability is crucial for separating organic waste (e.g., food scraps, manure, specific crop residues) destined for biological valorization pathways like composting or anaerobic digestion from potentially recyclable inorganic materials (e.g., plastics, metals from packaging or equipment) or contaminants, thereby significantly enhancing the purity of feedstock streams and enabling higher recycling rates or more efficient downstream processing [61, 70]. Concurrently, **robotics is being integrated into processing facilities** to perform tasks that are repetitive, physically demanding, hazardous, or require a high degree of precision [71]. This includes robotic arms equipped with specialized grippers for sorting specific items or removing contaminants from a conveyor belt, automated systems for shredding or chopping fibrous agricultural residues, robotic platforms for precise mixing or packaging of processed materials (e.g., compost, biochar), and automated guided vehicles (AGVs) or drones for internal logistics and material transport within large facilities [72, 73]. Robots often incorporate sensors and AI capabilities that allow them to adapt to variations in the waste stream – for instance, adjusting grip strength based on the size or fragility of an object, or modifying shredding parameters based on the density of the biomass [74, 75]. This integration of robotics into processing workflows ensures consistent quality and throughput, reduces human exposure to potentially harmful materials or environments, and allows for continuous operation, significantly augmenting or replacing manual labor for demanding tasks [76].

The implementation of automation and robotics in agro-waste management yields a multitude of tangible benefits that contribute to operational excellence and sustainable outcomes. A primary advantage is enhanced labor efficiency, directly reducing the reliance on manual labor, which can be scarce, costly, and physically demanding for tasks like sorting, handling heavy materials, or working in unpleasant environments [13]. By automating these tasks, operational costs associated with labor can be reduced, and the workforce can be potentially reallocated to supervisory, maintenance, or more specialized roles [28]. Furthermore, automating hazardous tasks, such as handling potentially contaminated materials or operating heavy machinery, significantly improves workplace safety, reducing the risk of accidents, injuries, and exposure to harmful substances for human workers [46, 59]. Another critical benefit is the improvement in quality control and consistency of waste-processing [33]. Automated sorting systems perform segregation with a higher degree of accuracy and consistency compared to manual sorting, leading to purer streams of recyclable or organic materials [69]. In processing facilities, automated systems ensure consistent feeding rates, mixing ratios, and operational parameters, which are essential for maintaining optimal performance in valorization processes like anaerobic digestion or composting, resulting in higher product yields and more uniform product quality (e.g., consistent nutrient content in compost, stable biogas production) [62]. This enhanced purity and consistency of processed materials are vital for supporting closed-loop recycling systems and ensuring that the output from valorization processes meets the required standards for market applications or safe reintroduction into agricultural cycles (e.g., using digestate as fertilizer) [77]. Finally, automation solutions offer increased scalability and operational resilience [2]. Automated systems are typically designed to handle larger volumes of waste and accommodate fluctuations in feedstock supply or operational demands more effectively than manual operations [30]. The inherent robustness of robotic and automated equipment enables continuous operation with potentially less downtime than manual labor, enhancing the overall system resilience and ensuring a steady flow of materials to processing facilities [78]. This scalability allows for expanding the capacity of waste management operations as needed, supporting larger agricultural enterprises or aggregating waste from multiple smaller farms.

Despite these significant benefits, the adoption of automation and robotics in agro-waste management is hindered by notable challenges, primarily related to initial investment and ongoing support. The most prominent challenge is the high initial investment or upfront cost required for acquiring, installing, and integrating complex automated systems and robotic equipment [21, 22]. Specialized sorting lines with advanced sensors and AI, robotic arms, or automated material handling systems represent a significant CapEx, particularly when compared to the relatively low costs of manual labor or simpler mechanical equipment [23]. This high CapEx can be prohibitive for smaller agricultural facilities, individual farms, or rural cooperatives operating with limited financial resources and less access to capital compared to large industrial entities [24]. The cost includes not only the purchase of hardware and software but also the expense of site preparation, infrastructure modifications to accommodate the equipment, and initial integration and calibration efforts [25]. Beyond the initial outlay, the long-term viability of automated systems depends on consistent performance, which necessitates ongoing maintenance and technical support [36]. Automated and robotic systems, being complex pieces of machinery and software, require regular preventative maintenance, calibration, and occasional repairs by skilled technicians to ensure optimal performance, reliability, and longevity [79]. Accessing qualified technical support, particularly in remote agricultural areas, can be challenging and costly, requiring either hiring specialized in-house staff or relying on potentially expensive external service providers or maintenance contracts [80]. The technical expertise required for troubleshooting complex robotic or AI-driven systems is significantly higher than that needed for manual equipment, contributing to the operational expenditure and posing a potential barrier if local technical capacity is insufficient [71]. Ensuring the long-term availability of spare parts and updates for proprietary software systems can also be a concern [34]. These combined costs and technical requirements for ongoing maintenance and support represent a critical challenge that must be factored into the total cost of ownership and addressed through robust support models, training programs, or potentially shared service centers to make automation and robotics a sustainable and economically viable option for widespread adoption in agro-waste management.

12.12 Conclusion

In summary, the exploration of digital solutions in agro-waste management reveals significant potential for innovation, efficiency, and sustainability. Digital technologies such as IoT, data analytics, and digital platforms play a pivotal role in optimizing agro-waste management. They enable real-time monitoring, data-driven decision-making, and efficient resource allocation, transforming traditional practices into more sustainable and economically viable solutions. Examples include smart farming practices in the Netherlands, biogas production initiatives in India, and CE models in Europe, all showcasing how digitalization enhances productivity, reduces waste, and promotes environmental stewardship. Although challenges persist, lessons learned emphasize the importance of standardization, data governance, inclusive policies, and effective communication in overcoming them and maximizing the impact of digital initiatives. To capitalize on the potential, strategic approaches include capacity building, public–private partnerships, innovation funding, supportive policies, and robust monitoring and evaluation frameworks. These strategies aim to foster innovation, address barriers, and ensure sustainable outcomes across agricultural value chains. A concerted effort from governments, private sector entities, research institutions, and communities is essential to scale successful digital solutions, promote knowledge sharing, and drive continuous improvement. Embracing digital innovation offers a pathway toward achieving global sustainability goals while unlocking economic opportunities in agricultural technology.

In conclusion, the journey toward effective agro-waste management through digitalization is dynamic and multifaceted. By building on successes, learning from challenges, and implementing strategic initiatives, stakeholders can pave the way for a more resilient, resource-efficient agricultural sector that contributes positively to environmental stewardship and economic development.

The management and recycling of waste materials present a significant challenge globally, in both developed and developing nations. Traditional waste collection methods have often been inadequate and financially burdensome. Addressing this substantial volume of waste optimally requires employing technology and digital tools. The CE has emerged as a crucial concept in recent years, focusing on sustainable development in urban areas. As cities face increasing waste volumes, sustainable waste management practices are paramount for ecological resilience of smart cities.

References

1 G.A. Kamilaris, A., and Fonts, A. The role of big data analytics in the agricultural sector: a systematic review. *Comput. Electron. Agric.* vol. 142, pp. 4, 55, 56, 61, 2017.

2 O.D. Corcoran, P., and Leahy, K. The role of smart city technologies in the waste sector: a review. *Waste Manag. Res. J. A Sustain. Circ. Econ.* vol. 108, pp. 4, 10, 34, 54, 58, 61, 66, 79, 2020.

3 Wang, Y., Kung, L., and Byrd, T.A. Big data analytics: understanding its capabilities and potential benefits for healthcare organizations. *Technol. Forecast. Soc. Chang.* vol. 130, pp. 4, 44, 33, 38, 53, 56, 2018.

4 W.P. Dinh, H.T., Lee, C., and Niyato, D. Offloading in mobile edge computing: a survey. *IEEE Commun. Surv. Tutorials* vol. 19, no. 4, pp. 4, 10, 38, 63, 2017.

5 Y.M. Chen, D., Wang, S., Liu, X., and Hu, S. Multi-source data integration for smart city applications: a review. *J. King Saud Univ. Inf. Sci.* vol. 32, no. 5, pp. 4, 38, 64, 65, 2020.

6 Batini, C. and Scannapieca, M. *Data quality: Concepts, Methodologies and Techniques*. Springer Science Business Media, 2008.

7 Saikia, K., Rajkumari, K., Moyon, N.S., Basumatary, S., Halder, G., Rashid, U., et al. Sulphonated biomass-based catalyst for solketal synthesis by acetalization of glycerol – a byproduct of biodiesel production. *Fuel Process. Technol.* vol. 238, pp. 107482, Dec. 2022. https://doi.org/10.1016/J.FUPROC.2022.107482.

8 Agrawal, P.S., Belkhode, P.N., and Rokhum, S.L. Biofuel production technologies – an overview. In: *Sustainability in Biofuel Production Technology*. John Wiley & Sons, Ltd, 2022. https://doi.org/10.1002/9781119888864.ch4.

9 Changmai, B., Sudarsanam, P., and Rokhum, L. Biodiesel production using a renewable mesoporous solid catalyst. *Ind. Crops Prod.* vol. 145, Mar. 2020. https://doi.org/10.1016/j.indcrop.2019.111911.

10 M.R. Ghiani, G., Laganà, D., and Manni, E. Logistics management of municipal solid waste: a literature review. *Waste Manag.* vol. 34, no. 12, pp. 15, 39, 43, 44, 67, 2014.

11 X.J. Qu, S., Ma, H., Guo, Y., and Jiang, S. Literature review on supply chain optimization: research framework and knowledge gaps. *Int. J. Prod. Res.* vol. 58, no. 23, pp. 11, 39, 40, 48, 68, 73, 2020.

12 R.R. and Klerkx, L. Dealing with digital divides in agriculture and rural areas: pegging a moving target. *NJAS-Wageningen J. Life Sci.* vol. 92, pp. 6, 39, 57, 2020.

13 D.A. Rotz, C.A., and Soni, P. Digital literacy in rural areas: a review. *J. Rural Stud.* vol. 64, pp. 7, 11, 17, 18, 40, 54,78, 2018.

14 K.B. Ijaz, F., Naeem, M., Iqbal, S., Khan, A.Z., and Ahmad, Z. Internet of things-based smart waste management: challenges and solutions. *Futur. Internet.* vol. 12, no. 4, pp. 8, 19, 57, 65, 2020.

15 M.P. Reyna, A., Chen, J., Sklar, E., and Martínez-López, F. On the use of IoT for the development of smart cities: a systematic review. *Appl. Sci.* vol. 8, no. 10, pp. 8, 31, 40, 54, 58, 65, 66, 2018.

16 Tiwari, R., Agrawal, P.S., Belkhode, P.N., Ruatpuia, J.V.L., and Rokhum, S.L. Hazardous effects of waste transformer oil and its prevention: a review. *Next Sustain.* 2024. https://doi.org/10.1016/j.nxsust.2024.100026.

17 G.V. and Dahiya, N. Adoption of smart waste management system (SWMS) using big data analytics. *Int. J. Inf. Manag. Data Insights* vol. 2, no. 2, pp. 19, 54, 66, 2022.

18 Ray, P.P. A survey on internet of things architectures. *J. King Saud Univ. Inf. Sci.* vol. 30, no. 3, pp. 291–319, 2018.

19 V.C. Mekikis, D.V., Riva, M., Antonopoulos, A., Kartsakli, E., and Alonso, L. Connectivity models for uplink and downlink in LoRaWAN. *IEEE Commun. Lett.* vol. 21, no. 11, pp. 22, 40,71, 2017.

20 D.S. and Shi, W. The promise of edge computing. *IEEE Comput.* vol. 49, no. 5, pp. 13, 41, 24, 2016.

21 Bohlmann, H.D. *The Economics of Digital Transformation.* Springer Nature, 2022.

22 J.N. Luthra, S., Mangla, S.K., Kumar, A., and Tripathi, N. Evaluating challenges to Industry 4.0 adoption in small and medium enterprises. *Prod. Plan. Control.* vol. 29, no. 16, 31, 65, 69, 80, 2018.

23 I.N. and Tjahjadi, M.E. Technical expertise and its influence on accounting information system success. *J. Manag. Bus. Environ.* vol. 1, no. 1, pp. 13, 41, 69, 80, 2019.

24 D.R. Schuh, G., Potente, T., Wesch-Potente, C., and Hauptvogel, A. Collaboration maturity levels for industrial companies in production networks. *Prod. Eng.* vol. 6, no. 7, pp. 13, 41, 69, 81, 2013.

25 Dastan, A. Cost optimization in cloud computing. *Int. J. Eng. Adv. Technol.* vol. 8, no. 4, pp. 14,42,45,60,61,69,81, 2019.

26 R.J. Syafrudin, M., Alfian, G., and Fitriyani, N. Performance analysis of IoT-based waste management using smart bin: affected by distance and obstacle. *Int. J. Autom. Smart Technol.* vol. 9, no. 4, pp. 14,42,64,68,71, 2019.

27 S.N. Suraj, P., Bhangale, A., Jadhav, P., and Raut, N. Smart waste management system using IoT and machine learning. *Int. J. Eng. Res. Technol.* vol. 9, no. 6, pp. 14, 42, 68, 71, 2020.

28 Koundinya, V. Farmers' perception of digital agricultural tools: a review. *J. Agric. Educ. Ext.* vol. 26, no. 4, pp. 14, 24,25,42,72,75,78, 2020.

29 P.D. and Davenport, T.H. Data scientists. *Harvard Bus. Rev.* vol. 90, no. 5, pp. 15, 65, 75, 2012.

30 F.S. Boysen, N., and Clausen, U. Production logistics: setting the stage for a digital transformation. *Int. J. Prod. Res.* vol. 57, no. 1, pp. 15, 43, 68, 80, 2019.

31 L.D. Ghiani, G., and Improta, G. The periodic vehicle routing problem with dynamic scheduling. *J. Sched.* vol. 22, no. 5, p. 15, 2019.

32 L.G. Demir, E., and Bektas, T. A review of richer vehicle routing problem variants. *Eur. J. Oper. Res.* vol. 257, no. 1, pp. 15, 43, 66, 73, 2017.
33 S.G. and Stock, T. Opportunities of circular economy in manufacturing companies. *Procedia Cirp.* vol. 40, pp. 16, 43,64,79,79, 2016.
34 K.R. Rajput, M., Haider Ansari, A., and Jawed, M.I. Legacy system migration: challenges and solutions. *Int. J. Softw. Eng. Appl.* vol. 3, no. 6, pp. 43, 16, 63, 69, 2012.
35 Vigo, D. and Toth, P. Vehicle routing: problems, methods, and applications. *siam* 2002.
36 K.A. Demirel, E., Onur, N., and Alan, H. Digitalization and digital applications in waste recycling: an integrative review. *Sustainability* vol. 16, no. 1, pp. 17, 44, 67, 81, 2024.
37 G.Y. and Wang, P. Deep learning for dynamic vehicle routing. *Eur. J. Oper. Res.* vol. 269, no. 3, p. 44, 2018.
38 Jordan, M.I. and Mitchell, M. Machine learning: trends, perspectives, and prospects. *Science* vol. 349, no. 6245, pp. 44,75, 2015.
39 Davenport, T.H. Data science and analytics: an overview. *Int. J. Forecast.* vol. 30, no. 4, p. 44, 2014.
40 Bakindi, A., Wiberg, A.H., Norman, J., Marsh, E., and Allen, S. Hybrid life cycle assessment (H-LCA) for buildings and construction materials: a systematic review and meta-analysis. *Build. Environ.* vol. 272, pp. 112630, Mar. 2025. https://doi.org/10.1016/j.buildenv.2025.112630.
41 Norman, D.A. *The Design of Everyday Things.* Basic books, p.45, 2013.
42 N.A. Wadhwa, B., and Das, S. Data sharing in smart cities: a review. *Int. Conf. Comput. Power Commun. Technol*, pp. 45, 67, 68, 2019.
43 P.-B.F. and Kamilaris, A. Deep learning in agriculture: a survey. *Comput. Electron. Agric.* vol. 147, pp. 45, 53, 56, 66, 67, 73, 74, 76, 2018.
44 B.M. Wolfert, S., Ge, L., and Verdouw, C. Big data in smart farming–a review. *Agric. Syst.* vol. 153, pp. 46, 74, 2017.
45 B.T. Tzounis, A., Katsoulas, N., Kavallieratos, N., Panagakis, P. Internet of things in agriculture, recent advances and future prospects. *Biosyst. Eng.* vol. 164, p. 46, 2017.
46 L.J. Roman, R., and Zhou, J. Mobile application security: vulnerabilities and countermeasures. *Comput. Commun.* vol. 36, no. 3, pp. 46, 74, 78, 2013.
47 M.M. Zafar, M.A., Khan, A.A., and Hanif, M. A systematic review of cybersecurity challenges in smart grids. *Sustainability* vol. 12, no. 21, pp. 46, 58, 63, 2020.
48 T.R. and Queiroz, M.M. Blockchain and supply chain management: a review of theoretical and practical aspects. *Int. J. Prod. Econ.* vol. 220, p. 46, 2020.
49 L.Y. and Xu, L. Blockchain technology in agricultural production. *IET Blockchains* vol. 3, no. 1, p. 46, 2020.
50 A.M. Adebayo, A.A., and Adedeji, O.M. Data security and privacy in smart cities: issues and solutions. *J. Phys. Conf. Ser.* vol. 1734, no. 1, pp. 46, 58, 62, 2020.
51 G.M. Sisinni, E., Saifullah, A., Han, S., and Jennehag, U. Industrial internet of things: challenges, opportunities, and directions. *IEEE Trans. Ind. Informatics.* vol. 14, no. 11, pp. 55,58, 2018.

52 C.G. Buyya, R., and Srirama, S.N. *Cloud Computing: Principles and Paradigms*, pp. 55, 62, 2021.

53 C.S. Wamba, S.F, Akter, S., and Dubey, R. The potential of blockchain technology for data sharing in the supply chain: a literature review and future research directions. *Int. J. Prod. Econ.* vol. 225, p. 47, 2020.

54 A.Q. Minhas, N.A., Lu, Y., Song, Z., and Imran, M.A. A review on blockchain-based security and privacy for smart agriculture. *Comput. Electron. Agric.* vol. 209, p. 47, 2023.

55 Z.K. Wan, J., and Cai, H. Smart farming: challenges and opportunities. *Int. J. Autom. Smart Technol*, pp. 54, 55, 2018.

56 A.N. Frank, A.G., and Dalenogare, L.S. The impact of industry 4.0 and digitalisation on the circular economy: a literature review and future research agenda. *Int. J. Prod. Res.* vol. 58, no. 23, p. 56, 2020.

57 Koberstein, J.T. Regulatory policy issues in chemical engineering. *Chem. Eng. Sci.* vol. 63, no. 24, p. 59, 2008.

58 Nabgan, W., Abdullah, T.A.T., Mat, R., Nabgan, B., Gambo, Y., Ibrahim, M., et al. Renewable hydrogen production from bio-oil derivative via catalytic steam reforming: an overview. *Renew. Sustain. Energy Rev.* 2017.

59 Borchard, R., Zeiss, R., and Recker, J. Digitalization of waste management: insights from German private and public waste management firms. *Waste Manag. Res. J. a Sustain. Circ. Econ.* vol. 40, no. 6, pp. 775–792, Jun. 2022. https://doi.org/10.1177/0734242X211029173.

60 Bogner, J.E. Landfill gas migration and emissions. *Environ. Monit. Assess.* vol. 15, no. 3, p. 70, 1990.

61 F.E. Boldrin, A., Andersen, J.K, Møller, J., and Christensen, T.H. Composting and compost utilisation: accounting of greenhouse gas emissions and global warming potentials. *Waste Manag. Res. J. a Sustain. Circ. Econ.* vol. 27, no. 9, pp. 70, 77, 2009.

62 Liu, X., Li, C., Yao, Z., Zhang, S., Liu, Z., Liu, L., et al. Zeolite and silica fume as enhancers for cementitious solidification of borate radioactive waste. *Nucl. Eng. Technol.* vol. 57, no. 9, pp. 103609, Sep. 2025. https://doi.org/10.1016/j.net.2025.103609.

63 Tian, Z., Niu, Y., Fan, D., Sun, L., Ficsher, G., Zhong, H., et al. Maintaining rice production while mitigating methane and nitrous oxide emissions from paddy fields in China: evaluating tradeoffs by using coupled agricultural systems models. *Agric. Syst.* vol. 159, pp. 175–186, Jan. 2018. https://doi.org/10.1016/j.agsy.2017.04.006.

64 L.G. Demir, E., and Bektas, T. An adaptive large neighborhood search heuristic for the pickup and delivery problem with multiple time windows. *Eur. J. Oper. Res.* pp. 73, 2012.

65 Y.L. Pan, S., Zhu, S., Li, Q., and He, Q. Digital twin-driven smart composting for agricultural waste. *J. Clean. Prod*, p. 74, 2021.

66 M.D. Liakos, K.G., and Busato, P. Machine learning in agriculture: a review. *Sensors* vol. 18, no. 10, 2018.

67 P.N. and Shmueli, G. Predictive analytics in information systems research. *MIS Q*, p. 75, 2014.

68 Agrawal, P.S., Belkhode, P.N., and Rokhum, S.L. Modelling and simulation to predict the performance of the diesel blends. In: *Sustainability in Biofuel Production Technology*. John Wiley & Sons, 2022. https://doi.org/10.1002/9781119888864.ch10.

69 Z.X. Guo, T., Zhong, Z., Xu, F., Guo, F., and Li, Y. Automated sorting system for construction and demolition waste based on deep learning. *Autom. Constr.* vol. 120, pp. 46, 76, 2020.

70 G.M. Zamorano, M., Pérez-Fortes, M., and Carreño, M.P. Automated waste sorting systems: a review. *J. Environ. Sci. Heal.* vol. 54, no. 7, pp. 46, 77, 2019.

71 K.D. and Bicchi, A. Trends and new challenges in robot manipulation. *IEEE Int. Conf. Robot. Auto*, pp. 46,77, 2008.

72 W.T. Mittal, S., Khan, M.A., and Romero, D. A knowledge-based framework for implementing industrial robotics and automation. *Prod. Plan. Control.* vol. 30, no. 16, pp. 46, 77, 2019.

73 AGVs: Automated Guided Vehicles. In: *The International Encyclopedia of Robotics*, pp. 46,77, 1988.

74 Siciliano, B. and Khatib, O. (Editors) *Springer handbook of robotics*, p. 77, 2016.

75 Kusiak, A. Smart manufacturing. *Int. J. Prod. Econ.* vol. 187, p. 77, 2017.

76 Parasuraman, R. Automation and robotics in the home. In: *The Handbook of Automation and Robotics in the Home*. 2019.

77 U.S. Ghisellini, P., and Cialani, C. From linear to circular economy: development of European countries. *J. Clean. Prod*, p. 79, 2016.

78 G.F. and Verbong, G.P. The ongoing energy transition: lessons from a socio-technical perspective. *Energy Policy* vol. 35, no. 12, p. 80, 2007.

79 K.F. Moulijn, J.A., and Van Diepen, A.E. Catalyst deactivation: is it predictable? *Appl. Catal. A Gen.* vol. 212, p. 81, 2001.

80 K.M. Abbasi, M.S., Qureshi, M.A., Alvi, A.R., and Qamar, A.M. Technical challenges of robotics applications and their solutions in agriculture. *Int. J. Adv. Res. Comput. Eng. Technol.* vol. 3, no. 11, p. 81, 2014.

Index

Note: Page numbers in *italics* and **bold** refers to figures and tables respectively.

b

d

e

f

g

u

v

w